Cryogenic Technologies at the European Spallation Source

A big science case study

Online at: https://doi.org/10.1088/978-0-7503-3223-1

Cryogenic Technologies at the European Spallation Source

A big science case study

Edited by
J.G. Weisend II
European Spallation Source and Lund University, Lund, Sweden

IOP Publishing, Bristol, UK

ISBN 978-0-7503-3223-1 (ebook)
ISBN 978-0-7503-3221-7 (print)
ISBN 978-0-7503-3224-8 (myPrint)
ISBN 978-0-7503-3222-4 (mobi)

DOI 10.1088/978-0-7503-3223-1

Version: 20240801

IOP ebooks

British Library Cataloguing-in-Publication Data: A catalogue record for this book is available from the British Library.

Published by IOP Publishing, wholly owned by The Institute of Physics, London

IOP Publishing, No.2 The Distillery, Glassfields, Avon Street, Bristol, BS2 0GR, UK

US Office: IOP Publishing, Inc., 190 North Independence Mall West, Suite 601, Philadelphia, PA 19106, USA

Dedicated to the Memory of

Sebastien Bousson

1972–2022

Irene Julie-Curie Laboratory

Orsay, France

Scientist, Collaborator, Friend

Contents

Preface

Detailed work on the cryogenic system of the European Spallation Source (ESS) started in 2009. During the past 14 years, this effort has involved more than a hundred people in dozens of laboratories and firms throughout Europe. The total system, including cryogenic plants, cryomodules containing superconducting radio-frequency cavities, a hydrogen moderator system, and sophisticated distribution lines, represents more than 50 million Euros of value. It is the second largest helium cryogenics system, after that of the Large Hadron Collider at CERN, in Europe[1].

The ESS cryogenics system is noteworthy for more than its size. First, the ESS is a completely new laboratory built on a green field site. The development of the ESS cryogenics system went hand in hand with the development of this new laboratory; both in terms of physical infrastructure and in terms of organizational development. Due to the time required to procure and build some of these cryogenic components, the ESS cryogenics system was frequently one of the first subsystems to access new organizational procedures and rules. In some cases, the cryogenics system drove the establishment of new processes and rules. For example, the accelerator cryoplant procurement was the second large procurement (after the civil construction contract) in the history of ESS. A second feature is the dependence on in-kind contributions by member states for the construction and operation of ESS. In the cryogenics system, in-kind contributions provided the superconducting radio frequency cavities and associated cryomodules (chapters 8 and 9), the accelerator cryogenic distribution line (chapter 4), and the cryogenic moderator system (chapter 6). In-kind contributions assisted in the testing of the elliptical cavity cryomodules (chapter 10). Testing of the spoke cavity cryomodules was carried out at Uppsala University (chapter 11). Successfully communicating and coordinating work with in-kind contributors, as well as with vendors and subcontractors, took significant effort and was key to the success of the ESS cryogenic system.

The goals of this book are to:

- Describe in detail the final installed cryogenic system at ESS and its performance to date.
- Describe the processes by which the system was designed, procured, and installed, including collaborations with in-kind partners.
- Provide lessons learned that are useful for future projects.
- Ensure that proper credit is given to everyone that contributed to the ESS cryogenics system.

The ESS cryogenics system is nearing completion (in 2024) and meets all its performance requirements. Along the way, we learned quite a bit. This book describes that journey.

[1] It will be surpassed eventually by the ITER Fusion Project and the FAIR Accelerator Project, which were both under construction in 2024.

Acknowledgements

Just like the construction of the ESS, this book has required many people working together to be successfully completed.

I first wish to thank all the contributing authors who wrote the detailed chapters that follow while still building and commissioning the ESS cryogenics system.

This book was first proposed by Joe McEntee of IOP Publishing in 2020. His support, along with that of others of the IOP Team (including John Navis and Phoebe Hooper) has been invaluable.

The staff at ESS, in particular my management (Mats Lindroos and Kevin Jones), have also been very supportive of this work.

My family Shari, Rachel, Nick, Alex, Patrick, and Lovisa have been through a number of book projects with me and as always their support makes this book possible.

Editor biography

John Weisend

John Weisend is a senior accelerator engineer at the ESS and adjunct professor at Lund University in Sweden. A specialist in cryogenic engineering, he has worked at the Superconducting Supercollider Laboratory, the Centre D'Etudes Nucleaires Grenoble, the Deutsches Elecktronen-Synchrotron Laboratory (DESY), the Stanford Linear Accelerator Laboratory (SLAC), the US National Science Foundation, and Michigan State University. He is chief technical editor of *Advances in Cryogenic Engineering* and the chair of both the International Cryogenic Engineering Conference Committee and the Cryogenic Society of America. His other books include: *He is for Helium, The Handbook of Cryogenic Engineering* (ed), *Cryostat Design* (ed), *Cryogenic Safety* (with T Peterson), *Going for Cold: A Biography of a Great Physicist, Kurt Mendelssohn* (with G T Meaden), and *Superfluid: How a Quantum Fluid Revolutionized Modern Science.*

Contributor biographies

Philipp Arnold

Philipp Arnold (standing fourth from right) and members of the ESS cryogenics Team. Other chapter authors pictured are Hideki Tatsumoto (standing second from left), Jianqin Zhang (standing fifth from right), and Nuno Elias (standing sixth from right)

Philipp Arnold is currently group leader and workpackage manager of the cryogenics system at the ESS AB in Lund, Sweden. He received his Master's degree in Mechanical and Thermal Engineering from the Technical University of Dresden, Germany. He has worked first as process engineer and later as sales manager for Linde Kryotechnik AG in Pfungen, Switzerland where he contributed in leading roles to small and large scale helium plants worldwide.

Pierre Bosland

Pierre Bosland is an engineer graduated from Ecole d'Electrochimie et Electrométallurgie de Grenoble, and PhD graduate on thin films deposited by magnetron sputtering. After his studies, he became the leader of the laboratory of the sputtered cavities at CEA Saclay. He then joined the CEA team in charge of the development of cryomodules for synchrotron light sources: SOLEIL (France), SLS (Switzerland), and ELETTRA (Italy). He was in charge of the design of the cryomodules and became the project leader of these projects. Then he was appointed project leader for the 12 SPIRAL2 low beta cryomodules produced by CEA for GANIL.

He has been involved with ESS for 10 years. He started as the project leader for the implementation of Irfu in-kind contribution to the ESS Accelerator and is now the Scientific Responsible and WP5 CEA work package leader.

Sébastien Bousson

Sébastien Bousson, who passed away in November 2022, was an internationally recognised expert in the scientific community of accelerators. He was a PhD who graduated from Paris-Saclay University in 2000 on the topic of thermal effects in stiffened superconducting cavities. He spent his entire career directing R&D on superconducting accelerator cavities and worked especially on the development of so-called 'spoke' cavities, which offer now major performance gains in particle acceleration. For more than 20 years, he held a number of senior positions at the institute (IN2P3), including project manager of the SUPRATech technology platform, coordinator of major work packages in major European programmes (EURISOL, TIARA) and project manager of the CNRS contribution to the ESS till 2020. He had been the Director of the Accelerator Division and Associate Scientific Director of IJCLab. He was awarded the CNRS Crystal Medal in 2015 and was one of several international experts defining the European Accelerator R&D strategy for the future of particle physics.

Christine Darve

Christine Darve is an engineering scientist at the ESS, Sweden. She obtained her PhD from Northwestern University on the topic of superfluid helium, following a diplome d'ingenieur on thermo-mechanics of systems and materials at the UTBM (Institut Polytechnique de Sevenans) in 1998.

She has worked at CERN and Fermilab designing, testing, and operating cryogenic systems for the US-LHC, the Tevatron, the Muon Collider and other large-scale infrastructures.

She moved back to Europe in 2011 and has been in charge at ESS of the technical coordination of the superconducting linear accelerator. In that role, she coordinated the preparation, construction, and testing of the spoke and elliptical cryomodules, with in-kind partners: CEA, CNRS, INFN, STFC, and UU.

Since 2020, she has bridged technical and quality challenges of the ESS in-kind partner projects.

In addition to those assignments, she has been involved in numerous educative and outreach programs since 1996. She has co-founded the *African School of Fundamental Physics and Applications* (ASP) in 2010 and the *Nordic Particle Accelerator Program* (NPAP) in 2015. Currently, she is the Chair of the *Forum on International Physics* (FIP) at the *American Physical Society* (APS) and has received an APS Fellowship in 2016.

Patxi Duthil

Patxi Duthil is a Cryogenist Engineer and head of the cryogenic group of the Accelerator Department at the Laboratoire de Physique des 2 Infinis Irène Joliot-Curie (IJClab). After graduating with a diplôme d'ingénieur on thermo-mechanics of systems, he obtained his PhD from the Université Pierre-et-Marie-Curie (Sorbonne University, Paris) and from the Hong Kong University of Science and Technology in energetics and fluid mechanics. He began to work on thermoacoustic devices for refrigeration and electricity production purposes at the LIMSI and IPNO laboratories. His activities then focused on cryogenics for accelerators: he participated in the developments and tests of the Super Proton LINAC cryomodule for CERN and of the spoke cryomodules for ESS. From 2014, he led the French in-kind contribution for the cryogenic distribution system for the spoke LINAC of ESS (WP11).

Nuno Elias

Nuno Elias is currently the Lead Cryogenics Engineer in the superconducting RF section within the accelerator division at ESS, holding a pivotal role in the rigorous testing, commissioning, and cryogenic operational aspects of cryomodules.

With a decade-long tenure at ESS, Nuno has been instrumental in shaping the cryomodule designs, in collaboration with in-kind partners while integrating with stakeholders from various engineering disciplines.

Before joining ESS, Nuno developed his skills at CERN as an applied fellow contributing in the development of superconducting magnets utilizing Nb_3Sn technology.

Nuno's academic journey culminated in the attainment of a PhD from the Technical University of Lisbon in Technological Physics. His doctoral work intersected with his tenure at CERN, where he played a pivotal role in the CAST experiment. Here, he conceptualized, constructed, and successfully operated of a cryogenics Helium-3 gas system, a critical component in the pursuit of dark matter particle detection, specifically axions.

He also holds a Master's degree in mechanical engineering with specialization in applied thermodynamics from the Technical University of Lisbon.

Jaroslaw Fydrych

Jaroslaw Fydrych is a Cryogenics Engineer at ESS ERIC. He received his PhD in technical sciences with a specialization in fluid mechanics from the Wroclaw University of Science and Technology in Poland. He was involved in WUST's collaborations in the field of cryogenics with XFEL, ITER, and FAIR projects, and he also spent 2 years at CERN as a Project Associate at Cryogenics Group. Jaroslaw joined ESS in 2013 as the Project Engineer for the ESS Cryogenic Distribution System.

Wolfgang Hees

Wolfgang Hees is a senior cryogenic engineer at the ESS in Lund, Sweden. He has a degree in Engineering Physics and worked for many years at the European Laboratory for Particle Physics (CERN) in Geneva, Switzerland. After some years in industry as a program manager for R&D in carbon capture technologies, he came back to accelerators. At ESS, Wolfgang built up the cryogenics group from scratch and lead the conceptual design of the cryogenic systems for both accelerator, target, and neutron instruments. Since then he has led projects as varied as building ESS' cryomodule test stand and installing a backup compressor for the accelerator cryoplant.

John Jurns

John Jurns is a senior engineer with over 40 years of experience in cryogenic system and component design, development, research, testing, and operations. He received his BS from SUNY at Buffalo and MS from Cleveland State University. He has extensive experience in both private industry and government research facilities. His expertise includes design and integration of bulk cryogenic storage and high pressure gas systems for industry, cryogenic system research, program development, design, build-up, and operation of cryogenic test facilities for NASA, specifying, procuring, and managing installation of a cryogenic helium refrigeration system for the ESS project, and leading a team responsible for the design, construction, and operation of liquid hydrogen and liquid deuterium cold neutron sources at the NIST Center for Neutron Research.

Rocio Santiago Kern

Rocio Santiago Kern has been part of the FREIA team since its start in 2012. Her main tasks were related to the liquefier's and cryostats' development, purchase, and operation. In March 2023, she became the project leader of the FREIA Laboratory.

Guillaume Olry

Guillaume Olry is a research engineer at the Laboratoire de Physique des 2 Infinis Irène Joliot-Curie (IJClab), expert in superconducting cavities and cryomodules. He graduated his PhD from Paris-Saclay University in 2003 on the topic of the design end test of a new superconducting spoke cavity and has been working especially on spoke cavities for more than 20 years now. Guillaume Olry has been involved in several Europeans R&D programs dedicated to superconducting cavities developments (XADS, CARE, EUCARD, EURISOL, SPL). After being in charge until

2014 of the seven high-beta cryomodules for the SPIRAL2 LINAC at GANIL, he was the project deputy leader of the CNRS contribution to the ESS, in charge of the in-kind contribution for the spoke cryomodules (Work Package 4) and then in 2020 was appointed as the project leader. He is now leading the RF group and is the deputy director of the accelerator division at IJCLAB.

Paolo Pierini

Paolo Pierini started working on the design of superconducting cavities and accelerators from the late-1980s, for a variety of applications, ranging from free-electron lasers, nuclear waste transmutation systems, to high intensity proton beams. He contributed to several design iterations for the TESLA/TTF cryomodules, which later became the baseline for the design of the ILC and XFEL accelerator cryomodules. He was responsible for the design and fabrication of the special cryomodule of the XFEL 3.9 GHz third harmonic system for beam phase space linearization, which was successfully operated in 2015 during the XFEL injector commissioning phase. In 2017, he joined ESS to lead the superconducting RF section of the accelerator division, in charge of testing, installing, and preparing the ESS cryomodules to operation. He presently acts as machine section coordinator for the SCL beamline.

Roger Ruber

Roger Ruber studied applied physics at Eindhoven University of Technology and obtained his PhD in high energy physics from Uppsala University. He worked at KEK and CERN on the development of superconducting magnets; contributed to several international accelerator development projects such as CLIC, ESS, and HL-LHC; and established the FREIA Laboratory for accelerator and instrumentation research at Uppsala University. Currently, he is at Jefferson Lab developing new technologies for superconducting accelerating cavities.

Hideki Tatsumoto

Hideki Tatsumoto has been a cryogenic engineer at ESS ERIC since 2018 and has responsibility for the cryogenic moderator system. Hideki received a PhD from Kyoto University in Japan and has 21 years of experience working in cryogenics (especially liquid hydrogen).

Xilong Wang

Xilong Wang is currently division head of superconducting and cryogenic technologies at the Dalian Advanced Light Source (DALS) project, proposed by Dalian Institute of Chemical Physics (DICP), Chinese Academy of Science(CAS) in Dalian, China. He received his PhD in cryogenic engineering from Technical Institute of Physics and Chemistry (TICP), CAS, Beijing, China. He has worked at the BEPCII/IHEP, China, the Deutsches Elecktronen-Synchrotron Laboratory (DESY), Germany and ESS ERIC, Sweden. Dr Wang's research interests include large scale accelerator cryogenics, cryomodule design and test facilities, and related superconducting RF components and equipment R&D.

Jianqin Zhang

In 2015, Jianqin received her PhD in Institute of High Energy Physics (IHEP), China Academy Sciences (CAS), her major is nuclear technology and its application with the direction of cryogenics engineering. From 2015 to 2019, Jianqin worked as a cryogenics engineer in IHEP. Since 2019, Jianqin joined ESS as a cryogenics engineer. Jianqin is mainly responsible for the ACCP operation and process analysis.

List of contributors

Philip Arnold
European Spallation Source, Lund, Sweden

Pierre Bosland
CEA-Université Paris-Saclay, DRF/IRFU/DACM/LISAH, Gif sur Yvette cedex, France

Sebastien Bousson
Deceased

Christine Darve
European Spallation Source, Lund, Sweden

Patxi Duthil
Laboratoire De Physique Des 2 Infinis Irène Joliot-Curie, Orsay, France

Nuno Elias
European Spallation Source, Lund, Sweden

Jaroslaw Frydrych
European Spallation Source, Lund, Sweden

Wolfgang Hees
European Spallation Source, Lund, Sweden

John Jurns
Retired

Guillaume Olry
Laboratoire de physique des 2 infinis - Irène Joliot-Curie, Orsay, France

Paolo Pierini
European Spallation Source, Lund, Sweden

Roger Ruber
Uppsala University, Uppsala, Sweden

Rocio Santiago Kern
Uppsala University, Uppsala, Sweden

Hideki Tatsumoto
European Spallation Source, Lund, Sweden

Xilong Wang
Dalian Institute of Chemical Physics, Dalian, China

John Weisand II
European Spallation Source, Lund, Sweden

Jianqin Zhang
European Spallation Source, Lund, Sweden

Contributors

The authors of this book are by no means the only contributors to the ESS cryogenics system. Many people from ESS, in-kind partner institutions, and industrial firms have contributed to the success of this system. As complete as possible a list of the people that contributed to the success of the ESS cryogenics system follows. Apologies in advance for anyone who was missed.

ESS:
John Weisend
Philipp Arnold
Jianqin Zhang
Xiaotao Su
Xilong Wang
Jaroslaw Fydrych
John Jurns
Hideki Tatsumoto
Wolfgang Hees
Christine Darve
Paolo Pierini
Per Nilsson
Piotr Tereszkowski
Bjorn Rundcrantz
Kent Wigren
Matthew Conlon
Saeed Haghtalab
Emilio Asensi
Wojtek Fabianowski
Marcelo Juni Ferreira
Laurence Page
Niklas Cerovina
Fabien Rey
Tomasz Zawierucha
Pawel Garsztka
Joakim Meyer
Emil Lundh
Marc Kickulies
Attila Horvath
Miklós Boros
Mats Segerup
Niklas Ivanov
Jaime Arriagada
Theodoros Vasilopoulos
Mattias Olsson
Romain Goncalves

Johnny Råström
Iris Haag
Tobias Lexholm
Alexander Hermansson
Piotr Tereszkowski
Mikael Beise
Jarich Koning
Jonathan Moberg
Magnus Jönsson
Luis Ortega
Meredith Shirey
Mirko Menninga
Malcolm De Silva
Ohad Graber-Soudry
Paolo Pierini
Nuno Elias
Cecilia Maiano
Henry Przybilski
Muyuan Wang
Phlippe Goudket
Felix Schlander
Miklos Boros
Eirikur Magnusson
Peter van Velze
Artur Gevorgyan
Morten Jensen
Delphine Hardion
Adalberto Fontoura
Luca Sagliano
Cédric Lombard
Andrea Bignami
Daniel Piso

Accelerator and Cryogenic Systems:
Jean-Pierre Thermeau
Philippe Bujard
Tomás Junquera
Olga Kochebina
Florian Dieudegard

CryoDiffusion:
Adrien Striebig
Christine Lhuillier
Benjamin Brule
Astrid Morisse

Christophe Adam
Anthony Oury

CEA Saclay:
Christian Arcambal
Florence Ardellier
Stéphane Berry
Quentin Bertrand
Pierre Bosland
Adrien Bouygues
Philippe Brédy
Enrico Cenni
Christelle Cloué
Michel Desmons
Guillaume Devanz
Gaël Disset
Fabien Eozénou
Michel Fontaine
Alexis Gaget
Françoise Gougnaud
Thibault Hamelin
Xavier Hanus
Vincent Hennion
Tom Joannem
Stéphane Jurie
François-Paul Juster
Catherine Madec
Claude Marchand
Christophe Mayri
Franck Peauger
Olivier Piquet
Juliette Plouin
Anthony Raut
Sylvie Régnaud
Bertrand Renard
Patrick Sahuquet
Thierry Trublet
David Vurpillot
Audrey Saliou
Claire Simon

DEMACO:
Peter Rijneveld
Ana Perez
Rossi Mendez

Marco van der Walle
Siebe Rooseboom de Vries
Ryan Rotteveel

FREIA (Uppsala University):
Lars Hermansson
Konrad Gajewski
Johan Eriksson
Åke Jönsson
Esat Pehlivan
Iaroslava Profatilova
Carl Svanberg
Anders Wirén
Rocio Santiago-Kern
Roger Ruber
Kjell Fransson
Han Li
Akira Miyazaki
Magnus Jobs
Vitaliy Goryashko
Anirban Krishna Bhattacharyya
Tor Lofnes
Rolf Wedberg
Rutambhara Yogi

Forschungszentrum Jülich:
Beßler, Yannick
Rosenthal, Eberhard
Grates, Carsten
Claudio Weber, Tania

IFJ-PAN:
Michal Sienkiwvicz
Wawrzyniec Gaj
Pawel Halczynski
Marek Skiba
Filip Skalka
Marcin Wartak

IJCLab:
Sébastien Bousson
Guillaume Olry
Patxi Duthil
Sylvain Brault
Denis Reynet
Nicolas Gandolfo
Matthieu Pierens

Jean Nsimaketo
Gilles Olivier
Patricia Duchesne
Emmanuel Rampnoux
Véronique Poux
Thierry Pépin-Donat
Ludovic Renard
Lê My Vogt
Sylvain Berthelot
Frédéric Chatelet
David Le Dréan
Natacha Bippus
Thibault Gérardin
Alice Thiébault
Olivier Frossard
Fetrarimanga Rabehasy
Richard Martret
Sylvie Durand
Virginie Quipourt
Soukaïna Gueddani

INFN-LASA:
P Michelato
D Sertore
C Pagani
A Bosotti
L Monaco
T Grimaldi
M Bertucci
R Paparella
G Fornasier
A d'Ambros
E del Core
A Bignami
M Chiodini
L Sagliano
M Bonezzi
M Fusetti
F Fiorina
J F Chen
S Pirani

KrioSystem:
Piotr Grzegory
Jacek Podolski
Maciej Matkowski

Konrad Zawada
Bartosz Kopania
Grzegorz Orłowski
Mateusz Maciązka
Grzegorz Michalski

Linde Kryotechnik:
Claudio Böni
Andreas Rüegge
Nikolay Kolev
Roger Harrer
Jan Hildenbeutel
Pietro Ferretti
Paolo Selva
Manuel Messmer
Jens Rother
Adrian Zenklusen

iSA GmbH:
Frank Lex

Aerzen:
Roland Wischöfer

Air Liquid Advanced Technologies:
Pierre Roux
Sarah Bigeard
Pierre Gourlet
Anne Barbier
Emmanuel Felix
Regis Caratis
Pascale Dauguet

Oak Ridge National Laboratory:
Lyngh, Daniel
Yong Joong Lee

Power Heat:
Greger Samuelsson
Magnus Hammarstedt
Sebastian Malmqvist
Bengt Persson

STFC:
Mike Ellis
Simon Gregory
Peter McIntosh
Alan Wheelhouse

Joseph Diakun
Philippe Goudket
Keith Middleman
Niklas Templeton
Mark Pendleton
Shrikant Pattalwar
Ayomikun Akintola
Louis Bizel-Bizellot
Rachael Buckley
Gary Collier
Phil Davies
Keith Dumbell
Andy Goulden
Sean Hitchen
Phil Hornickel
Conor Jenkins
Matthew Jones
Mike Lowe
Dave Mason
Andrew Blackett-May
George Millar
Jennifer Mutch
Adrian Oates
Peter Owens
Ninad Pattalwar
Paolo Pizzol
Oliver Poynton
Gareth Roberts
Paul Smith
Stuart Wilde
James Wilson

Technische Universität Dresden, Institute of Power Engineering:
Hans Quack
Marcel Klaus

Université Paris Diderot:
Jean-Pierre Thermeau

Wroclaw University of Science and Technology:
Maciej Chorowski
Jaroslaw Polinski
Agnieszka Piotrowska
Janusz Skrzypacz
Przemysław Jaszak
Maciej Grabowski

Ziemowit Malecha
Adam Ratajczak
Piotr Felisiak

Consultants:
Guy Gistau Baguer
Hans Quack

Chapter 1

Introduction to the European Spallation Source

J.G. Weisend II

The European Spallation Source (ESS) is a new facility designed to be a world class center for neutron science. This chapter describes the motivations for ESS, the basic layout of the facility and the in-kind funding structure. The role of cryogenics in the ESS project is also described.

1.1 Motivations for ESS

Neutron science is generally defined as the use of neutrons to understand matter. Neutron science contributes to a wide range of technical disciplines, including materials science, condensed matter physics, biology and medicine, chemistry, and engineering [1]. The neutron science community is very active. In Europe alone, there are roughly 5000 active researchers in this field.

Neutrons have long been recognized as a valuable probe for investigating matter. They possess a number of useful features, including the ability to penetrate deeply into materials and interact with the atomic nucleus rather than the surrounding electron cloud; the ability to interact with the lighter elements, such as hydrogen (and thus water), lithium, and oxygen; and the possession of a magnetic moment, which makes neutrons ideal for studying paramagnetic and superconducting materials. These features are complementary to the other common probe of matter—x-rays produced by light sources.

Neutrons for science come from two different sources: nuclear reactors and accelerator-driven spallation sources. In reactors, the neutrons are produced as a byproduct of nuclear fission. In spallation sources, a metal target is struck by a beam of protons. These protons interact with the nuclei of the metal and, in a nuclear process known as spallation, large amounts of neutrons are emitted. In both reactor and spallation sources, the initial energy of the neutrons created is too high to be useful as a probe into matter. Thus, these sources almost always include a moderator through which the neutrons scatter, reducing their energy to a value more useful for material studies. These moderators typically contain liquid hydrogen operate at near

doi:10.1088/978-0-7503-3223-1ch1

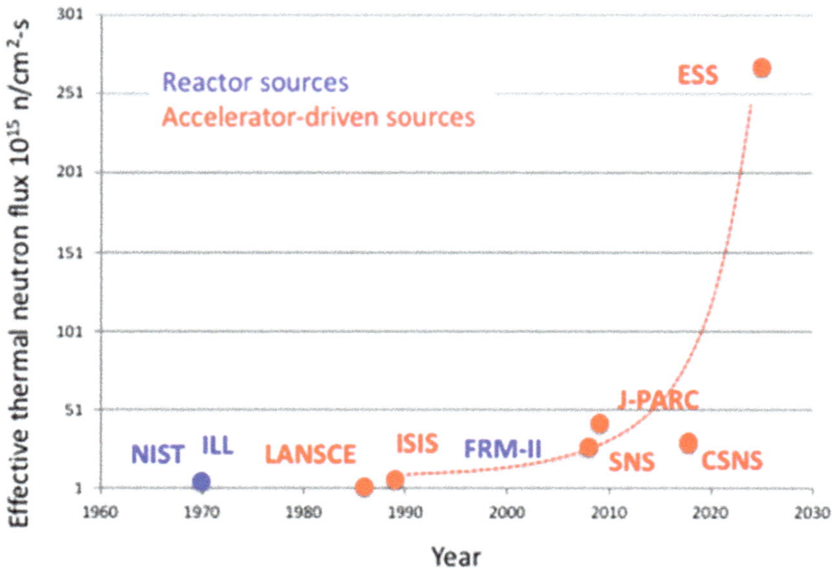

Figure 1.1. Expected ESS performance in terms of neutron flux compared to existing neutron sources (Courtesy of ESS).

20 K. Scattering through this very low temperature medium produces neutrons with an energy of most interest to neutron scientists.

An important parameter in neutron science is the brightness of the neutron source. This refers to the number of neutrons of the appropriate energy available to interact with the sample being studied. High brightness means that experiments can be completed in less time and that rare interactions can be studied, making new areas of research possible. In spallation sources, brightness is in part a function of the proton beam power and the ability of the moderator to cool the neutrons down to the desired energy. The ESS design optimizes both these parameters.

When the ESS project was first discussed, neutron sources in Europe were reactor based, along with two spallation sources: SINQ (Switzerland) and ISIS (UK). Many of these reactors are approaching the end of their lifetime and will be decommissioned. One of the motivations for the construction of ESS is to replace the soon to be lost neutron science capability associated with the decommissioned reactors. However, from the start, the ESS project was not only interested in replacing the reactors with similar capability. ESS uses advances in technology to build a much brighter neutron source. Figure 1.1 compares the expected ESS performance with that of other neutron sources (both reactor and accelerator based) around the world.

1.2 The ESS facility

Figure 1.2 is a schematic of the ESS facility. ESS may be divided into three primary technical systems: the accelerator, the target, and the neutron instruments. A fourth system, the integrated controls system, links the first three systems through data

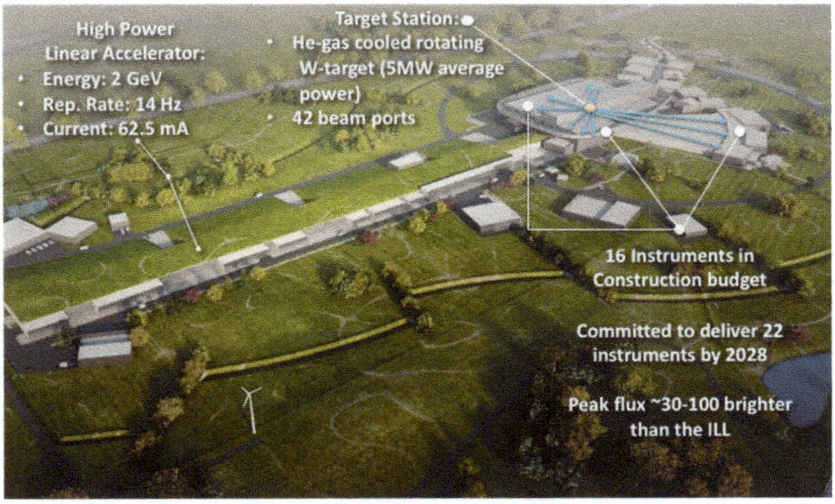

Figure 1.2. Schematic of the ESS project (courtesy of ESS).

acquisition and control. In the ESS design, the accelerator creates and accelerates the proton beam that strikes the target. The target creates, through spallation, neutrons, which then pass through the 20 K hydrogen moderator that is also part of the target system. The now lower energy neutrons move out to an array of neutron instruments where the science of ESS takes place. A full, detailed description of ESS may be found in reference [1]. A newer document reflecting the most recent changes to the accelerator and target systems is given in reference [2].

An important feature of the ESS facility is that the project started with a green field site. Site construction began in 2014 on what was previously agricultural land. All systems had to be constructed, including buildings, roads, and utilities. As will be seen, the order in which some facilities were built was driven by the need for early installation of the cryogenic plant components. Figure 1.3 compares the green field site with the current, near final, state of construction.

The organization of the ESS laboratory was also started from scratch. This meant that ESS had to develop its own policies and procedures for activities such as procurement, engineering reviews, documentation, and permission to energize equipment. The accelerator construction, in general, and the cryogenics system, in particular, were frequently driving the need for these policies. For example, the procurement contract for the Accelerator Cryoplant was one of the first large procurements made by ESS; second only to the civil construction contract.

1.2.1 Accelerator overview

Once fully powered, the ESS linear accelerator (LINAC) will be the most powerful proton accelerator in the world with an average beam power of 5 MW. Figure 1.4 is a schematic of the accelerator design showing components and beam parameters. The accelerator beam is pulsed at a frequency of 14 Hz. Each beam pulse is 2.86 ms

Figure 1.3. Comparison of the ESS site at start of construction (2014), top panel, and near final construction (2022), bottom panel (picture used with the permission of ESS ERIC).

in length. This long beam pulse is one of the unique features of ESS. The final energy of the beam will be 2 GeV.

The LINAC is contained in a tunnel whose roof is at ground level and covered with 5 m of earth for radiation shielding. Running parallel to the tunnel is a building known as the klystron galley, which contains all of the instrumentation racks, radio-frequency (RF) power sources, and other ancillary systems required by the accelerator. The LINAC consists of a room temperature section followed by a cryogenic section that uses superconducting cavities. Throughout the LINAC, RF

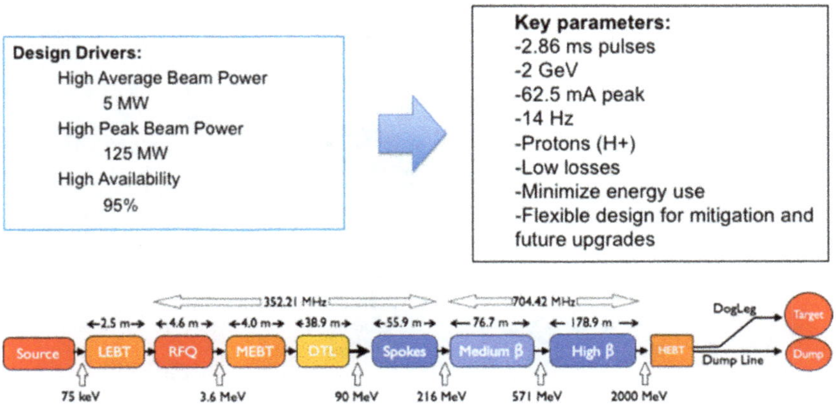

Figure 1.4. Schematic of the ESS accelerator and key parameters. Reproduced from [2]. © IOP Publishing Ltd. CC BY-NC-SA 3.0. Courtesy of ESS.

energy is fed into the various structures that provide acceleration of the beam. In the warm section of the LINAC, these structures are a radio-frequency quadrupole (RFQ) and a drift tube LINAC (DTL). They are connected to each other by a medium energy beat transport (MEBT) section. The RFQ is connected to the ion source that creates the protons from a hydrogen plasma by a low energy beam transport (LEBT) section. The LEBT and MEBT contain components such as magnets to compress and steer the beam, beam instrumentation to measure the beam characteristics, and bunchers and choppers to generate the correct beam pulse shape.

After the DTL, the beam is accelerated by three different types of superconducting radio-frequency (SRF) cavities. These cavities, made from pure niobium, operate at a temperature of 2 K. Note that the cryogenic (superconducting) section of the LINAC provides more than 95% of the total acceleration of the beam. The three types of SRF cavities used in the LINAC are double-spoke cavities and two styles of elliptical cavities: medium beta and high beta. Here, beta (β) refers to the ratio of the proton speed to the speed of light. In the ESS LINAC, medium beta refers to a β of 0.67 and high beta refers to a β of 0.86. The SRF cavities are contained within cryomodules. Figure 1.5 shows cross sections of the spoke and elliptical cavity cryomodules. The design and performance of these cavities and cryomodules is described in more detail in chapters 8 and 9 of this book. Between adjacent cryomodules are LINAC warm units (LWUs). These room temperature components contain beam instrumentation, beam tube vacuum systems, and magnets to shape and steer the beam. Figure 1.6 is a photo of an elliptical cavity cryomodule within the ESS tunnel.

All of the cryomodules are fully tested at cryogenic temperatures and full RF power prior to their installation in the tunnel. The double-spoke cavity cryomodules are tested in the FREIA facility at Uppsala University (see chapter 11), while the elliptical cavity cryomodules are tested at the ESS site (see chapter 10).

After the last cryomodule in the LINAC comes the high energy beam transport (HEBT) section. This section connects the accelerator with the target or in parallel to

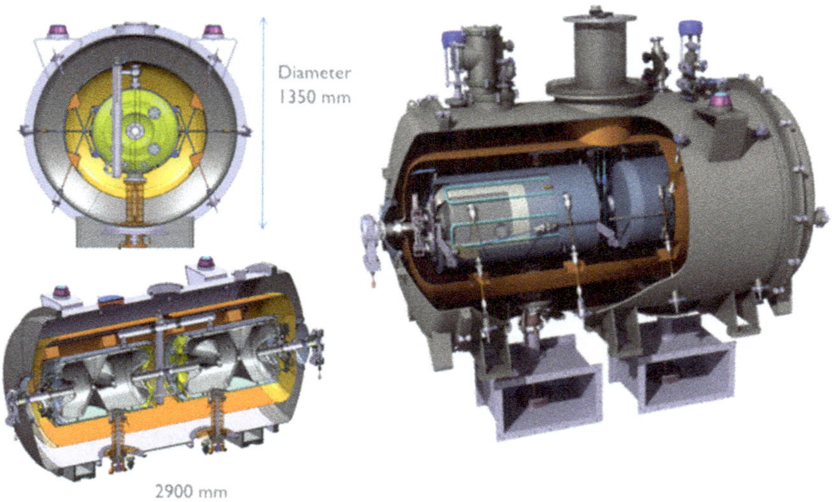

Figure 1.5(a). Cutaway view of a spoke cavity cryomodule (courtesy of CNRS & CEA, France).

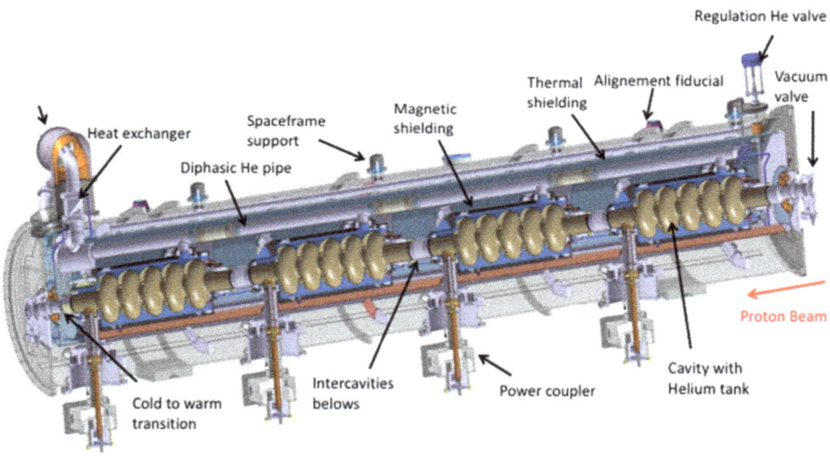

Figure 1.5(b). Cutaway view of an elliptical cavity cryomodule, the cryomodule is roughly 1 m in diameter and 8 m long (courtesy of CNRS & CEA, France).

a beam dump for low power beam tuning. The HEBT section is also at room temperature and contains LWUs at the same interval spacing as in the cryogenic section of the LINAC.

In order to be conservative, it was decided to design the accelerator and its supporting systems from the start to allow for the addition of 14 more high beta cryomodules if some or all of these become necessary to reach a final energy of 2 GeV. This decision meant that the klystron gallery was extended to support these additional cryomodules and that parts of the cryogenic system (see chapters 2 and 3) were sized for the heat load from these additional cryomodules.

Figure 1.6. Elliptical cavity cryomodule installed in the ESS tunnel (courtesy of H Przybilski, ESS).

The project's costs rose significantly during construction. To adjust to this, it was decided that while the entire LINAC and most of its systems are designed and installed to produce a 2 GeV beam, only enough RF power sources should be purchased to accelerate the beam up to 0.8 GeV, which is equivalent to an average beam power of 2 MW. In practice, this means that only the first five high beta cryomodules will accelerate the beam, with the remaining ones installed and operated at cryogenic temperatures but not powered by RF systems. Additional RF power sources will be purchased in the first years of ESS science operations, eventually returning the accelerator to its designed 5 MW average beam energy.

Table 1.1 is a summary of most of the components of the ESS accelerator. Further details on the accelerator may be found in references [1, 2].

1.2.2 Target overview

The ESS target converts the high energy proton beam produced by the accelerator into neutrons via a nuclear process known as spallation. In this process, the protons are absorbed by the nuclei of the target material (in this case tungsten), which causes an emission of a large number of neutrons.

The ESS target consists of a wheel of tungsten 2.5 m in diameter and rotating at 25 revolutions per minute. The total target weight is about 2 tonnes. The target is cooled by a flow of room temperature helium gas. Note that while the average beam power of the completed accelerator is 2 MW, the proton LINAC is pulsed at 14 Hz. During a single 2.86 ms pulse, the beam deposits up to 125 MW on the target. In

Table 1.1. Summary of accelerator sections.

Accelerator section	Beam energy at end of section (MeV)	Number of modules	Number of cavities per module	Operating temperature (K)	Length of cryomodule (m)
Ion Source	0.75	1		~300	
LEBT	0.75			~300	
RFQ	3.6	1	1	~300	
MEBT	3.6		3	~300	
DTL	90	5	1	~300	
Spoke	220	13	2	2	4.14
Medium β	570	9	4	2	8.28
High β	2000	21	4	2	8.28
HEBT	2000			~300	

order to spread this energy out on the target, a set of magnets at the end of the accelerator moves the beam across the surface of the target during the beam pulse. The lifetime of the target during full power operations is estimated to be 5 years.

The neutrons produced by the spallation process are generally at too high an energy to be useful for neutron science. Thus, the neutrons are reduced in energy by passing them through a moderator. In the moderator, the neutrons scatter off hydrogen atoms, losing energy in the process. In order to both maximize the amount of scattering and to achieve the desired final energy of the neutrons, the moderator at ESS is mainly composed of supercritical hydrogen operating at 20 K. There is also a smaller 300 K water moderator. One of the innovations of ESS is that the hydrogen moderator has been optimized to produce a 'cold source brightness that is three times greater on a per-proton than existing neutron sources' [2]. This innovation, coupled with the high power of the proton LINAC, results in ESS being between 10 and 100 times brighter than existing neutron sources. The moderator is expected to have roughly a 1-year lifetime at full beam power. Its mechanical design allows for straightforward removal and replacement using remote handling equipment.

The neutrons scattering through the 20 K hydrogen moderator deposit up to 30 kW of heat, which must be removed by the cryogenic system. This system consists of two parts. The cryogenic moderator system (CMS), which is a closed loop of hydrogen that circulates through the moderator absorbing heat from the neutrons (figure 1.7). The CMS transfers this heat via a heat exchanger to a helium flow at between 16 and 17 K that is produced by the target moderator cryoplant (TMCP). More details on the CMS may be found in chapter 6, while additional details on the TMCP can be found in chapter 5.

The target, moderator, and many supporting systems are contained within a shielding monolith (figure 1.8). Neutron beam ports at the periphery of the monolith allow up to 44 instruments to be installed at ESS. Note that there is no beam dump

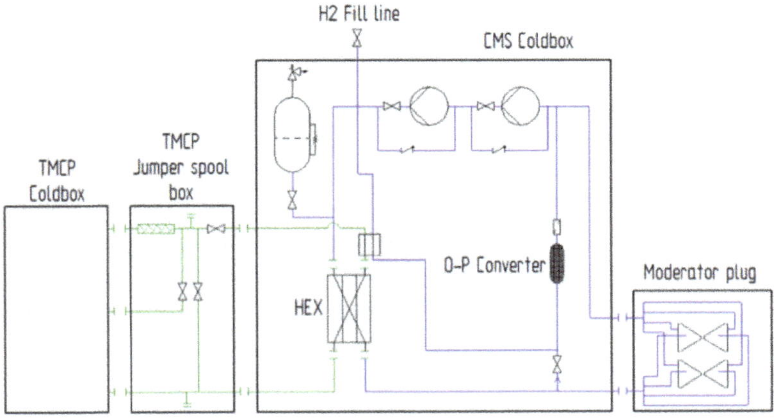

Figure 1.7. Simplified schematic of the CMS showing the connection to the TMCP. Reproduced from [2].

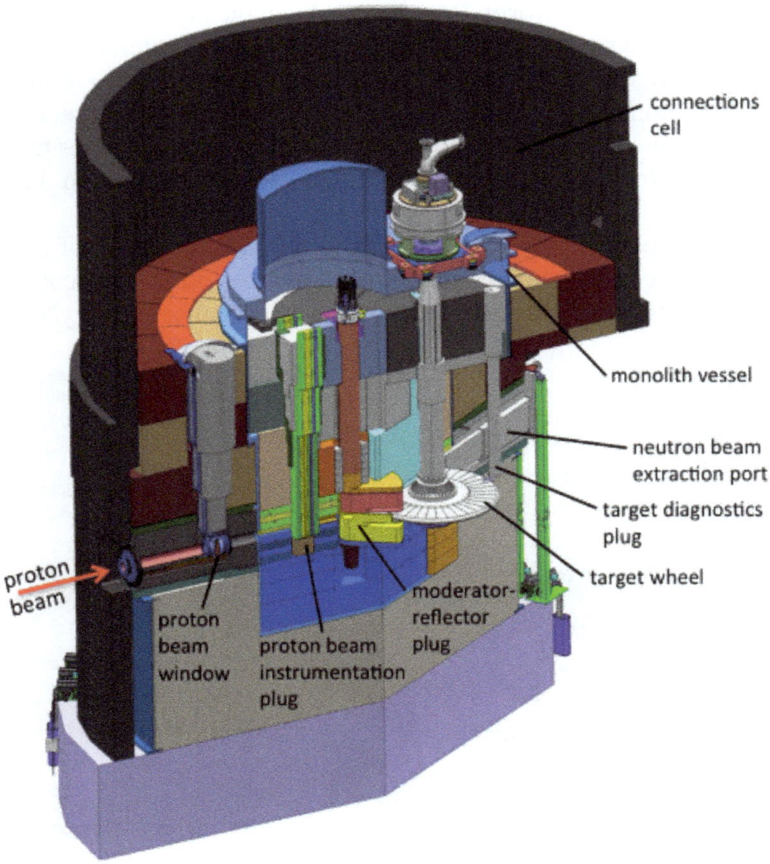

Figure 1.8. Cutaway view of the ESS target monolith. Note the target, moderator, and typical neutron beam port. Reproduced from [2].

in the ESS design, all the beam power entering the monolith is either absorbed by the target, moderator, and other monolith systems or removed by the neutrons transmitted to the instruments. A lower power (12 kW) beam dump is installed at the end of the accelerator tunnel to allow for beam commissioning and tuning.

1.2.3 Neutron science systems

The science at ESS takes place in the instruments; more formally the neutron science systems. All of the activities described so far exist only to ensure that copious amounts of neutrons with the right parameters (e.g., energy and pulse length) reliably arrive at the instruments. The ESS construction project funds 15 initial instruments with an additional seven to be built shortly afterwards using operations funds.

The instruments are connected to the monolith by neutron beam lines and are located as shown in figure 1.9 in both near and far experimental halls, depending on their size and timing requirements.

Each instrument is an international collaboration of scientists, with components coming in from all over the world. Broadly speaking, the neutron instruments can be divided into diffractometers, which study the structure of materials, and spectrometers, which study the dynamics of materials. There are also other types of instruments and some instruments serve multiple functions. Figure 1.10 lists the first instruments, along with details of their type and the research areas to which they can contribute. More information on the ESS's instruments can be found in reference [3].

Cryogenics appears in the ESS instrument suite in a number of applications. Some instruments contain superconducting magnets or other equipment that must be cooled down to cryogenic temperatures. The instruments may also contain

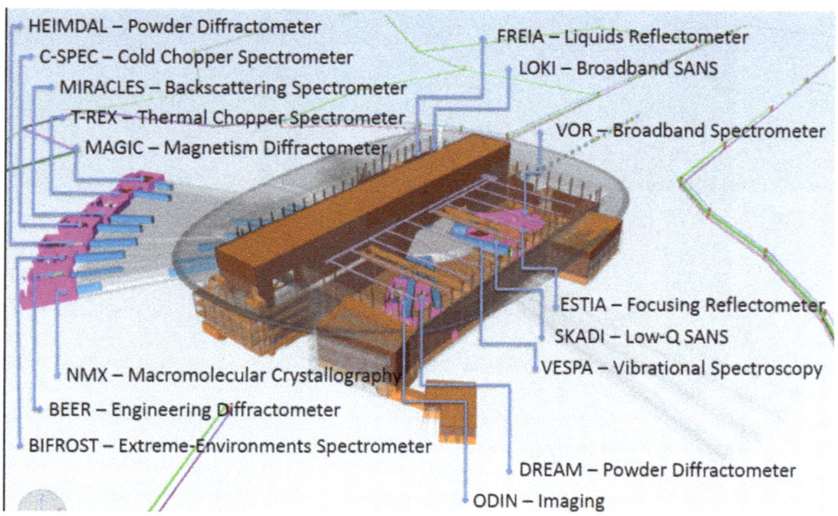

Figure 1.9. ESS neutron instruments and their locations (Courtesy of ESS).

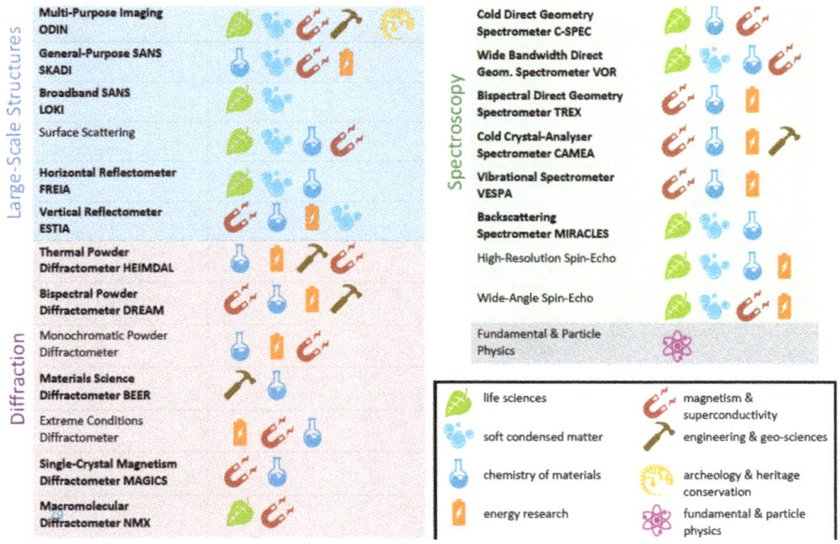

Figure 1.10. Science drivers for ESS's instruments (Courtesy of ESS).

sample environments that operate at cryogenic temperatures, some well below 1 K. Sample and test laboratories that support the instruments may also require cryogenics. Cryogenic temperatures can be produced by liquid helium supplied to the instruments from the test and instrument cryoplant (TICP—see chapter 7) or by discrete cryocoolers and dilution refrigerators procured as part of the instrument project. An ESS-wide helium recovery system is set up to recover released helium gas and return it to the ESS cryogenics facility for purification and liquefaction (chapter 7).

1.3 Sustainability and heat recovery

A major aspect of the Swedish/Danish bid to host ESS was the commitment to build a sustainable laboratory. This included use of sustainable materials where possible, minimization of environmental impact, power via renewable energy resources, and, significantly, recovery of waste heat.

From the beginning, ESS committed to recover 50% of energy used on the site. This is accomplished by recovering the heat removed by various ESS cooling systems and using it to heat the district hot water system in Lund. In fact, there are no large cooling towers or cooling ponds on the ESS site. The heat sink for the ESS laboratory is the district hot water system.

This level of heat recovery is unprecedented in scientific facilities and is made possible by a number of factors, including a serious commitment in terms of money and effort by ESS, the advantage of starting from a green field site with no preexisting infrastructure, and the presence of district heating in Lund.

Lund, like many cities in Scandinavia, produces hot water for building heat, washing, and other domestic needs at central plants and then distributes this water

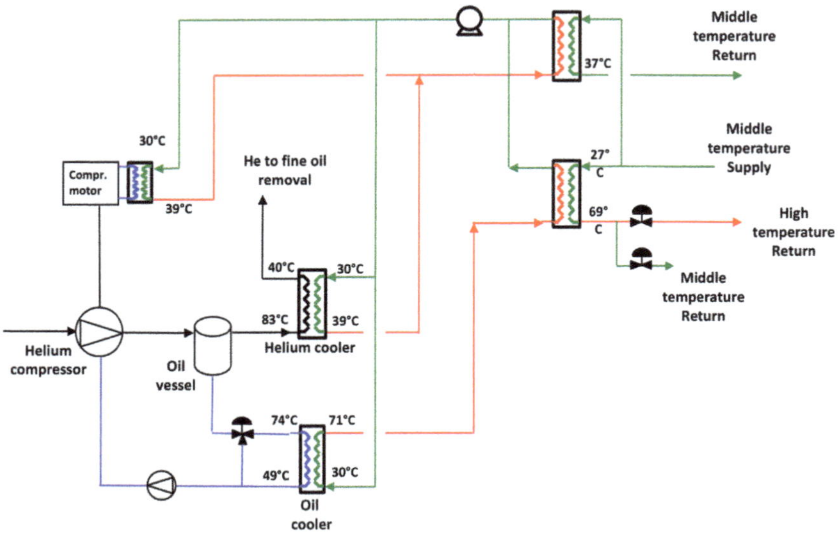

Figure 1.11. Example of a heat recovery scheme from an ESS cryoplant helium compressor (Courtesy of ESS).

via an extensive piping system. The existence of this system provides a ready heat sink for ESS's waste heat.

The cryoplants are a significant source for heat recovery. Heat removed from the motor and the oil, helium heat exchangers in the main compressors, and from the turbine brakes is sent to the domestic heating system. A maximum of 4.5 MW of heat is recovered from the cryogenic systems compared to a total input power of 5.1 MW.

Figure 1.11 is a schematic of the heat recovery from an ESS helium compressor. This recovery is accomplished without raising the helium or oil temperatures outside the specifications of the compressor's manufacturer.

1.4 Availability

ESS is a user facility. Neutron scientists whose experiments have been approved by a scientific advisory committee bring their material samples to ESS and make measurements using one of the instruments. These individual experiments are generally short-lived; taking from a few hours to perhaps a week or so, with hours to days being the most common duration. Thus, to be effective, the ESS neutrons must be available to researchers as scheduled. Not only must the neutrons, and by extension the accelerator and target, be available a certain fraction of the year but they must also be available when scheduled during the year. This is in contrast to high energy physics laboratories where experiments last months to years and the total integrated amount of beam time is more important. In effect, a visiting scientist is only interested in the neutrons being available during their scheduled experiment.

High availability was an ESS goal from the beginning of the project. The goal is that ESS should have an availability of 95% after 5 years of operation. Here availability is defined as the amount of time that neutrons are available as scheduled divided by the amount of scheduled time. Note that 95% is the total goal for ESS. This means that subsystems must achieve a higher availability, with the accelerator goal being 98%.

This requirement drove a number of decisions for the cryogenic system, including use of conservative design choices, operating margin, provisions for in situ repair of cryomodules, testing of all cryomodules prior to installation, redundant systems where possible, and extensive spares inventory. More details on ESS availability may be found in reference [4].

1.5 Project funding and in-kind contributions

The ESS is funded by a group of 13 European countries. Aside from some small grants for communication and research, very little of the funding comes from the European Union. Instead, the funding comes from the member countries that make up the ESS. Sweden and Denmark are considered the host states. The main ESS site is located in Lund, Sweden and the data management and science center is located in Copenhagen, Denmark. The other member states are the Czech Republic, Estonia, France, Germany, Hungary, Italy, Norway, Poland, Spain, Switzerland, and the United Kingdom.

During the construction of ESS the host states provide 47.5% of the costs, of which almost all is provided as cash. The non-host states provide 52.5% of the construction costs; 70% of this is in the form of In-kind contributions and 30% is in cash.

The use of in-kind contributions is one of the special features of the ESS project. In-kind refers to countries providing needed equipment or expertise of staff in lieu of cash during the construction phase. This approach has been taken at a smaller scale for the European XFEL [5] and FAIR [6] projects in Germany, and to a much larger extent for the ITER Project [7] in France. Roughly half of all the equipment needed for the ESS accelerator, including almost all the equipment in the beamline, was provided as in-kind contributions. Figure 1.12 shows in-kind contributions to the ESS accelerator. The target and instruments portions of ESS project also contain large amounts of in-kind contributions. Altogether, more than 100 European institutions made in-kind contributions to the ESS project.

Employing in-kind contributions to such an extent means that the ESS team had to devote much of their time to coordinating the in-kind effort. This work included negotiating with the in-kind partners to determine what they would contribute, working these contributions into the project schedule, ensuring that the interfaces between components match up, carefully tracking progress and dealing with issues as they arise, addressing schedule delays and conflicting priorities at in-kind partners, holding sufficient reviews to ensure that components will meet their performance requirements, deciding how in-kind components will be accepted and turned over to ESS, and collecting appropriate design and operating documentation.

As will be seen, this was accomplished, successfully for the most part, by establishing very detailed communication links between ESS and the in-kind

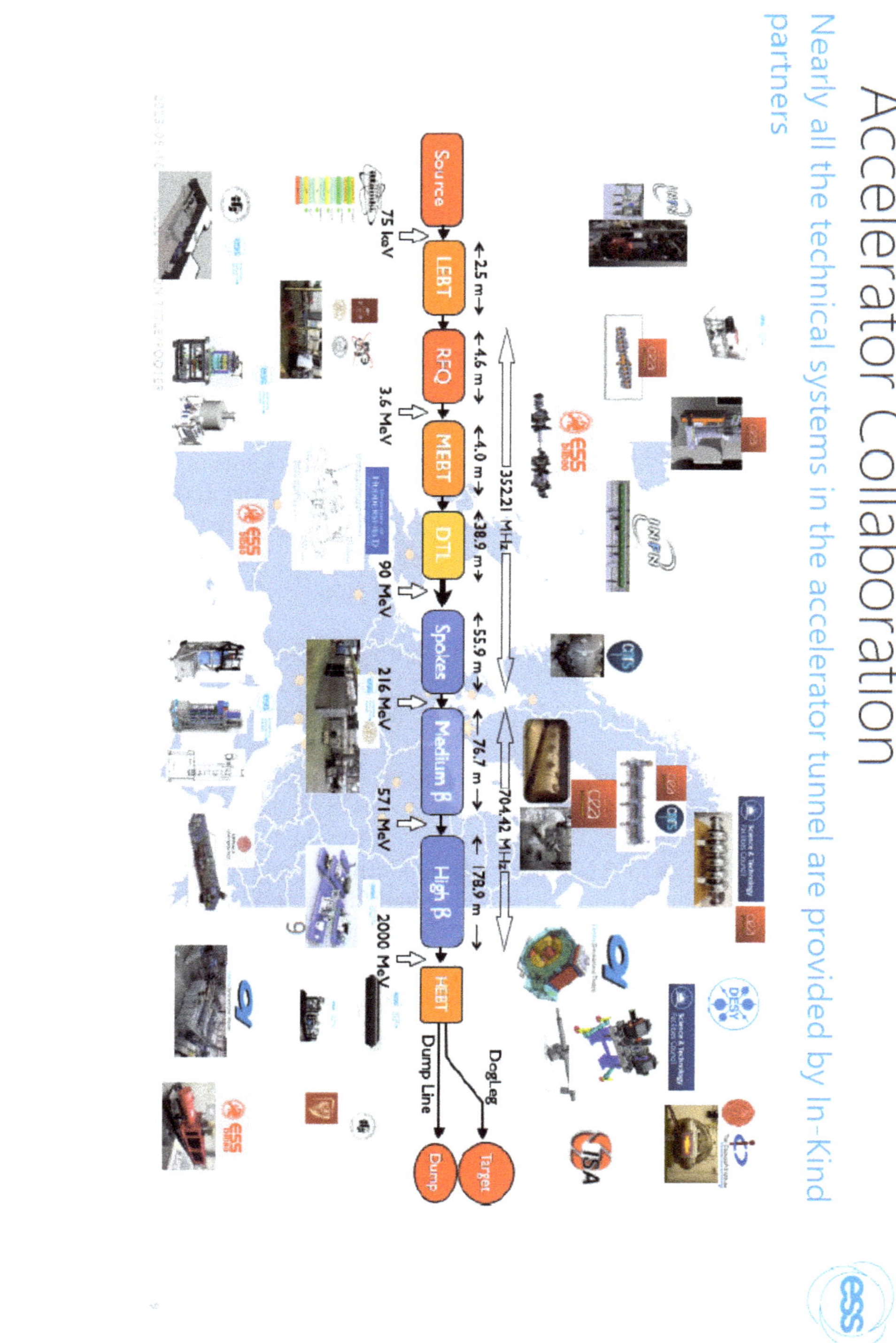

Figure 1.12. In-kind contributions to the ESS accelerator. Institution logos are shown near the provided equipment. (Courtesy of ESS).

partners at all levels, from the laboratory directors down to the working engineers and scientists. Most importantly, we regarded our in-kind partners as scientific collaborators rather than subcontractors. Good personal relationships have been shown to be critical for success.

Setting aside the fact that no other funding model was likely to be agreed upon, there are advantages to the in-kind approach. It allowed ESS to take advantage of the existing expertise in other European laboratories, which jump started the detailed design of components early in the project. For example, a separate Swedish–French contract resulted in the early (2009) start of the SRF cavity and cryomodule design in France. Cavities and cryomodules have a long development time and this early start aided the project schedule. Overall, the efforts in in-kind partner laboratories effectively doubled the available resources for use in accelerator construction. In-kind work also maintained, and in some cases strengthened, the capabilities of existing European Institutions, which had a side benefit of keeping member states supportive of the project.

In the case of the ESS cryogenic system, in-kind contributions played a lessor role with all three cryoplants, and the purification and storage system specified and purchased directly by ESS. The three notable exceptions to this were the cryogenic distribution system (chapter 4), the SRF cavities and cryomodules (chapters 8 and 9), and the CMS (chapter 6).

References

[1] Peggs S 2013 *ESS Technical Design* Report ESS-2013-001. https://europeanspallationsource.se/sites/default/files/downloads/2017/09/TDR_online_ver_all.pdf

[2] Garoby R *et al* 2018 The European Spallation Source design *Phys. Scripta* **93** https://iopscience.iop.org/article/10.1088/1402-4896/aa9bff

[3] https://europeanspallationsource.se/science-instruments

[4] Bargalló E, Andersson R, Nordt A, De Isusi A, Pitcher E and Andersen K H 2015 ESS reliability and availability approach *IPAC 2015*

[5] *The European XFEL Technical Design Report* 2007 https://bib-pubdb1.desy.de/record/77248/files/european-xfel-tdr.pdf

[6] https://fair-center.eu/

[7] https://iter.org/

Chapter 2

An overview of the ESS cryogenic system and its design evolution

J.G. Weisend II and W Hees

Superconductivity and cryogenics were recognized as necessary to the European Spallation Source (ESS) project from its earliest days. This chapter provides a high-level overview of the ESS cryogenics system and describes the reasoning behind early conceptual design choices. Some requirements for the cryogenic system did change over time and that evolution is discussed. In addition, the recruitment of staff to support the ESS cryogenic system is discussed. The overall safety strategy for the cryogenics system is also described. Best practices and lessons learned from this work are given.

2.1 Overview of the ESS cryogenics system

Figure 2.1 is a block diagram of the as-built ESS cryogenic system. The system is centered on three separate cryoplants [1]: the accelerator cryoplant (ACCP) for cooling the superconducting radio-frequency (SRF) cavities in the accelerator, the target moderator cryoplant (TMCP) for cooling the 17 K hydrogen moderator, and the test and instruments cryoplant (TICP), which provides cooling to the ESS cryomodule test stand and liquid helium to the instruments. Each of these plants is connected to the equipment it cools via a distribution system. These systems are vacuum insulated cryogenic transfer lines that vary in complexity from multiple pipe transfer lines with distribution boxes in the case of the ACCP (figure 2.2) to a two-pipe transfer line connecting the TMCP to the cryogenic moderator system (CMS). In the case of the accelerator, the distribution system is further divided into the cryogenic transfer line that runs from the ACCP to the tunnel where it connects to the cryogenic distribution system (CDS), which is located parallel to the accelerator; connecting to each of the cryomodules (figure 2.3). Liquid helium is supplied to the instruments and other laboratories via smaller, portable, helium dewars. Both the ACCP and TICP have large, external, and fixed liquid helium dewars incorporated

doi:10.1088/978-0-7503-3223-1ch2

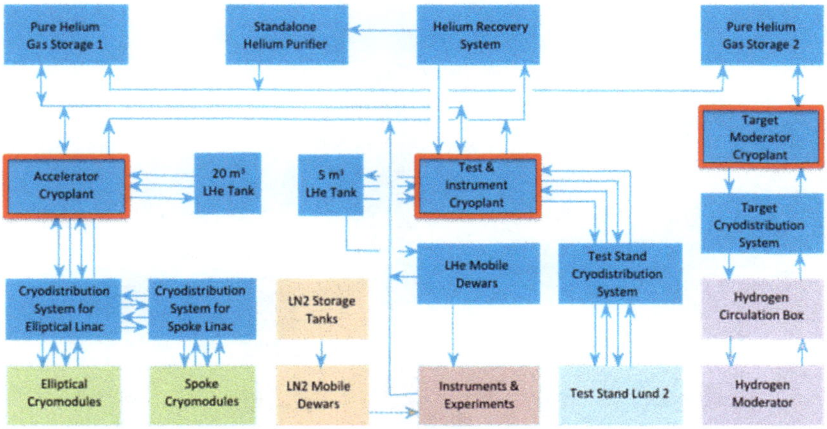

Figure 2.1. Block diagram of the ESS cryogenic. Reproduced from [2]. © IOP Publishing Ltd. CC BY-NC-SA 3.0. Courtesy of ESS.

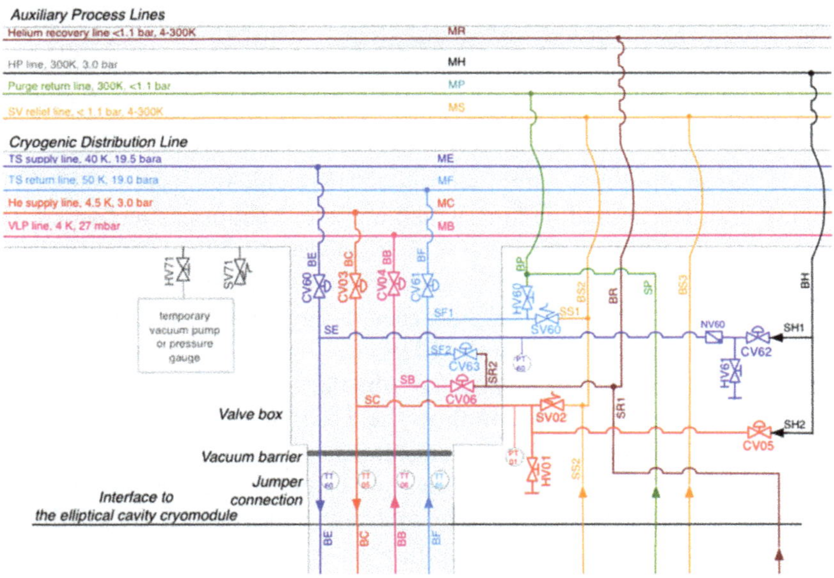

Figure 2.2. Piping and instrumentation diagram of the ESS ACCP CDS showing the valve box and the connection to the cryomodules. Reproduced from [2]. © IOP Publishing Ltd. CC BY-NC-SA 3.0. Courtesy of ESS.

into their design. The TMCP, which does not produce liquid helium, does not include such a dewar.

Each cryoplant has its own warm compressor system, cold box, and controls, and each is operated independently. There are no provisions for sharing loads between the plants. For example, the ACCP cannot be used to cool the moderator. The plants are connected only to their individual loads. The cryoplants do share a common helium gas recovery, purification, and storage system. Pure helium gas is

Figure 2.3. View of the CDS in the ESS tunnel prior to installation of the cryomodules (courtesy of ESS.)

stored at 20 bar in the pure helium storage (PHS), which consists of 19 70 m^3 gas tanks (figure 2.4). Recovered helium, which includes helium gas recovered from the instruments via a dedicated recovery line, is collected first in a gas bag and then compressed to 200 bar for storage in high-pressure cylinders as dirty helium. This helium is then cleaned via the 80 K adsorber in the TICP or in some cases via the external purifier and then stored in the PHS.

All the PHS tanks may be connected together. However, more typically, we assign 14 tanks to the ACCP, one tank to the TICP, and four tanks to the TMCP. Helium is a limited resource and as part of our sustainability efforts we have enough storage space to store the entire ESS helium inventory as 300 K gas at various pressure levels, both in the storage system and in the linear accelerator (LINAC) and distribution system. The total helium inventory at ESS is typically 3.5 tonnes. Due to leaks and operating errors, we expect to have to replace 30% of the helium inventory annually during the first 5 years of operation. After that the expectation is that we will only have to replace 25% or less of the inventory each year. These goals are in line with best-in-class facilities, such as CERN.

2.2 Conceptual design and early decisions

2.2.1 Introduction

The earliest conceptual designs of the ESS cryogenics system were discussed in an internal technical note written in 2010. This was expanded in the 2012 ESS Conceptual Design Report [3] and presented with significant detail in the 2013

Figure 2.4. The ESS helium compressor facility with the PHS storage tanks (19 in total) (courtesy of W Hees and P Arnold ESS).

ESS Technical Design Report [4]. Reference [2] contains the final design of the system. As will be seen, with the exception of important changes in some of the cryoplants' capacities, the system built is very close to that described in the Technical Design Report.

Simultaneous to the early conceptual designs (2010–13) were discussions between ESS and member states on which systems would be suitable for in-kind contributions. From the beginning, it was known that the SRF cavities and cryomodules would be in-kind contributions. The cryoplants were identified early as purchases from commercial venders and did not excite any real interest from the member states as an in-kind contribution. However, the unique CDS was selected by Poland as an in-kind contribution. Originally, the entire CDS was to be provided by Poland but as the design of the spoke cavity cryomodules developed (chapter 9) the JT valve and JT heat exchanger associated with the spoke cavity cryomodules was moved from the cryomodule to the CDS valve box adjacent to it. This change resulted in much closer coupling in both operations and schedule of the spoke cavity cryomodules and the section of the CDS connected to the spoke cavity cryomodules. Thus, it was decided that the French Laboratory (Laboratoire de Physique des 2 Infinis Irène Joliot-Curie) providing the spoke cavity cryomodules would also provide the spoke portion of the CDS as an in-kind contribution. The remainder of the distribution system, including the Cryogenic Transfer Line linking the ACCP with the tunnel was provided as an in-kind contribution by the Wroclaw University of Science and Technology in Poland.

The original ESS project schedule had a milestone of 'Ready for Beam on Target' in 2019. In order to accomplish this goal, we would have to complete the ESS cryogenic system before then. Given the complexity of developing new SRF cavities and cryomodules, and that it takes between 4 and 5 years to design, order, construct, install, and commission a large-scale helium cryoplant, development of the cryogenics system was given significant priority in funding and staffing. The schedule influenced a number of design choices and also meant that the cryogenics system would frequently face challenges as being the first part of ESS to require items such as energization of high-power equipment, shift work, and oxygen deficiency hazard polices. In a number of cases, the requirements of the ESS cryogenics system resulted in laboratory-wide solutions and policies. The ESS cryogenics team played a significant role in both raising these questions and finding solutions to them.

2.2.2 Number of cryoplants

An early choice involved the number of cryoplants required to meet all the requirements: accelerator, target, and instruments. Separating the large 2 K cryoplant needed for the accelerator from the large 16 K cryoplant needed for the target moderator was an easy choice. Such a separation allowed us to procure cryoplants optimized to their respective operating temperatures and the fact that we would procure multiple plants potentially increased commercial competition. Having two cryoplants also gave us important flexibility in operating the accelerator cryogenic system independent of the target cryogenic system.

Deciding on a separate cryoplant for the cryomodule test stand and the production of helium for instruments was a less clear choice. Both the test stand and instrument needs could be met by increasing the capacity of the ACCP. In fact, a key design requirement of the cryomodule test stand was that it should operate at the same temperatures and pressures as the accelerator itself. This permits an easier interpretation of the test stand results (see chapter 10). The choice to use a separate cryoplant was driven by schedule and availability concerns.

Since each cryomodule has to be fully tested prior to tunnel installation, the cryomodule test stand had to be operational long before the start of accelerator operations. Given the size and complexity of the ACCP, we felt that it was better for the schedule to procure and commission a smaller separate TICP. This proved to be true, the TICP was the first cryoplant to be fully commissioned at ESS and was available when cryomodule testing started. The ACCP was not yet fully commissioned at that time.

The other deciding factor for a separate TICP was the availability of the accelerator. If the ACCP shutdown suddenly, then it would result in at least 2 h of lost accelerator operation while the ACCP was restarted and steady state operation resumed. This would in turn affect the overall ESS availability and a high availability is one of the top requirements for ESS. Both cryomodule testing and transferring of liquid helium to portable dewars are potentially disruptive activities. Separating these activities from the ACCP was seen as one way to reduce the likelihood of the ACCP shutting down and increase availability.

Availability was also a motivation behind not connecting the three cryoplants together in such a way that they could share cooling duties. While at first glance such an approach would appear to provide redundancy, it was decided that the complexity in controls and operations needed for the cryoplants to be interconnected in this way would make it more likely for an unplanned shutdown to occur, affecting availability. Instead, redundancy was added to individual cryoplants by building in a reasonable excess capacity. This meant that a cryoplant might be able to operate at reduced performance and still meet its cooling requirements.

2.2.3 Location of the cryoplants

Once the number of cryoplants was decided, the next choice to be made was where to locate them. Since ESS was a green field site, there was a significant amount of flexibility in this decision. It was decided to locate all three cryoplants together. This would allow them to share common services and permit easier operation, inspection, and maintenance. It was also decided to locate the helium compressors of all three cryoplants in a separate building from that of the three cold boxes. This choice, consistent with that of a number of other large cryogenic facilities, separated the high noise of the compressors from the cold boxes and dewars of the cryoplants. The compressor facility (building G04) also contained the external purifier and the helium recovery system (gas bag, recovery compressor, and high-pressure storage). The medium pressure (20 bar) pure helium storage tanks are located adjacent to the compressor facility. In order to reduce the possible influence of vibration and noise on the accelerator operation, the compressor facility was located across the street from the accelerator tunnel.

The three cold boxes and their associated dewars were located in the cold box hall, which is adjacent to the klystron gallery that runs parallel and next to the accelerator tunnel. The cold box hall is located just at the end of the superconducting portion of the ESS accelerator. This location in the klystron gallery just after the end of the superconducting portion of the accelerator minimizes the distance between the ACCP and the accelerator, as well as between the TICP and the cryomodule test stand. Underground helium gas lines connect the compressors in the compressor facility with the cold boxes in the cold box hall. Figure 2.5 shows the relationship of these buildings.

This arrangement does mean that the target moderator cryoplant is a significant distance (roughly 335 m) from the CMS located in the target building. The long distance and significant mass of helium in the transfer line connecting the TMCP and CMS leads to some unique control challenges, as described in chapter 5. An alternative solution, which was considered, would have been to locate the TMCP cryoplant in or much closer to the target building. This idea was rejected because it was felt that the advantages of placing the cryoplants in one location outweighed the advantages of a shorter transfer line. Circumstances resulted in this being a fortuitous decision for another reason. The target building construction was greatly delayed. Given the significant time required for installation and commissioning of large cryoplants, had the TMCP been located in or near the target building, it is very

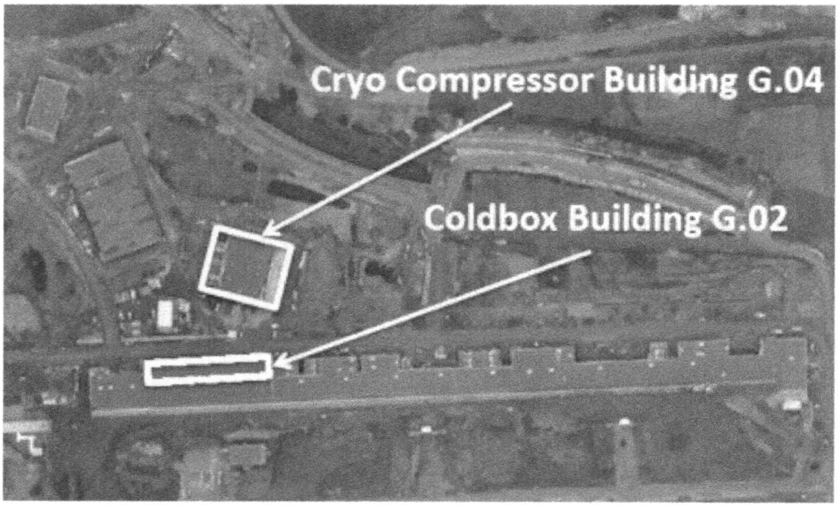

Figure 2.5. Aerial view of ESS site showing the relative location of the compressor building and the cold box building. The building below the cold box building is the klystron gallery. immediately below the klystron gallery is the earthen berm covering the accelerator tunnel (courtesy of ESS).

unlikely that the cryoplant would have been ready in time to support scheduled beam operations. Instead, the TMCP was commissioned in 2019, well ahead of beam operations.

2.2.4 Building sizes

In order to keep to the project schedule, we had to supply the required sizes of the compressor facility and cold box hall to our conventional facilities team before we knew the sizes of the actual equipment to be installed in these spaces. We did, however, know from experience the approximate space required and based on that created conservative requirements. The size of the compressor facility (figure 2.6) was based on buildings at CERN which held similar sized compressor systems. In the case of the cold box hall (figure 2.7), we selected the size based on the expected size of cold boxes, laying them out in simple solid model (figure 2.8) to check for spacing and then added some additional height and length to be conservative. As a secondary check, we did request the expected equipment sizes as part of an industry study that we commissioned for the ACCP (see chapter 3).

The basic parameters of the compressor building are:

- Access door: 5.0 m wide and 5.0 m high.
- Crane: 7.0 m height under hook, lifting capacity of 15 t.
- Maximum floor load: 10 000 kg m^{-2}.
- Space available: two halls, both 15 m wide, one 32 m long, the other 40 m long.

While the basic parameters of the Cold Box Hall are:

- Access door: 5.0 m wide and 5.0 m high.

Figure 2.6. Plan view of ESS compressor building (courtesy of ESS).

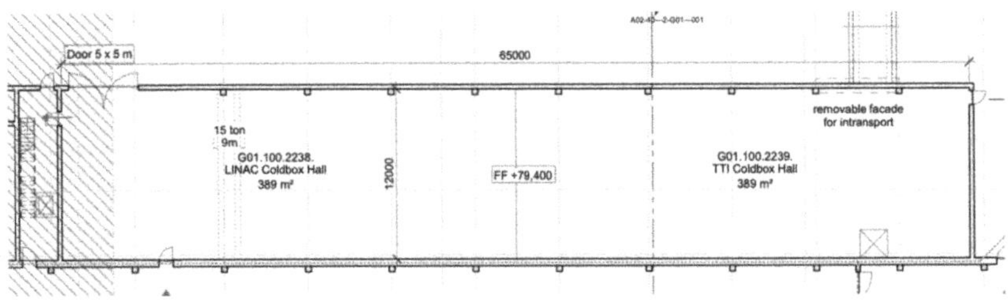

Figure 2.7(a). Plan view of ESS cold box hall (courtesy of ESS).

- Crane: 9.0 m height under hook, lift capacity 15 tons.
- Maximum floor load: 20 000 kg m^{-2}.
- Space available: 12 m width and 50 m length.

In addition, we defined maximum vibration and noise limits for the helium compressors and used these in the technical specification for the ACCP.

The maximum allowable vibration levels for the compressor station are as follows:

- At the top of the compressor skid foundations, less than 1.0 mm s^{-1} RMS (10–1000 Hz).
- On the motors, less than 4.5 mm s^{-1} RMS (10–1000 Hz).
- On the compressors, less than 7.5 mm s^{-1} RMS (10–1000 Hz).
- On piping or equipment around the compressors, less than 30 mm s^{-1} RMS (10–1000 Hz).

Figure 2.7(b). View of the interior of the cold box hall showing TMCP cold box (left-hand side) and ACCP cold box (right-hand side) (courtesy Ulrika Hammarlund/ESS).

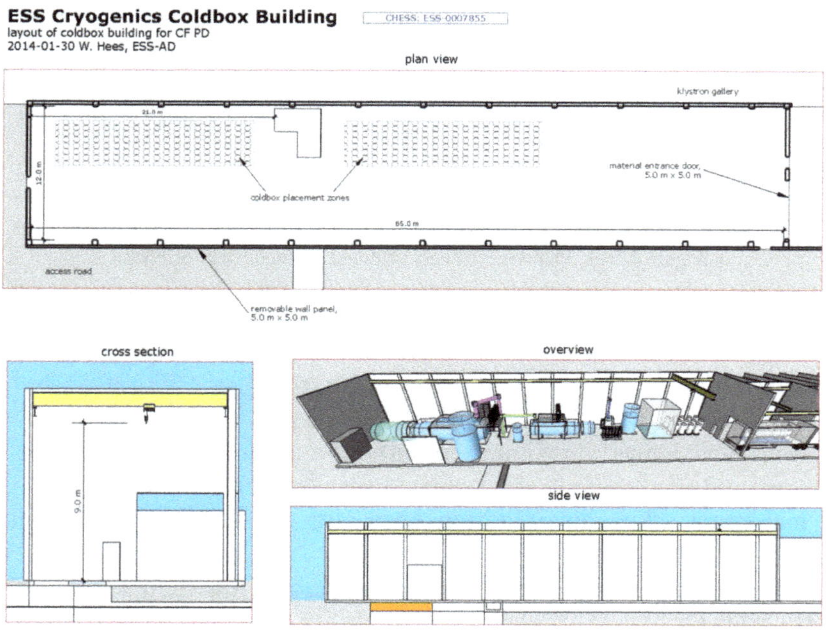

Figure 2.8. Preliminary sizing and layout of the cold box hall (Courtesy of ESS).

The noise produced by all helium compressors combined cannot be greater than 110 dB (A) anywhere in the compressor room.

The results of this approach were successful. The compressor building had sufficient space for the final systems and allows for the installation of a future backup compressor for the ACCP. However, the cold box hall turned out to be just barely wide enough, we were limited here by cost and site conditions, but is tall enough. Once we had the final sizes of the cold boxes from the vendors, it was clear that the cold box hall was longer than needed but by then detailed design of the building had progressed to the stage where reducing the length would actually cost more than proceeding with the original design. The extra space was put to good use during the project, serving as a klystron assembly area. Other uses for this space will be found during ESS operations.

Early in the project, it was decided to locate the main control room for the ESS in the target building. Given the distance from the target building to the cryogenic facilities, we designed a cryogenic control room and some associated office space adjacent to the cold box hall. The motivation here was to make commissioning of the cryoplants, which experience has shown to require frequent movement by staff between the control room and the cryoplants, easier. The commissioning experience at ESS showed this to be a good choice. Moreover, we also used this control room to operate the adjacent cryomodule test stand and, due to the delay in the target building construction, for early beam commissioning of the normal conducting portion of the LINAC. Without this cryogenic control room (now renamed the Local Control Room), significant additional project delays would have occurred.

With the completion of the target building, all ESS systems (including the cryogenic systems) are controlled from the main control room. The local control room is only used as a backup.

2.2.5 Plant capacities and safety margins

A typical challenge of cryogenic systems for large-scale scientific projects such as ESS is that the cryogenic plant needs to be specified and ordered before all the cryogenic heat loads are fixed. If the cooling capacity of the procured cryoplant is too small, then the facility will not function but a cryoplant with too much capacity will be more expensive than needed and also may not be as efficient as an optimally sized plant. Waiting until all the cryogenic loads are completely known may not be an option because this approach may delay the project.

The solution to this problem is to explicitly define and keep track of safety factors applied to the plant capacity. Clarity is important here because one does not want to inadvertently apply safety factors to numbers that already have safety factors applied. This would result in too large a capacity requirement. Our application of safety factors differed between the three cryoplants and also varied with the stages of the project.

There were four separate occasions in which we estimated the required cooling capacity of the ESS cryoplants, which were a technical note produced in 2011, the Conceptual Design Report (CDR) [1] produced in 2012, the Technical Design

Report (TDR) produced 2013 [2], and the actual technical specification with which we procured each plant. Both the technical note and the CDR used top-down estimates of cooling requirements scaled from similar systems. There was significant uncertainly in these estimates because we did not have fully detailed designs for the components being cooled. In both these cases, we applied a safety factor of 1.8 for the accelerator and target cryoplant capacities. This number was chosen based on experience with other large cryoplants where safety factors of near 2 were used in early project stages. In the CDR, the maximum expected requirement for liquid helium to be used in the neutron instruments was 7500 l/month. This was based on the experience of similar facilities and because this number was given as a maximum requirement, i.e., it already contained a safety factor, no additional safety factor was applied.

The cryoplant cooling capacity numbers in both the TDR and the cryoplant specifications were based on detailed bottom-up calculations of the various heat loads at each temperature level of interest. This was possible because the design of components such as the CDS, cryomodules, and hydrogen moderator were much more advanced. As a result, the overall accuracy of the heat load estimates was judged to be much better than in the earlier cases. Here again, different safety factors were used for different cryoplants.

In the case of the ACCP, the capacity was calculated by the following formula:

$$C = F_o(F_{ud}Q_d + Q_b + F_{us}Q_s) \tag{2.1}$$

where:

- C is the total capacity of the plant at a given temperature.
- F_o is the operational safety factor. This value, set to 1.15, gives us some margin within which to control the plant, as well as providing some margin for suboptimal plant performance.
- F_{ud} is the safety factor for the dynamic heat loads associated with the SRF cavities. In the case of the ACCP, this value is set to 1.0 as the SRF cavity head loads are calculated assuming a fairly poor performing cavity. That is, the safety factor is already built into the estimated cavity dynamic heat load.
- Q_d is the SRF cavity dynamic heat load estimate.
- Q_b is the heat load caused by beam losses in the accelerator. This is fixed at 1 $\mathrm{W\,m^{-1}}$ in order to limit activation of materials in the accelerator by the beam. The accelerator is not allowed to operate at higher beam losses, and thus no additional safety factor is required.
- F_{us} is the safety factor associated with static heat loads. This is set at 1.5. Note that this safety factor is less than the 1.8 applied earlier in the project. This difference reflects the greater confidence in the bottom-up heat load calculation based on a more mature design.
- Q_s is the static heat load estimate.

The capacities at the cryoplants at each project stage are presented in section 2.3.

2.2.6 Procurement of the cryoplants

Another set of early decisions in the project was the manner by which we would procure the cryoplants. We decided that we would create detailed statements of work and technical specifications for each cryoplant and then procure them via competitive bid from industry. We did not require that all the cryoplants came from a single vendor. This was done to increase ESS's competitive advantage and in recognition that ESS is funded by multiple member states, all of whom should have the ability to bid on procurements. Our statements of work did require that any successful bidder have previous experience with similar size cryoplants. In the end, both of the two largest suppliers of large helium cryoplants (Air Liquide and Linde) won contracts to provide cryoplants. Linde provided the ACCP and TMCP, while Air Liquide provided the TICP along with the external purifier and the helium recovery system.

We did standardize certain aspects of the cryogenics system. The controls systems of all three plants were required to use the Experimental Physics and Industrial Control System (EPICS) to connect to the human–machine interface (HMI) and to connect the HMI to the broader ESS control system. In addition, ESS specified that all programmable logic controllers had to come from the Siemens S7-300, and/or S7-400, and/or S7-1500 families. This requirement extended through the majority of the control systems in the ESS project.

The procurement of the ACCP cryoplant was the second biggest procurement at that time in the ESS project, with only the ESS civil construction contract being bigger. Thus, again, the cryogenics system was at the forefront of helping to develop processes that later became ESS wide. The ESS procurement department and cryogenics section worked closely together and all three cryoplant procurements were accomplished successfully on schedule and within budget.

2.3 Evolution of cryogenic system design

Between the publication of the ESS TDR in 2013 and the start of the procurements of the cryogenic plants in 2014, two developments resulted in significant changes in the cryogenic system design.

The first of these was driven by cost. Simply put, the cost of the accelerator described in the TDR was too large to fit within the overall project cost of 1843 million Euro. During 2013, the accelerator was redesigned to reduce its cost. It was realized that a major cost driver was the expensive radio-frequency (RF) power sources and their associated SRF cryomodules. In order to reduce cost, the proton beam energy was reduced from 2.5 to 2.0 GeV, while the beam current was raised from 59 to 62.5 mA, allowing the average beam power to remain at 5 MW. Reducing the beam energy meant that the number of cryomodules and their RF power supplies could be reduced. The number of spoke cavity cryomodules went from 14 to 13, the number of medium beta cryomodules went from 15 to 9, and the number of high beta cryomodules went from 30 to 21. Removing the cryomodules meant less cooling was required by the ACCP, and thus its capacity could be reduced, further saving costs.

However, the new design meant that parts of the cryomodules, particularly the RF power couplers, were operating very near their maximum design capabilities. This added technical risk to the project in that underperforming cryomodules could result in the accelerator not meeting its specifications of 2 GeV and 5 MW. In order to reduce this risk, space was kept in the accelerator design to add an additional 14 high beta cryomodules, known as contingency cryomodules. These cryomodules could be added later in the event that the original cryomodules did not meet their specifications so that the final 2 GeV/5 MW beam parameters would still be met. It was judged that expanding the capacity of the ACCP later to cool these additional cryomodules was both difficult and expensive, so the ACCP was designed from the start to cool both the nominal set of cryomodules or the nominal set plus the 14 additional contingency cryomodules. This also meant that the ACCP would have two very different operating capacities at 2 K. The solution to this problem is discussed in chapter 3. The CDS was also designed so that it could be easily expanded to serve the contingency cryomodules and the Cryogenic Transfer Line that connects the ACCP with the tunnel was sized to allow the higher flowrates needed for the additional cryomodules. Details of the CDS are given in chapter 4.

The second change to the ESS cryogenic system was driven by innovation. A new design for the hydrogen moderator was developed that resulted in the moderator being significantly more efficient [3] at cooling the spallation neutrons down to the energies needed by the neutron scientists. This meant that for the same accelerator and target parameters, the neutron source was now three to five times brighter than at the time of the TDR. The more efficient moderator also meant that more heat would be deposited into the hydrogen and in turn removed by the TMCP. This led to a significant increase in the required TMCP cooling capacity. Fortunately, this change occurred just before the procurement of the TMCP and we were able to adjust the technical specifications of the cryoplant prior to releasing the request for bids.

Table 2.1 shows the evolution of the required cooling capacity of the ESS cryoplants over time. Note that the earliest estimates were scaled from existing plants and only discussed capacity in terms of 4.5 K equivalent cooling. Once more detailed cryomodule and distribution designs were available, we were able to better estimate cooling needs at specific temperatures.

Additional details on the evolution of the ESS cryogenics system can be found in reference [5].

2.4 Staffing and recruitment

The recruitment and development of a highly skilled cryogenics team was an early priority. In August of 2012, there were only three cryogenics engineers at ESS, one of whom departed for CERN shortly afterwards. We expended considerable effort in recruiting the people that ESS would need to design, procure, install, commission, and operate the ESS cryogenic system. The vision was to have a dedicated project engineer for each of the cryoplants and one for the CDS. Later, a much-needed additional cryogenic engineer was added to work on all the interfaces between the

Table 2.1. Evolution of ESS cryogenic plant capacities.

Document and year	ACCP	Target cryoplant	TICP	References
Technical Note 2011	13.4 kW @ 4.5 K equivalent	11 kW @ 4.5 K equivalent	31 l h^{-1}	
Conceptual Design Report 2012	14 kW @ 4.5 K equivalent	None given	45 l h^{-1}	3
Technical Design Report 2013	1.7 kW @ 2 K + 1.2 kW @ 5–8 K + 7.9 kW @ 40 K + 7.2 g s^{-1} liquefaction	25 kW @ 16 K	50 l h^{-1}	4
Technical specification	2.23 kW @ 2 K + 0.83 kW @ 2–4 K + 11.4 kW @ 33–53 K + 9.0 g s^{-1} liquefaction (2014)	30.3 kW @ 15 K (2015)	76 W @ 2 K 387 W @ 40–50 K 0.2 g s^{-1} liquefaction (2015)[1]	

[1] These parameters for the TICP refer to its operation as a refrigerator for cryomodule testing. When operated in pure liquefaction mode, a plant with these parameters will exceed the 7500 l/month requirement of ESS.

cryogenic system components and to provide support as needed to the various cryogenic projects. Additional design support was provided by experienced consultants.

From the start, we knew that we wanted to have cryogenic operators and technicians in place once equipment started to arrive on site. This was driven by the desire to have ESS technicians heavily involved in the installation and commissioning of the systems that they would eventually operate and maintain. Thus, in 2014, we started to recruit technicians and operators to the cryogenics team. These were some of the first technicians hired by ESS and some of the first to perform on-call and shift work.

We were able to successfully attract a group of talented and experienced engineers and technicians from both industry and academia, and the strength of these people is a major reason for the success of the ESS cryogenic system. Their efforts, were of course, augmented by the contributions of our in-kind partner laboratories in Poland, France, the United Kingdom, and Sweden, which had their own talented staff members.

An early and important decision in the development of the cryogenics team was that one team would be responsible for almost all cryogenic engineering throughout ESS. This differs from other institutions in which there are separate cryogenic groups for accelerator, instruments, and target systems. We specifically wanted a single team so that common approaches could be used, cross training could be carried out, and an overall common purpose could be developed.

Note that design, procurement, testing, installation, and commissioning of the cryomodules is the responsibility of the separate group of people. They are in the SRF section, which has its own talented staff and in-kind partners. The cryogenics and SRF teams work very closely together. Many of the authors of this book work in one of these teams.

2.5 Cryogenic safety at ESS

Attention to safety issues was paramount from the start of the cryogenic system design. ESS is required to meet all European and Swedish safety regulations. All of the pressure systems at ESS are built in compliance with the European Pressure Equipment Directive (PED) and applicable Swedish regulations. Electrical equipment and vacuum systems are built consistent with Swedish regulations.

Safety was a topic at all equipment reviews and dedicated HAZOP reviews were held for each of the cryoplants. Testing, sizing and number of pressure relief valves, and inspections were carried out consistent with the PED. When required, third party inspection and testing was observed by certified firms such as TÜV Nord and Dekra.

One challenge was the parallel installation of some cryogenic equipment while other equipment nearby was under commissioning or operation. This led to the real possibility of the inadvertent opening of a pressurized system by installation workers. In order to prevent this, ESS developed a system of documentation, work permits, and Lock Out/Tag Out [6].

Oxygen Deficiency Hazards (ODHs) arose early at ESS with the first commissioning of the cryoplants. The ESS cryogenics group drove the need for ESS to develop site-wide policies regarding ODH. These policies [7] were based on similar approaches taken at CERN and the SLAC National Accelerator Laboratory. An analysis of the tunnel and accelerator buildings [8] indicated that these areas were ODH Class 0, and thus required no additional technical mitigations. However, given the uncertainties in the analysis and the quantity of cryogens in use, a management decision was made to require installation of an extensive ODH detection and alarm system in all potential ODH areas in the accelerator complex.

More sophisticated computational fluid dynamic studies indicated that in the worst-case scenario (i.e., failure of the beam tube vacuum inside the cryomodules) the amount of helium released into the tunnel near the failure would result in a significant oxygen deficiency hazard. Thus, a helium collector system [9] was designed and installed so that even in this worst-case event all of the released helium would vent outside the tunnel.

2.6 Best practices and lessons learned

- People are the key to success. One should always start early to hire the best people possible and involve them in the design, procurement, and construction of the cryogenic system. While much can be done with consultants,

having a strong in-house capability is vital. It is best to bring in technicians and operators into the project early so that they can participate in the construction and commissioning, and truly learn all the details of the equipment.
- Good communications with vendors and in-kind partners are critical. Regular meetings, formal design reviews (with external experts), and factory visits are all best practices.
- Large helium cryoplants typically require between 3 and 5 years to specify, design, build, install, and commission. Time must be allowed for this in the schedule and work should be started early.
- Be very explicit about, and document the use of, safety factors on cryoplant capacities to avoid having too much or too little capacity.
- Safety systems should be included in the design from the beginning of the project. Be sure to consider the safety issues of integrated systems, not just isolated components.
- Work closely and early with the civil construction team to ensure that building sizes and utilities will meet the needs of the cryogenic system.
- Be careful not to underestimate the amount of work required for the engineering, construction, and commissioning of ancillary systems such as warm gas or water cooling.

References

[1] Weisend J G, Arnold P, Fydrych J, Hees W, Jurns J M and Wang X L 2015 Cryogenics at the European Spallation Source *Phys. Proc.* **67** 27–34

[2] Garoby R *et al* 2018 The European Spallation Source design *Phys. Scr.* **93** 1

[3] Peggs S 2012 *Conceptual Design Report* ESS-2012-001

[4] Peggs S 2013 *ESS Technical Design Report* ESS-2013-001

[5] Hees W, Arnold P, Fydrych J, Jurns J, Wang X L and Weisend J G 2015 The evolution of the cryogenic system of the European Spallation Source *IOP Conf. Ser.: Mater. Sci. Eng.* **101** 012073

[6] Arnold P, Gous H, Phan D, Xiaotao S and Weisend J G 2019 Challenges of parallel ESS cryoplants installation and commissioning activities *IOP Conf. Ser.: Mater. Sci. Eng.* **502** 012108

[7] 2016 *ESS Guideline for Oxygen Deficiency Hazards* US Particle Accelerator School, Report No. ESS-0038692

[8] 2019 *ODH Assessment of Accelerator Buildings* US Particle Accelerator School, Report No. ESS-0063324

[9] Fydrych J, Lundh E, Moberg J, Phan D, Tereszkowski P and Weisend J G 2019 The safety helium collectors for the ESS superconducting linac: functional specification and detailed design *IOP Conf. Ser.: Mater. Sci. Eng.* **502** 012132

Chapter 3

The Accelerator Cryoplant

X L Wang and J Q Zhang

The accelerator cryoplant (ACCP) is the largest and most complex of the ESS cryoplants. This chapter reviews the conceptual design and procurement process of the ACCP, and gives details on the final plant design, construction installation and commissioning. Topics such as laboratory visits, industry studies, bid evaluation, control systems, and the importance of vendor communication are also covered.

3.1 Early work

The ESS Technical Design Report [1, 2] was published in early-2013. By that time, we knew that we would have three separate cryoplants and knew at least nominally what the temperature levels and cooling capacities of these plants would be (see chapter 2). We also knew that we would have to place the order for the ACCP in 2014 in order to ensure that it was commissioned consistent with the project's schedule.

At the beginning of 2013, the ESS cryogenics team consisted of W Hees, X L Wang and J Weisend. X L Wang was appointed the Project Engineer for the ACCP and we were soon joined by P Arnold, who was made Work Package Leader for Cryogenics, responsible for the scope, cost, and schedule of almost all ESS cryogenic systems. We also had available to us as consultants G Gistau (formerly of Air Liquide) and H Quack (formerly of Linde and the Technical University of Dresden).

We needed to fully understand the requirements and interfaces of the ACCP, and how best to meet these requirements. From this, we could develop a conceptual design and a detailed technical specification. This early work of information gathering and conceptual design took most of 2013 and was vital to having a successful outcome. These efforts included laboratory visits, discussions with users, definitions of interfaces and operating modes, and industry studies.

doi:10.1088/978-0-7503-3223-1ch3

3.1.1 Laboratory visits

Large helium cryoplants such as ESS's ACCP generally belong to the big science facility community, which supports and encourages free and open exchange of scientific and technical knowledge, expertise, engineering designs, experiences, and lessons learned.

We took any possible opportunities to visit similar projects at various laboratories via conferences (International Cryogenic Engineering Conference, Cryogenic Engineering Conference), workshops (Cryogenic Operations, TESLA Technology Collaboration), and contacting previous colleagues and experts directly. These included LHC (CERN); European XFEL; the Accelerator Module Test Facility at DESY; the Cryomodule Test Facility at Fermilab; the Spallation Neutron Source (SNS) at ORNL; Continuous Electron Beam Accelerator Facility (CEBAF) I and II at Jlab; the W-7X Stelarator at IPP, North Germany; FAIR at GSI; the KSTAR Tokamak at NFRI, South Korea; and the Institute of High Energy Physics (IHEP) in Beijing, China. A few pictures and a summary of our laboratory visits are shown in table 3.1 and figure 3.1. Each visit started with a short meeting in person where the

Table 3.1. Laboratory visiting summary. (Courtesy of ESS)

Name	Machine type	Laboratory	Performance	Comments
LHC	Accelerator	CERN	2.4 kW @ 1.8 K 0.3 kW @ 4.5 K 33 kW @ 50–75 K	8 plants. 18 kW @ 4.5 K Eqiv. 3CC+SP compressor (ALAT) 4CC+SP compressor (LKT/IHI)
XFEL	Accelerator	DESY	2.4 kW @ 2 K 4.0 kW @ 5–8 K 30 kW @ 40–80 K	Two HERA Plants: 12 kW @ 4.5 K Eqiv. Upgrade to: 8 kW @ 4.5 K Eqiv. Each 4CC to LP
AMTF	Accelerator	DESY	0.8 kW @ 2 K 0.5 kW @ 5–8k 3 kw @ 40–80 K	Vacuum pump
CMTF	Accelerator	FERMI	0.5 kW @ 2 K 0.6 kW @ 5–8k 5 kW @ 40–80 K	3CC + SP compressor/pumps
SNS	Accelerator	ORNL	2.4 kW @ 2.1 K 15 g s^{-1}@ 4.5 k liq. 8.3 kW @ 35 K	4CCs to LP

CEBAF II	Accelerator	JLAB	4.6 kW @ 2.1 K 10 g s^{-1} @ 4.5 k liq. 8.3 kW @ 35 K	Two plants 5CC to LP
W-7X	Stellarator	IPP	3 kW @ 3.4 K 25 g s^{-1} @ 4.5 k liq. 14 kW @ 80 K	Two cold circulators
FAIR	Aocderator	GSI	35 kW @ 4.5 K 33 g s^{-1} @ 4.5 K liq. 24 kW @ 50 K	40 kW @ 4.5 k Eqiv. Magnets
KSTAR	Tokamak	NFRI	5.3 kW @ 4.3 K 17 g s^{-1} @ 4.5 k liq. 270 g s^{-1} @ 50–70 K	9 kW @ 4.5 K Eqiv. Two cold circulators Magnets

project and the facility overview, the design consideration, technical issues, challenges, experiences, and lessons learned were introduced by our host colleagues. Following a site visit, our staff had straightforward impression of what worked and, possibly more importantly, what did not work in each of these projects. Our discussions and communications continued beyond these visits during the whole ACCP project period. We received a lot of great help, very useful advice and comments, and friendship from cryogenic experts from all over of the world. Some of these people later served on design reviews for the ACCP and other ESS cryogenic installations.

There is another reason why we think that laboratory visits are necessary and essential. it is extremely beneficial, not only for the cryogenic staff but also for our new and fresh colleagues from other departments in ESS such as controls, conventional facilities, radiation licensing, quality control, safety, etc (as ESS is a new institution) to get early and direct impressions about the cryogenic system and facilities. We would like to express our deep thanks to all our host colleagues and laboratories.

3.1.2 User discussions

The users of the ACCP are the ESS accelerator and associated cryomodules (CMs). We held repeated, detailed discussions with the people responsible for these components during the pre-study phase. The ACCP will serve the superconducting section of the accelerator, which consists of 43 CMs and cryogenic distribution system (CDS) [2] with a length of about 300 m at optimum design scenario. There is a possibility to install 14 additional CMs in the future as a design contingency. The bulk of the acceleration is carried out by three types of superconducting radio-frequency (SRF) cavities. There are a 13 double spoke cavity CMs, nine medium

(a)

(b)

(c)

Figure 3.1. (a) KSTAR/NFRI; (b) W-7X/IPP, Germany; (c) LHC/CERN, Switzerland/France; (d) CEBAF, Jlab/USA; (e) European XFEL, Germany; and (f) IHEP, China. (Courtesy of ESS)

(d)

(e)

(f)

Figure 3.1. (Continued.)

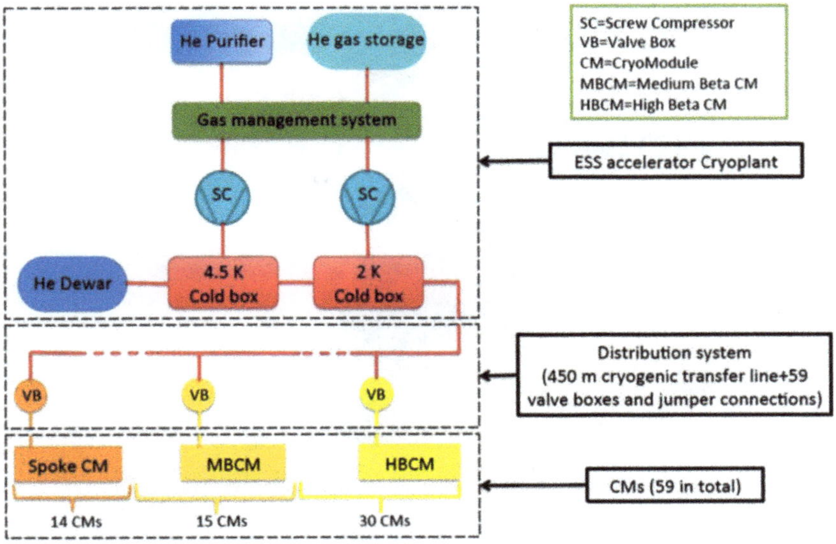

Figure 3.2. Block diagram of ESS accelerator cryogenic system (2013 version). Reprinted from [3]. Copyright © 2014 ASHRAE, reprinted by permission of Taylor & Francis Ltd, hWp://www.tandfonline.com on behalf of 2014 ASHRAE.

beta elliptical CMs, and 21 (plus 14) high beta elliptical CMs. All the SRF cavities operate in saturated 2 K He II baths. Each of the CMs also contains a 40 K thermal shield (TS) and requires helium cooling of the radio-frequency (RF) power couplers. The power coupler cooling returns the helium at ambient temperature, and thus is a liquefaction load. There are no superconducting magnets in the ESS SRF accelerator. The CMs will be connected with the ACCP by a CDS, which will be mainly composed of a cryogenic distribution line with 43 plus 14 valve boxes. The ACCP is used to provide the cooling for all circuits. The ACCP will provide the cooling capacities at three temperature levels: 2 K bath cooling for cavities, 4.5 K forced helium flow for the power couplers, and 40–50 K helium flow for the TS. This is summarized in figures 3.2 and 3.3.

Unlike other big science projects, the ESS cryomodule (see also chapters 8 and 9) had been under development for a long time and the CM and Cryogenics groups are in the same division. This resulted in a general common understanding about the cryogenic system and the cryomodule. ESS CMs are based on an in-kind contribution. Since our partner and ESS staff had various background and experiences, it is very important to have decent discussions at the beginning to bring them onto the same page.

3.1.3 Interfaces

Together with our in-kind partners, we carefully clarified the process and physical interfaces as shown in the below figures. The ACCP is connected to the CMs by the CDS (see also chapter 4), which consists of cryogenic transfer line, the related warm

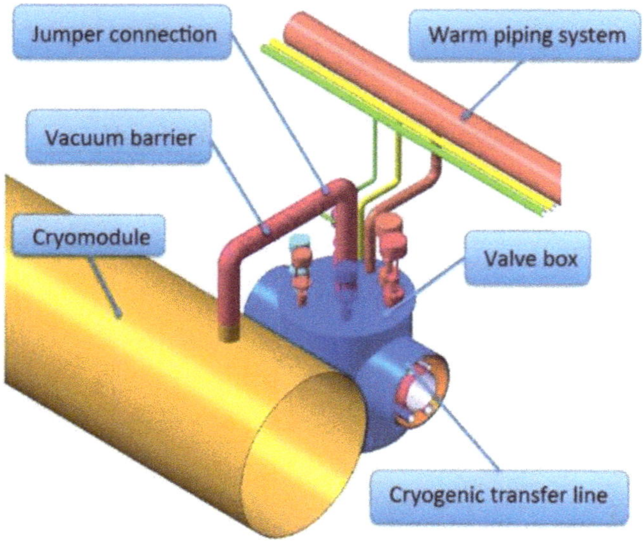

Figure 3.3. Draft physical interface between the CM and ESS cryogenic system (2013 version). Reprinted from [3]. Copyright © 2014 ASHRAE, reprinted by permission of Taylor & Francis Ltd, hWp://www.tandfonline.com on behalf of 2014 ASHRAE.

piping system, and valves boxes. The ACCP and the CDS are in the scope of ESS. CMs are provided by our in-kind partners.

Each of the CMs requires three levels of cryogenic process cooling. The CM Piping and Instrumentation diagram (P&ID) is given in figures 3.4 and 3.5. The primary circuit (3 bar, 5 K) uses a heat exchanger to precool the forward flow, followed by a Joule-Thomson (J-T) valve to expand to saturation pressure at 2 K within the CM. The second circuit supplies a forced flow of 4.5 K helium to cool the RF power couplers. The helium warms up from 4.5 K to ambient temperature toward the warm end of the power coupler and returns to the cryoplant via a warm gas return line. The appropriate mass flow is regulated at ambient temperature with a flow controller in the return connection at each power coupler. The distribution system will deliver the helium to the CM for coupler cooling at 3 bar. The coupler design calls for an operating pressure of 1.4 bar. That reduction in pressure must be accomplished inside the CM. The third circuit provides helium at 19.5 bar and 40 K to cool the thermal radiation screens. It will return to the ACCP at 19 bar and 50 K. Each CM will be connected to the cryogenic transfer line via a valve box and jumper connection in U-shape design. The draft design of the valve box and jumper connection is shown in figure 3.3. A vacuum barrier in the jumper connection separates the vacuum of the transfer line from the individual insulation vacuum of the CM. The system permits cool down and warm up of the individual CM while keeping the remaining CMs at operating temperature. Major subsystems in the cryoplant include gaseous helium storage vessels, helium purification, helium compressors, cold boxes, liquid helium dewar, and cryogenic transfer line.

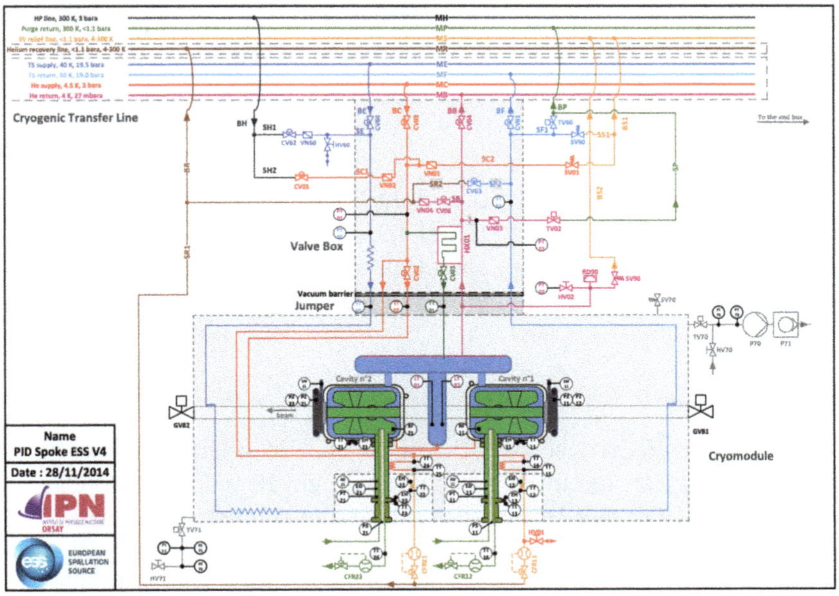

Figure 3.4. Spoke CM P&ID. Courtesy of ESS, IPN Orsay

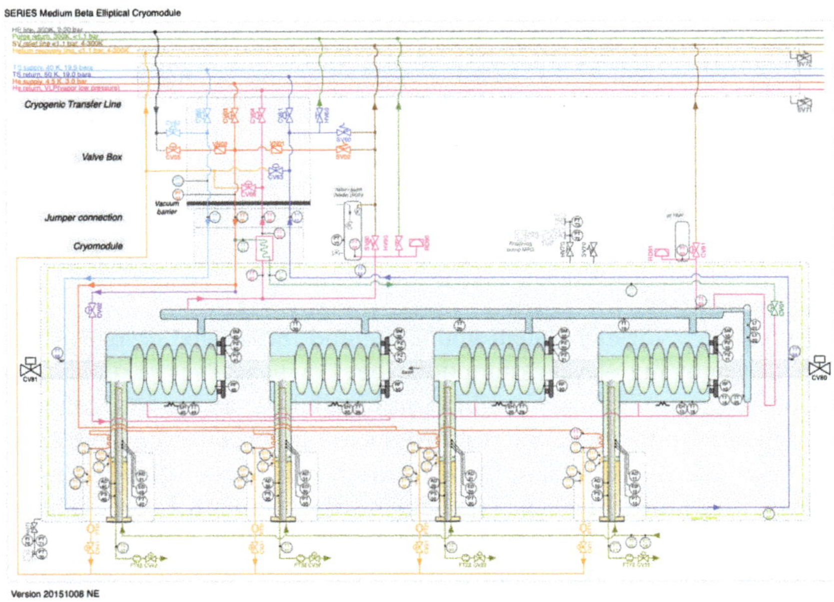

Figure 3.5. Elliptical CM P&ID. Courtesy of ESS, CEA

The process and physical dimension clarification become important factors for the ACCP process and economic optimization and tunnel size and the conventional facility design.

3.1.4 CMs and CDS database

Based on the database from the CMs and the CDS, we collected the cold mass and the 2 K volume as shown in tables 3.2 and 3.3. This information has a big impact on the ACCP process optimization, operation modes, cool-down time estimation, cryogenic safety (helium release in the case of worst scenario), and helium inventory, etc.

3.1.5 Project staging

Recall from chapter 2 that the ACCP will be built to operate in two stages; with Stage 2 containing up to an additional 14 High Beta Elliptical Cavity CMs. Accordingly, the ACCP will be designed and manufactured to ultimately fulfill the requirements at Stage 2. However, to optimize the plant operation and efficiency in the first years of operation, the respective Stage 1 configuration will be fulfilled as well.

3.1.6 Operation modes

An important step in detailing the design of the cryogenic system is to describe the expected cryogenic operating modes. This step allows us to ensure that the correct number of connections between the CMs and the distribution system are made, and allows us to determine the number and type of valves and instrumentation. It is also important in determining the number and type of connections between the distribution system and the ACCP and in developing the technical specification of the ACCP.

The high-level cryogenic operating modes for the ESS accelerator are described below. These modes are distinct from the accelerator operation modes, which are related to the beam status.

Operating assumptions

The design of the cryogenic system and CMs, and the desired cryogenic operating modes are driven by high-level assumptions that are themselves linked to the project goals, such as reliability, availability, sustainability, and cost reduction. The assumptions are also linked to the planned operating schedule for the ESS in which there is expected to be a contiguous two-month down time each year for maintenance with little to no scheduled down times during the rest of the year.

Key assumptions that affect the accelerator cryogenic system are:

1. Each cryomodule shall be able to be individually cooled down and warmed up without affecting the temperatures of the other CMs or the distribution system.
2. The isolation vacuum space of each cryomodule shall be able to be evacuated or raised to room pressure without affecting the isolation vacuum space of other CMs or of the distribution system.

Table 3.2. The whole LINAC cold mass in kg. (Courtesy of ESS)

The whole LINAC cold mass (kg)						
Stage 1: Optimum design without contingency						
Circuits	Materials	Spoke CMs	MB-CMs	HB-CMs	CDS	Total
		13	9	21		
2 K	SS	949	499.5	1166	10 262	12 876
	NB	1950	2160	4620		8730
	Ti	1040	1485	3507		6032
	Ni	936	972	2268		4176
40–50 K	AL	676	1080	2520		4276
	SS	117	12.6	29.4	3430	3589
	Cu				15 488	15 488
Stage 2: Optimum design with contingency						
Circuits	Materials	Spoke CMs	MB-CMs	HB-CMs	CDS	Total
		13	9	35		
2 K	SS	949	500	1943	13 332	16 723
	NB	1950	2160	7700		11 810
	Ti	1040	1485	5845		8370
	Ni	936	972	3780		5688
40–50 K	AL	676	1080	2520		4276
	SS	117	12.6	49	4503	4682
	Cu				20 369	20 369

Table 3.3. The whole LINAC 2 K volume in m^3. (Courtesy of ESS)

2 K Volume in the whole LINAC (m^3)						
Stage 1: Optimum design without contingency						
Circuits		Spoke CMs	MB-CMs	HB-CMs	CDS	Total
		13	9	21		
2 K	Liquid	2.2	2.0	4.7		8.9
	Gas	0.4394	0.36	0.84	15.7	17.3
Stage 2: Optimum design without contingency						
Circuits		Spoke CMs	MB-CMs	HB-CMs	CDS	Total
		13	9	35		
2 K	Liquid	2.2	2.0	7.9		12.1
	Gas	0.0338	0.04	0.04	20.1	20.2

3. High power RF operation of the SRF cavities will only occur at the nominal operating temperature of 2 K. There is no requirement for full power RF operation at 4.5 K.
4. The replacement of an operating cryomodule outside of the scheduled two-month maintenance period shall be a rare event (less than once every 5 years). This will be accomplished via proper design, prototyping, testing, and QA/QC procedures.
5. Cryomodule failures that occur at a frequency greater than once every 5 years are expected to be addressed by either fixing the cryomodule while in the tunnel (depending on the failure, this may be accomplished with the cryomodule at operating temperatures or at room temperature) or by leaving the cryomodule in place and taking it partly or completely out of service but not removing it until the next scheduled two-month maintenance period.
6. Assumptions 4 and 5 together mean that the replacement of a cryomodule in the tunnel while the distribution system is at cryogenic temperatures, while possible, is a rare event (occurring less than once every 5 years). However, sufficient piping and valving will be installed (see figures 3.4 and 3.5) to allow for such a removal.
7. Initial installation of the CMs during the construction of ESS will occur while the CDS is warm.
8. All ESS cryogenic equipment, including the CMs, is designed to allow repeated thermal cycling between 300 K and cryogenic temperatures with no impact on performance.
9. To conserve helium and prevent safety issues, all helium will be recovered, purified, and reused. Only in the event of catastrophic failures will helium be released into the tunnel via parallel plate or burst diskF relief valves.

Operating modes

- Purging and purification
  Starting with a warm cryomodule, 300 K helium gas from the high-pressure supply is circulated through all the cryomodule helium circuits and returned to the purifier via the purge return line. This process continues until the measured purity of the gas returning to the purifier meets ESS standards.
- Cool down and warm up
  Cold helium is brought from the distribution line, mixed as appropriate with warm helium gas, and sent into the various circuits of the cryomodule. The warmer gas leaving the cryomodule is returned to the cryogenic plant via the cool-down/warm-up line. Currently, there are no limits on the cool-down rate driven by either Q disease issues or cryomodule design issues. Once the cryomodule circuits reach their nominal operating temperatures, the flows will be returned to the cryoplant via the dedicated return lines rather than the cool-down/warm-up line. Once the cavity circuit has reached a nominal temperature of 4.5 K, the cold compressors (CCs) will be started, and the cavity circuit will be pumped down to the nominal 2 K level.

In the warm-up phase, 300 K helium gas is brought into the cryomodule circuits warming up the cryomodule. The return flow is brought back to the plant via the cool-down/warm-up line. Currently, there are no limits on the warm-up rate driven by either Q disease issues or cryomodule design issues.

- Nominal 2 K operations
  Operation with the SRF cavities at 2 K is the nominal operating mode and should be the most common mode seen at ESS. This mode can occur with both the beam and RF power on and with the beam and RF power off. In the case of RF and beam off, the total heat deposited into the cryogenic system is much less than with the beam and RF on. For short duration interruptions of the RF and beam, electric heaters in the cryogenic circuits will turn on to allow the heat load to the cryogenic plant to remain constant. For longer duration interruptions, the cryogenic plant capacity will be 'turned down' via the control system to allow for this reduced heat load requirement. Based on current technology and experience, the time to turn down the plant will be on the order of an hour or more.
- 4.5 K standby mode
  Under certain circumstances, such as multiple day interruptions to the RF power and beam, it may be energy efficient to stop pumping with the CCs and let the helium return to the plant at roughly 4.5. The valve settings are the same as in the nominal 2 K case, but the CCs are switched off and the returning He gas is sent directly to the suction side of the 4 K cold box (CBx).

3.1.7 Heat load estimation

Detailed studies and analyzes have been performed to estimate the heat loads imposed on the CMs, CDS, and due to beam losses, and are addressed in several internal technical notes [4–8] The heat loads, including safety factors, will be summarized in this section.

Safety factors

Compared with Technical Design Report (TDR), it was commonly decided to decrease the uncertainty safety factor for static loads on the CDS from 1.5 to 1.3 while keeping the operational safety factor of 1.15, as well as the safety factors used for the constant CM loads. The predicted heat loads per CM are extracted from [5] and given in table 3.4. Accordingly, the safety factors for sizing the ACCP are listed in table 3.5. See chapter 2 for a broader discussion of the safety factors.

Heat loads at Stage 1 and nominal design mode

In optimum design, the linear accelerator (LINAC) lattice includes 13 spoke CMs, 9 medium beta CMs, and 21 high beta CMs. The distribution system consists of the cryogenic transfer line with a length of 385 m, as well as 43 valve boxes and jumper connections.

With the above safety factors, the predicted heat loads, and installed capacity for the ACCP at stage 1 and nominal design mode are summarized in table 3.6. As a

Table 3.4. Predicted heat loads per CM, W (extracted from [5]). (Courtesy of ESS)

T levels	2 K							4.5 K Liq.	40–50 K
	Static				Dynamic			Static	Static
CM types	Others	Valves	Coupler	Total	Beam	Cavity	Total	$g\ s^{-1}$	
Spoke CM	3.3	0.2	3.5	7.0	1.5	5.0	6.5	0.092	30.0
MBCM	6.3	0.2	6.8	13.3	3.3	20.0	23.3	0.092	46.5
HBCM	6.3	0.2	6.8	13.3	3.3	24.4	27.7	0.092	46.5

Table 3.5. Safety factors for sizing the ACCP. (Courtesy of ESS)

	For CMs			For CDS	
Safety factors	F_o	F_{ud}	F_{us}	F_o	F_{us}
2 K	1.15	1	1.5	1.15	1.3
4.5 K liquefaction	1.15		1.5		
40–50 K	1.15		1.5	1.15	1.3

Recall from chapter 2 $Q = F_o(F_{ud} Q_d + F_{us} Q_s)$.

Table 3.6. Heat loads at Stage 1 and nominal design mode. (Courtesy of ESS)

	2 K		4.5 K Liq.	40–50 K
	Static	Dynamic	Static, $g\ s^{-1}$	Static
Predicted CM in total	490	875	4.0	1785
Installed CM in total	845	1007	6.8	3079
Predicted CDS supply	108	—	—	111
Predicted CDS return	312	—	—	3549
Predicted CDS in total	419	—	—	3660
Installed CDS in total	627	—	—	5472
Installed in total	2478		6.8	8551

result, the ACCP shall have a capacity of about 2.5 kW at 2 K, 4.5 K liquefaction rate of 6.8 $g\ s^{-1}$, and 8.5 kW at 40–50 K in Stage 1.

Heat loads at Stage 2 and nominal design mode

In an optimum design with the contingency, the LINAC lattice includes 13 spoke CMs, 9 medium beta CMs and 35 high beta CMs. The distribution system consists

Table 3.7. Heat loads at Stage 2 and nominal design mode. (Courtesy of ESS)

	2 K		4.5 K Liq.	40–50 K
	Static	Dynamic	Static, g s^{-1}	Static
Predicted CM in total	676	921	5.2	2436
Installed CM in total	1166	1060	9.0	4202
Predicted CDS supply	142	—	—	147
Predicted CDS return	409	—	—	4654
Predicted CDS in total	551	—	—	4801
Installed CDS in total	824	—	—	7177
Installed in total	3050		9.0	11 380

of the cryogenic transfer line with a length of 505 m, as well as 57 valve boxes and jumper connections.

With the above safety factors, the predicted heat loads, and installed capacity for the ACCP at stage 2 and nominal design mode are summarized in table 3.7. As a result, the ACCP shall have a capacity of about 3.0 kW at 2 K, 4.5 K liquefaction rate of 9.0 g s^{-1}, and 11.4 kW at 40–50 K in Stage 2, which determine the maximum capacity of the ACCP.

Summary

The heat loads, helium properties and mass flow for all steady-state operation modes are summarized in tables 3.8 and 3.9. The definition of the operation modes is explained in a recently published paper [7]. These tables were used in the ACCP technical specification.

Including safety margins, the ACCP is specified with a capacity of 3050 W at 2 K, 9 g s^{-1} liquefaction rate at 4.5 K and 3 bar and 11,380 W at 40–50 K.

The heat load distribution along the LINAC is given as below with 28% from the CDS, 10% from the spoke CMs, 14% from the medium beta CMs, and 48% from the high beta CMs.

From cryoplant point of view, the exergy loss analysis along the different cooling circuits indicates that the real exergy loss in total includes 78% at 2 K circuit, 10% at 4.5 K liquefaction, and 12% at 40–50 K circuit.

3.1.8 Industry studies

An industry study is a common approach for the large plant procurement, and it is extremely valuable.

Several questions remained regarding the details of the ACCP, and we sought input into them by funding studies from the cryogenics industry. We competitively bid these studies and were able to fund two of them: one by Air Liquide and one by Linde Kryotechnik.

Table 3.8. Heat loads for all steady-state operations modes. (Courtesy of ESS)

Operation modes		2 K, W				4.5 K Liq., g s^{-1}	40–50 K, W
		$Q1$		$Q2$	$Q1+Q2$	$F1$	$Q3$
		Static	Dynamic	Static	Total	Static	Static
Stage 1 (2019–2023)	Nominal design	845	1007	627	2478	6.8	8551
	Nominal turn down	845	—	627	1472	6.8	8551
	4.5 K Standby	845	—	627	1472	6.8	8551
	TS Standby	—	—	—	—	—	8551
	Maximum Liquefaction	Loads in 4.5 K standby mode plus maximum liquefaction rate at rising level into the LHe storage tank					
Stage 2 (2023–)	Nominal design	1166	1060	824	3050	9.0	11 380
	Nominal turn down	1166	—	824	1990	9.0	11 380
	4.5 K Standby	1166	—	824	1990	9.0	11 380
	TS Standby	—	—	—	—	—	11 380
	Maximum Liquefaction	Loads in 4.5 K standby mode plus maximum liquefaction rate at rising level into the LHe storage tank					

Q1: Isothermal loads from CMs, Q2: Non-isothermal loads from CDS, F1: Liquefaction rate for coupler cooling, Q3: Heat loads from TS, Q1 and Q2 capacities in 4.5 K standby are at 4.5 K level.

Table 3.9. Helium properties and mass flow rates for all steady-state operation modes. The bold and black values listed in this table shall be strictly guaranteed. The blue italic values may be taken into account as references for the process design. (Courtesy of ESS)

				Mass flow, g s^{-1}		Pressure, bar		Temperature, K	
Operation modes		Levels	Heat load, W	Supply	Return	Supply	Return	Supply	Return
Stage 1 (2019–2023)	Nominal design	2 K	2478	_99.3_	_92.5_	⩾3	⩽0.027	4.5	_4.5_
		4.5 K Liq.	—		6.8	⩾3	1.05	4.5	_300_
		TS	8551	_79.3_	_79.3_	15–19.5	Ps- ΔP*	⩾33*	⩽53*
	Nominal turn down	2 K	1472	_49.0_	_42.2_	⩾3	⩽0.027	4.5	_6.1_
		4.5 K Liq.	—		6.8	⩾3	1.05	4.5	_300_
		TS	8551	_79.3_	_79.3_	⩾11.5	Ps- ΔP*	⩾33*	⩽53*
	4.5 K standby	4.5 K	1472	_51.7_	_44.9_	⩾3	1.2	4.5	_6.4_
		4.5 K Liq.	—		6.8	⩾3	1.05	4.5	_300_
		TS	8551	_79.3_	_79.3_	⩾11.5	⩾11	⩾33*	⩽53*
	TS standby	2 K	—	—	—	—	—	—	—
		4.5 K Liq.	—	—	—	—	—	—	—
		TS	8551	_79.3_	_79.3_	⩾11.5	Ps- ΔP*	⩾70	⩽90
	Maximum liquefaction mode	4.5 K	1472	_TBD_	_TBD_	⩾3	1.2	4.5	_TBD_
		4.5 K Liq.	—		6.8	⩾3	1.05	4.5	_300_
		TS	8551	_TBD_	_TBD_	⩾11.5	Ps- ΔP*	⩾50	⩽70
Stage 2 (2023-)	Nominal design	2 K	3050	_120.2_	_111.2_	⩾3	⩽0.027	4.5	_4.6_
		4.5 K Liq.	—		9.0	⩾3	1.05	4.5	_300_
		TS	11 380	_105.5_	_105.5_	15–19.5	Ps- ΔP*	⩾33*	⩽53*
	Nominal turn down	2 K	1990	_67.3_	_58.3_	⩾3	⩽0.027	4.5	_5.9_
		4.5 K Liq.	—		9.0	⩾3	1.05	4.5	_300_
		TS	11 380	_105.5_	_105.5_	⩾11.5	Ps- ΔP*	⩾33*	⩽53*
	4.5 K standby	4.5 K	1990	_70.9_	_61.9_	⩾3	1.2	4.5	_6.3_
		4.5 K Liq.	—		9.0	⩾3	1.05	4.5	_300_
		TS	11 380	_105.5_	_105.5_	⩾11.5	Ps- ΔP*	⩾33*	⩽53*
	TS standby	2 K	—	—	—	—	—	—	—

	4.5 K Liq.	—	—	—	—	—	—	—
	TS	11 380	*105.5*	*105.5*	⩾11.5	Ps- ΔP*	⩾70	⩽90
Maximum liquefaction mode	4.5 K	1990	*TBD*	*TBD*	⩾3	1.2	4.5	*TBD*
	4.5 K Liq.	—		9.0	⩾3	1.05	4.5	*300*
	TS	11 380	*TBD*	*TBD*	⩾11.5	Ps- ΔP*	⩾50	⩽70

* The contractor shall choose the TS flow and temperatures within the limits given in table 3.2 and the average shield temperature shall be ⩽43 K. Referred to the maximum supply pressure of 19.5 bar ($P_s = 19.5$ bar), the acceptable pressure drop of the TS is less than 0.5 bar. For any other supply pressure, this pressure drop shall be approached via the equation $\Delta P = 19.5/P_s \cdot 0.5 \cdot \frac{20^2}{(T_S - T_r)^2}$ bar, where T_s: TS supply temperature and T_r: TS return temperature.
TBD: To be defined.

The future accelerator cryogenic plant procurement will be based on a technical specification, which fully describes the work to be carried out and the required quality. The scope of supply comprises the following:

- The ESS accelerator cryogenic plant process design and layout.
- The design, construction, transportation, installation, instrumentation, testing, and commissioning of all specified components.
- All piping and cabling to the interfaces of components.
- Provide the cryoplant controls, including hardware and software, up to the programmable logic controller (PLC) level. ESS personnel will carry out connection of the PLC to the ESS integrated controls system, development of human–machine interfaces (HMIs), alarms, and data logging.
- Start-up of the whole system and execution of the acceptance tests, including simulation of all specified ESS operation modes.
- Documentation of all components and their interactions within the plant.
- Time schedule. The design and official specification will be finished in 2013. The official procurement process will start at the beginning of 2014. It is expected that 2.5 years are required for engineering and production, and 1 year for installation and commission. The cryoplant shall be ready for use at the first quarter of 2018.

Some general questions addressed in the studies included key design choice (process, machine, LN2 precooling, heat recovery, control, efficiency, etc), layout and footprint, utility requirements (electricals, cooling water, instrument air, and HVAC), schedule, laydown and staging area for installation, preventive maintenance plan, spare parts and redundancy strategy, interfaces, and rough budget. The specific questions for the ESS ACCP are listed below:

1. How should the 2 K cycle and machinery be optimized to provide the minimum investment plus operating costs?
2. Is the decision to split the subatmospheric pumping on the 2 K system between cold and warm compressors appropriate?
3. Is there a design approach that will allow the plant capacity to be reduced to 30% of its full capacity? What would such a design entail? Approximately how much time would such a reduction in capacity take?
4. If the plant shuts down suddenly, approximately how long will it take the plant to get back to its nominal operating point (with the cold He vapor return pressure at around 31 kPa)? Are there design choices that can minimize this recovery time?
5. Is a figure of merit of 26% Carnot at 4.5 K a reasonable requirement?
6. Is our approach regarding plant automation reasonable? Given this approach, how will the plant be commissioned?
7. It is likely that ESS will specify specific PLCs to be used in this plant automation. Are there vendor preferred PLC models that will optimize cost and performance?
8. Is liquid nitrogen precooling in the plant cycle recommended? Why? If needed, how much capacity?

9. Layout of the 2 K and 4.5 K cold boxes, division, or integration? How will the cold boxes be commissioned and tested?
10. What is the approximate total footprint of the cold boxes and compressors? How many electrical and instrument racks are required?
11. What are the expected voltage and current power requirements for the helium compressor motors?
12. What, approximately, are the expected cooling water requirements for the helium compressors?
13. What are the maximum allowable oil and helium temperatures at the compressors?
14. What is an appropriate amount of time for vendor response to the request for proposals for the procurement of the ACCP?
15. Once an order is placed, what is the approximate time to delivery of equipment on the ESS site? How will the performance of the sub-components be assessed, especially the key ones such as CCs?
16. How will the acceptance tests for the complete cryoplant be carried out? What is the foreseen installation and commissioning time?
17. Are the He flow pressures and temperatures reasonable? Or, are there optimizations that can be done? Are there optimizations to be made in view of the cryoplant design on the TS temperatures and pressures?
18. Are there decisions that can be made that will reduce the cost of the plant?
19. Is there an optimum scope of work for the vendor that will minimize the initial cost of the plant?
20. Estimate the amount of laydown and staging area required for plant installation.
21. Short description of how fault handling and recovery will deal with the following problems:
    - Sudden loss of helium or LN2;
    - Loss of instrument air;
    - Loss of electricity for power and instrumentation and control;
    - Failure of valves;
    - Loss of vacuum.
22. How should the components operating below atmosphere pressure be protected from air inleak? Are Helium guards an option?
23. Requirements for preventive maintenance. How many months of continuous operation can be anticipated? And, what is the minimum length of time that would be expected for a warm-up/cool-down cycle?
24. How many spare parts (i.e., CCs) and redundancies (i.e., oil pumps) are needed?
25. The preliminary schedule and the overall budget should be provided.

Based in part on the results of these studies, ESS made several decisions about the ACCP:

1. For the 2 K system, we would employ a mixed approach using a set of CCs for most of the pressure reduction but doing the final pumping via a room

temperature, subatmospheric screw compressor. We believed that this would allow us to both better turn down the plant capacity and have a simpler recovery from cryoplant upsets.

2. No liquid nitrogen (LN2) precooling: The ACCP shall be designed for the refrigeration performance without using LN2 precooling. One basic advantage of the refrigeration system with LN2 precooling is to boost the cool-down process in term of the massive cold masses, i.e., in a magnitude of thousand tons at LHC, CERN. The ESS accelerator has a total cold mass of around 20 metric tonnes, which could be cooled down within a reasonable time. The other fact is that LN2 is used to boost and increase LHe production capacity in 4.5 K liquefaction systems. It is noted from table 3.1 that the 2 K circuit dominates the ACCP heat loads. The industry study results confirm that the boosting effect of LN2 in a large 2 K helium refrigeration system is not as big as in a 4.5 K liquefier and the efficiency saving is relatively small. Other positive effects, such as increased reliability due to less utility interfaces, operating cost saving due to the Swedish electricity cost, traffic reduction and environment protection, all point towards a process without LN2 precooling.
3. We were originally going to have the vendor create the control system, including both software and hardware, up to the PLC level with ESS creating the HMI that linked to the PLCs and ran on the control room computers. However, feedback from these industry studies convinced us that it would be both more efficient and more reliable to have the cryoplant vendor create the complete control system, including the HMIs. We did specify that the control system had to be consistent with a variety of ESS standards, including use of the Experimental Physics and Industrial Control System (EPICS) and Control System Studio tools, and use of the one of the PLC types chosen by ESS for sitewide standardization. In the ACCP procurement, we specified PLCs from the Siemens S7–300, S7–400, and S7–1500 families. Further details on the ACCP control system, its commissioning, and lessons learned may be found in [10].
4. One integrated CBx: All ACCP cold components, including expansion turbines, CC system, heat exchangers etc, shall be integrated into one unique CBx. This solution brings advantages such as space saving and low investment cost.
5. Waste heat recovery: ESS has committed to an energy management strategy to recover a maximum of waste heat at the highest temperatures possible and to minimize the cooling water flow [13]. The recovery energy will be recycled from ESS to the city of Lund's district heating system and other possible users. To accomplish this in the ACCP, for the oil and helium coolers in the warm compressor system, the warm end temperature difference between oil/helium inlet and cooling water outlet shall be controlled to less than 10 K in normal operations.

Feedback from the industry studies also confirmed our planning of a roughly 4–5 year time from the release of tendering documents to a commissioned cryoplant at ESS.

Another valuable feature of these paid industry studies is that they made cryogenic firms, even those not paid to produce the studies, aware of the scope of what ESS was going to procure and the rough schedule for the procurement. This permitted potentially interested firms to plan their work and perhaps even to start thinking about what they would submit once the official request for bids came out. We felt that this maximized participation and allowed us to consciously reduce the time allowed for bid preparation in the procurement process.

3.2 Schedule

The project's schedule is an important management tool. It is the most visible proof of how the project will be executed, and thus provides the basis for follow-up and quality reviews during execution. In the schedule, the activity list is supplemented with a calendar that shows the start and the end points of activities alongside milestones and dependencies.

The high-level schedule and timing for the ACCP project is listed as follows:

- One year specification preparation, including drafting, external experts review, and internal procurement procedure.
- Half year procurement procedure, including vendor's proposal preparation, evaluation, and resolution of any challenges.
- Four years project execution, including design, manufacture, installation, and commissioning.

3.3 Budget and costs

The ACCP budget and cost estimation is based on the deep investigation of the contract cost similar projects in the recent five years, our cryogenic staff experiences, and the two comprehensive industry studies. It was fixed when the ESS TDR was released in 2013.

The baseline scope of the ACCP tender includes the warm compressor station (WCS), final oil removal system (FOR), a one-bad online full flow dryer, gas management panel (GMP), one CBx, and one ambient heater. There were two optional items for which vendors could propose a separate cost if interested: the first was a 20,000-l liquid helium storage dewar, and the second was the design and installation of warm piping connecting the vendor's equipment in the compressor hall together with the vendor's equipment in the cold-box hall. All other equipment, not mentioned above, will be separately ordered. This includes warm helium storage tanks, recovery system, and warm interconnecting pipes between the warm compressor hall and the cold-box building.

Eventually, the total contract value of the ACCP baseline scope was 18.5 MEuro (2015 price) plus two options of about 1.1 MEuro. There were about 20 changes of order (COR) during the contact execution, adding costs of about 1% of the main contract. These costs were within our budget for the ACCP.

3.4 Staffing

Staffing and resource acquisition are crucial for a project's success. Identifying needs and describing what is needed for project execution is necessary. You also need to determine which resources can be recruited internally and which must be found externally. This is true for all types of resources.

At the beginning of 2013, the ESS cryogenics team consisted of W Hees, X L Wang, and J Weisend. X L Wang was appointed as the Project Engineer and ESS Project Manager for the ACCP. We were soon joined by P Arnold, who was made Work Package Leader for Cryogenics and was responsible for the scope, cost, and schedule of almost all ESS cryogenic systems. J. Jurns joined the cryogenics team in 2014 and was the Project Engineer and ESS Project Manager for the Target Moderator Cryoplant (TMCP, chapter 5). G Gistau (formerly of Air Liquide) and H Quack (formerly of Linde and the Technical University of Dresden) were also available to us as paid consultants.

Once the procurement of the ACCP started, ESS prioritized the recruitment of cryogenic system operators and technicians. We wanted to have these people in place early to allow them to observe and participate in the ACCP's installation and commissioning. This was accomplished and allowed the ESS to quickly transition to operating the ACCP and other cryogenic equipment once they had been commissioned. The early recruitment of this staff is an important lesson learned.

The cryogenic team received strong support and involvement of the internal departments, such as procurement office and legal office for the contract management; quality office for quality control and quality assurance; integrated control group for the control and personal protection system; conventional facility department for the civil engineering and utilities; safety group for the general safety; oxygen deficiency hazard and reliability, availability, maintainability, and inspectability (RAMI); infrastructure team for designer and alignment; as well as external colleagues from the in-kind contribution labs.

3.5 Procurement strategy, technical specification, and bid evaluation

3.5.1 Procurement strategy

The procurement of the ACCP was the second large procurement in the history of ESS; with the first being the choice of Skanska as the civil construction contractor. As such, the ACCP procurement required ESS to develop new internal policies and procedures to release the request for quotes, evaluate the responses, and place the order. The ESS cryogenics team worked very hard with members of the ESS procurement office (Luis Ortega, Meredith Shirey, Mirko Menninga, and Malcolm De Silva) and the ESS Legal Office (Ohad Graber-Soudry) to ensure that the procurement went smoothly. In the end, the procurement of the ACCP occurred on schedule and without any significant challenges or issues.

ESS chose to procure the ACCP via competitive bids in response to a detailed technical specification and statement of work based on specific performance requirements for the ACCP. In this approach, we were procuring both the thermodynamic cycle design and the associated equipment to implement the design. The successful vendor would provide all the cryoplant equipment from the warm compressors up to the interconnect to the CDS. As mentioned earlier, the control system (both hardware and software) was also provided by the cryoplant vendor. There were two optional items for which vendors could propose a separate cost if interested. One was a 20,000 l liquid helium storage dewar, and the other was the design and installation of warm piping connecting the vendor's equipment in the compressor hall together with the vendor's equipment in the cold-box hall.

While there was no restriction on the bidder's country of origin, ESS did limit bidding to bidders that had previous experience in building large scale, automated helium cryoplants.

In chapter 1, the ESS use of in-kind contributions as a funding approach was described. Since all the cryoplants (i.e., ACCP, TMCP, and TICP) were planned as commercial purchases, the procurement of these plants was never seen as suitable for in-kind contributions and the cryogenic plants were funded directly by the ESS laboratory.

3.5.2 Technical specification

Creating a technical specification for such a procurement approach requires a fine balancing act. Too much detail may add unnecessary costs, limit the vendor's ability to innovate, and possibly even prevent some firms from submitting a bid. Meanwhile, too few details may result in a plant that does not reliably meet all its requirements. The ACCP Technical Specification, which was 117 pages in length, was principally written by X L Wang, with contributions by Phillip Arnold, Wolfgang Hees, and John Weisend. This document was heavily reviewed by our external consultants (G Gistau and H Quack), whose comments and suggestions were folded into the final draft. A sense of the specification may be seen in the specification's Table of Contents (appendix A).

Key topics in the ESS technical specification included:

- Codes and norms.
- ESS site condition/utility interfaces.
- Performance (heat load, operation modes, etc).
- Key components (WCS and CBx).
- Spare parts.
- Control.
- Instrumentation, interlocks, and electrical design.
- Requirements for mechanical design, manufacture, and mounting.
- Testing.
- Quality management.
- Documentation.

Some aspects of the technical specification are worth noting in more detail.

A series of operating modes were defined, and the capacities and helium parameters (temperature, pressure, and flow) for each of those modes were given in a table (table 3.9) in the specification. Note here, as described in chapter 2, that there are two distinct stages of operation. Stage Two, which is expected to occur later, includes the cooling required for the 14 additional CMs that may be added to the accelerator. To optimize the performance of the cryoplant in both stages, we specifically in the technical specification required:

- The warm compression from subatmospheric pressure (SP) to low pressure (LP) or middle pressure (MP) shall be equipped with a variable frequency drive (VFD) so that at least 25% of the stage two SP flow can be regulated continuously.
- The warm compression from LP to MP shall be equipped with VFD so that at least 25% of the stage two LP flow can be regulated continuously.
- All CC drives shall be equipped with VFD.
- Each CC shall have a spare cartridge comprising flow parts optimized for stage one operation that can easily be adapted with flow parts optimized for stage two operation.
- As many expansions turbine spare cartridges as are deemed optimal by the contractor in view of capital and operational cost shall be equipped with flow parts optimized for stage one operation, which should easily be adapted with flow parts optimized for stage two operation.
- The initial installation, commissioning, and acceptance testing shall be performed with all equipment optimized for stage two operation.
- After successful acceptance testing, the equipment shall be exchanged for equipment optimized for stage one operation, commissioned, and then acceptance tested.
- The cartridges with flow parts optimized for stage two operations shall be adapted with spare parts for stage one operation and the stage two flow parts to be stored for later use.

These requirements would clearly lead to a more expensive plant but were felt to be necessary to ensure that the plant could manage both operating phases and allow reasonable amounts of capacity turn down.

Given the large number of operations with subatmospheric pressure helium, we required that all non-welded connections (e.g., instrumentation ports, relief valves, or flanged connections) in the subatmospheric system be surrounded by atmospheric pressure helium guards. Thus, any leaks into the system will be helium and not air.

Accompanying the Technical Specification was a Statement of Work, which described the vendor's responsibilities, listed the two optional items for bidding (a 20,000 l liquid helium dewar, and the warm gas piping from the helium compressors and the CBx) and provided a list of mandatory design reviews and factory inspections that were to be conducted throughout the project.

3.5.3 Call for tender and bid evaluation

We formalized a set of tender documents, listed as follows:

- ACCP-A01 Tender Management (2 pages)

- ACCP-A02 Company Qualification (7 pages)
- ACCP-A03 Instruction to Bidders (7 pages)
- ACCP-A04 Statement of Work (23 pages)
- ACCP-A05 Technical Specification (117 pages)
- ACCP-A06 Evaluation Criteria (5 pages)
- ACCP-A07 Draft Supply Agreement and Performance Guarantee (27 pages)
- ACCP-A08 General Conditions of Tender (4 pages)

Along with the Technical Specification and the Statement of Work, we released the criteria, see appendix B, by which the bids would be evaluated. These criteria consisted of:

$$SC + SO + SQ \tag{3.1}$$

where:

SC was a score form 0–30 related to the capital cost of the plant

SO was a score from 0–30 related to the operating cost, defined as the electricity required to operate the warm helium compressors of the plant.

SQ was a score from 0–40 for the quality of the proposal, i.e., how well it was judged that the proposal would meet the technical requirements.

It is worth noting that the criteria gave equal weight to the initial capital cost of the plant and the plant's operating cost. Thus, a vendor could propose a more expensive plant and still win the bid if that plant would be more energy efficient and thus have a lower operating cost. This willingness of ESS to spend money up front to gain long term energy savings is consistent with ESS's commitment to sustainability.

In addition to the criteria weighting, there was another feature of the operating cost that drove potential vendors in the direction of energy efficiency. If, at the end of plant commissioning and performance testing, the actual operating cost of the plant is found to be less than that calculated by the vendor in their proposal, then 50% of the cost savings will be paid by ESS to the vendor. If, however, the measured operating cost of the plant was found to be more than that calculated by the vendor in the proposal, then the difference in cost will be paid by the vendor to ESS. We did not specify a Coefficient of Performance or Percent Carnot in the technical specification, but rather used the above approaches to drive the vendor design towards energy efficiency.

The formal Invitation to Tender for the ACCP was released by ESS on June 10, 2014. Responses were received on December 1, 2014. Multiple responses were received and on December 17, 2014, the contract was awarded to Linde Kryotechnik. The contract was signed on March 17, 2015. The relatively short time between release of the tender invitation and arrival of proposals was made possible by the previous industry studies. Without those, a reasonable time for proposal preparation would have been nine months. We also decided to give Linde the contract for the 20,000 l helium dewar and retained the warm piping work as an in-house ESS project.

3.6 Norms and rules

All components and equipment had to meet European and Swedish laws and regulations, as well as standards of good practice commonly accepted in the cryogenic community. A partial list of norms and rules applied for the ACCP is given below.

- ALPEMA: The standards of the brazed Aluminum Plate-Fin heat Exchangers Manufacturers' Association
- EN 10204: Metallic products—Types of inspection documents
- EN 12517-1: Non-destructive testing of welds—Part 1: Evaluation of welded joints in steel, nickel, titanium, and their alloys by radiography—Acceptance levels
- EN 12517-2: Non-destructive testing of welds—Part 2: Evaluation of welded joints in aluminum and its alloys by radiography—Acceptance levels
- EN 13445: Unfired pressure vessels
- EN 1435: Non-destructive examination of welds—Radiographic examination of welded joints
- EN 1779: Non-destructive testing—Leak testing—Criteria for method and technique selection
- EN 287–1: Qualification test of welders—Fusion welding—Part 1: Steels
- EN 288–3: Specification and approval of welding procedures for metallic materials—Part 3: Welding procedure tests for the arc welding of steels
- EN 30042: Arc-welded joints in aluminum and its weldable alloys—Guidance on quality levels for imperfection
- EN 439: Welding consumables—Shielding gases for arc welding and cutting
- EN ISO 14731: Welding coordination—Tasks and responsibilities
- EN ISO 15614-1: Specification and qualification of welding procedures for metallic materials—Welding procedure test—Part 1: Arc and gas welding of steels and arc welding of nickel and nickel alloys
- EN ISO 17635: Non-destructive testing of welds—General rules for metallic materials
- EN ISO 3834-2: Quality requirements for fusion welding of metallic materials —Part 2: Comprehensive quality requirements
- EN ISO 5817: Welding—Fusion-welded joints in steel, nickel, titanium, and their alloys (beam welding excluded)—Quality levels for imperfections
- EN ISO 9001: Quality Management Systems—Requirements
- EN13480: Metallic industrial piping
- IEC: International Electrotechnical Commission
- ISO 10438-1: Petroleum, petrochemical, and natural gas industries—Lubrication, shaft- sealing and control-oil systems and auxiliaries—Part 1: General requirements
- ISO 10438-2: Petroleum, petrochemical, and natural gas industries—Lubrication, shaft-sealing and control-oil systems and auxiliaries—Part 2: Special-purpose oil systems

- ISO 10438-3: Petroleum, petrochemical, and natural gas industries—Lubrication, shaft-sealing and control-oil systems and auxiliaries—Part 3: General-purpose oil systems
- ISO 10440-1: Petroleum, petrochemical and natural industries, rotary-type positive displacement compressor-Part 1: Process compressor
- ISO 2372 Group G: Mechanical vibration of machines with operating speeds from 10 to 200 rev s^{-1}—Basis for specifying evaluation standards
- ISO 2954: Mechanical vibration of rotating and reciprocating machinery—Requirements for instruments for measuring vibration severity
- ISO 3740: Acoustics—Determination of sound power levels of noise sources—Guidelines for the use of basic standards
- ISO 8573-1: Compressed air—Part 1: Contaminants and purity classes
- PED: Pressure Equipment Directive PED 97/23/EC with Annexes I to VII
- TEMA: The standards of Tubular Exchanger Manufacturers' Association

3.7 Quality assurance and control

Quality assurance and control were mainly managed by the vendor's internal QA/QC staff and the notify body (the third party). However, ESS established our own QA/QC department at the early stage of the project, who are deeply involved in the ACCP design phase, review meeting, factory inspection, and on-site installation. They had rich field experience, strong backgrounds, and well-understood the norm and rules, which become very valuable for the project's execution.

3.8 Naming convention

The ESS Naming Convention was agreed upon and approved at an early stage of the ESS project to ensure meaningful, short, and structured names of signals and devices. Given the millions of signals to control and thousands of devices to operate, clear communication is essential among operators, physicists, and engineers. Considerable efforts have been invested to make the ESS Naming Convention useful in a wider context to help enforce system integration activities across all divisions.

The ESS Naming Convention is based on a standard that was originally developed for the Super Superconducting Collider and later adopted by other large research facilities, e.g., the SNS, Facility for Rare Isotope Beams (FRIB), International Thermonuclear Experimental Reactor (ITER), CEBAF, Paul Scherrer Institute, and MAX-Lab (MAX IV). We gratefully acknowledge the use of ideas, nomenclature, and software, as well as conversations with personnel from these institutions.

All the signal and control device names were required to be named according to the ESS Naming Convention. ESS provided the tools and necessary assistance to enable the contractor to perform this task. The device naming and system or sub-system under the discipline CRYO was reviewed by the relevant stakeholders and approved officially. The in-kind contributors and contractors are obliged to give the device names accordingly. New devices are subject to review and approval.

3.9 System overview

The final Linde Kryotechnik (LKT) design of the ACCP is given below.

Overall process[1]

The simplified process design is shown in figure 3.6, which mainly consists of three oil-lubricated screw compressors, six gas dynamic bearing turbines, three CCs, and several heat exchangers blocks.

There are four main pressure levels: subatmospheric pressure (SP) in series with CCs, LP slightly above atmospheric pressure, MP, and high pressure (HP) connected to the inlet of the CBx. Following this classification, the WCS contains three stages. One stage after the three CCs compresses helium from SP directly to the MP level (SP stage). A second compressor (LP stage) is used to compress helium from LP level to the same MP level. At this pressure level, additional flow comes back from the CBx. The total flow is compressed in a single compression stage (HP stage) from MP to HP. The MP can be regulated consistent with the operation conditions. The adaptable MP ensures operation of turbines and screw compressors at highest efficiencies during all operation modes.

The screw compressors are chosen in such a way that the flow in each stage is compressed by a single screw compressor. Moreover, all three screw compressors are of identical screw block size, which allows a small number of spare parts, high redundancy, and reduced installation efforts and costs.

The 'mixed' compression cycle, which is based on a combination of CCs (CC1–CC3) in series with a warm subatmospheric compressor (SP compressor), is used to produce refrigeration at 2 K. The compressed helium from the SP circuit is directly discharged to the suction side of the HP compressor. The still low enthalpy of the gas at the outlet of the CCs is used for heat exchangers at corresponding temperature levels. The SP and LP compressors share the same bulk oil removal system, minimizing the investment cost of the system. Such a cycle can accommodate a large dynamic range without additional electrical heating. In addition, the SP compressor provides flexibility and advantages during transient operation modes, i.e., start-up of pumping down, in which the CCs are far from their design conditions and difficult to tune. The main drawback of this type of cycle is the risk of contamination from air leaks. Welded joints will be used wherever possible and all components interfacing subatmospheric circuits and ambient atmosphere, such as safety valves, dynamic seal of valve stems, and instrumentation, will be protected from air intake, either by a helium or vacuum guard.

The floating pressure cycle [11] is a common control strategy to provide flexibility and energy saving potential. Suction and discharge pressures of the HP compressor are varied proportionally. The mass flow of the compressor is thus adapted without a frequency converter keeping the volumetric flow rates unchanged. VFDs are applied to control the capacity of the SP and LP compressors to adapt the capacity

[1] Part of this section is reproduced from Wang X L *et al* 2015 *IOP Conf. Ser.: Mater. Sci. Eng.* **101** 012012, used with permission.

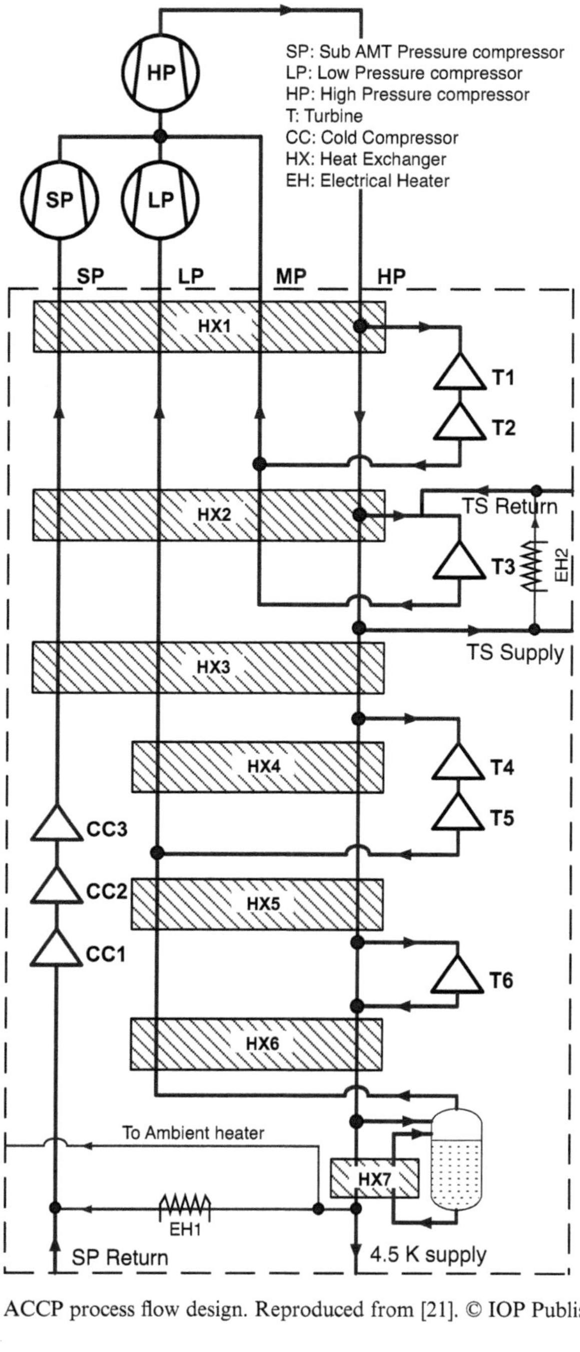

Figure 3.6. Simplified ACCP process flow design. Reproduced from [21]. © IOP Publishing Ltd. CC BY 3.0.

to the flow rate of returning gas. High efficiency and power savings are achieved with these arrangements in the turn-down modes.

The different capacity requirements for the two ACCP stages are matched by an exchange of CC wheels and flow parts and turbine flow parts, in combination with the capacity control of the warm compressors described above. High efficiency can be achieved in both stages of the project.

The floating pressure cycle and LKT's expander technology perform in highest process efficiency over a broad operation map, resulting in nearly constant process efficiency over a very large capacity range of the refrigerator and related power savings. These features have been proven by similar plants at XFEL-DESY [12, 13], FNAL [14], and JLab [15].

Two built in electrical heaters and one external ambient heater are used for the acceptance tests of the overall performance of the ACCP. The first test heater (EH1) is located upstream of the 4.5 K supply line and directed to the SP return line. This heater is used to simulate the cumulated static and dynamic heat loads in the 2 K or 4.5 K system depending on the mode tested. The second heater (EH2) is connected between the TS supply and TS return for shield circuit tests. Measuring with the flow meter downstream of the ambient heater located outside the CBx will test the liquefaction load of the coupler cooling. All cold equipment is incorporated in one CBx, which provides a compact, flexible, and cost-efficient setup.

The key machine and equipment parameters are listed in table 3.10.

- The total electrical power consumption of the WCS is 2.37 MW, which gives about 9 kW @ 4.5 K equivalent cooling power in total with an exergy efficiency of 25.9%.
- LKT selected the Aerzen WCS to circulate helium and achieve the performance of the ACCP. All warm compressors are oil-lubricated screw compressors and are designed for pressures from 0 bar (vacuum) to 25 bar abs. The WCS includes three compression stages: compressing helium from SP to MP, from the LP to MP, and from the MP to HP. One compressor in each stage is selected based on a technically safe and cost-effective operation. The compressor station is assembled and grouped on five skids in total: SP, LP and HP skids, SP and LP bulk oil removal (BOR) system skid, and HP BOR skid.
- The SP compressor is designed with an internal volume ratio of 4.0, mass flow of 117 g s^{-1}, shaft power of 303 kW, suction pressure and temperature of 0.607 bara and 33 °C, and discharge pressure and temperature of 4.25 bara and 71 °C. It is also equipped with a VFD.
- The LP compressor is designed with an internal volume ratio of 2.6, mass flow of 285 g s^{-1}, shaft power of 516 kW, suction pressure and temperature of 1.05 bara and 33 °C, and discharge pressure and temperature of 4.25 bara and 83 °C. It is also equipped with a VFD.
- The HP compressor is designed with an internal volume ratio of 2.6, mass flow of 735 g s^{-1}, shaft power of 1383 kW, suction pressure and temperature of 4.05 bara and 40 °C, and discharge pressure and temperature of 20.5 bara and 83 °C, and has no VFD.
- The SP and LP compressor skids share a common BOR skid, where a coalescer on the skid is equipped to keep the oil less than 10 ppm.

Table 3.10. Key machine and equipment parameters. (Courtesy of ESS)

Main electrical power	2.37 MW					
Power at 4.5 K equivalent	~9 kW with exergy efficiency 25.9%					
	V_i	P_{suc}, bara	P_{dis}, bara	$\dot{m}$ g s^{-1}	P_s, kW	VFD
SP compressor (33K/71K)	4.0	0.607	4.25	117	303	Yes
LP compressor (33K/83K)	2.6	1.05	4.25	285	516	Yes
HP compressor (40K/83K)	2.6	4.05	20.5	735	1383	NO
Impurity removal (ORS/dryer/cryogenic adsorbers)	Oil	BOR (1000 ppm)-Coalescer in skid (10 ppm)-FOR (0.5 ppm Aerosols and 10 ppb Vapor)				
	Water	One-bed online dryer (0.1 ppmv)				
	Air	Two 80 K adsorbers (73 K, 20 bar) and one 20 K adsorber (24 K, 20 bar)				
CCs	Ceramic ball motor bearing; exactly the same as DESY 1–3 stages, HS270 (CC1), and HS220 (CC2 and CC3); nominal mass flow rate:116.6 g s^{-1}					
Plate-fin heat exchanger	NTU<35 per block, calculations conducted for all modes, two installed horizontally					
Liquid He storage	20,000 dewar (2.2 tons) and subcooler in CBx (1500 l)					
Warm He storage tank	14 × 70 m^3 at 20 bar (2 7 tons) (CMs + CDS+ ACCP in total)					

- The HP compressor skids have their own BOR skid, where a coalescer on the skid is equipped to keep the oil less than 10 ppm after the warm compressor system.
- The FOR consists of a set of coalescing filters for the continuous separation of oil mist from the helium and a charcoal adsorber to adsorb oil aerosols. It keeps oil aerosols less than 0.5 ppm and oil vapor less than 10 ppb towards the dryer.
- A one-bed dryer downstream of the FOR is used to adsorb the water vapor molecules under the full flow conditions. It keeps water less than 0.1 ppmV before entering the CBx.
- The GMP includes valves and pipe terminals for process control.
- A CBx has all the equipment to achieve the required performance, mainly consisting of heat exchangers, expansion turbines, CC system, adsorbers, phase separator and subcooler, instrumentation, valves, and piping work, corresponding electrical cabinets, a working platform, etc.
- An ambient heater for the cool down/warm up of the ACCP alone or the ACCP with the whole accelerator.
- A liquid helium storage tank of 20,000 l, which has a capacity of 2.2 tons. This tank is intended to store the accelerator's helium inventory in liquid form as a second fill, support the ACCP when recovering from a trip, for an acceptance test for the maximum liquefaction mode, and to provide liquid helium to the system in the case of CM CD/WU.
- Warm helium storage tanks include 14 tanks and 70 m^3 each with a total capacity of 2.7 tons.

3.10 Meeting and reviews

During the ACCP execution it is vital to follow up on and report results and costs, analyze the situation, handle changes, and continuously update the plant. This is an interaction that requires solid communication between the stakeholders involved, including internal cryogenic teams, other internal ESS departments, and the vendor's project team. The project is tracked with heavy daily communications, monthly status reports, and the formal meeting and reviews.

Some of the design reviews (table 3.11) were linked to both progress payments and to permission for the vendor to order items such as heat exchangers and warm compressors. The reviews were organized by the vendor, but the review committees were made up of ESS staff and outside experts. In addition to our two external cryogenics experts (H Quack and G Gistau), we were fortunate to have available to us Andreas Thiel from the neighboring MAX IV Laboratory who is an expert in helium compressor technology.

The requirement for design reviews and inspections illustrates a very important aspect of this project. These are not the type of goods where you can place the order and simply wait for delivery. Monitoring this progress was a full-time job for the ACCP project engineer (X L Wang). His work included weekly phone meetings with the vendor, answering questions from both the vendor and ESS staff (e.g., from our

Table 3.11. List of ACCP meetings and reviews.

Date	Event
Mar 2015	• Contract signed
May 2015	• KOM
Aug 2015	• PDR-1
Oct 2015	• HAZOP
Nov 2015	• PDR-2
Jan 2016	• FMEA
Mar 2016	• CDR-1
June 2016	• IDR
Aug 2016	• LDR
Sept 2016	• CDR-2
Apr 2017	• IRR
Apr 2018	• WCS/CB start-up
Oct 2020	• SAT

civil facilities team regarding utilities), participating in design reviews, and inspections on the vendor's site. Dr Wang was aided by other ESS experts in welding, vacuum leak checking and controls. Only by investing a significant amount of staff resources into this project were we able to achieve the good result that we did. This is an important lesson learned from this project. It should be noted that the commercial vendors fully supported and were quite pleased with this level of involvement by ESS.

The actual meeting and reviews are summarized in table 3.11 and are briefly introduction as follows:

- The ACCP contract was signed in March 2015 and the kick-off meeting took place two months later. The project team from both sides, the communication plan, some open points in the proposal, and project management issues were addressed and discussed there.
- The first part of the preliminary design review (PDR-1) took place in August 2015. ESS received copies of order confirmation from the contract subsuppliers for warm compressor skids, CCs, and heat exchangers, and allowed the contractor to proceed with the order under some certain conditions, such as providing the remaining details of the CCs, verification of the sizing of the oil skid, marking the heat exchanger data sheets to show which are horizontal and which are vertical, etc.
- A hazard and operability (HAZOP) review took place in October 2015 after the basic process design was finished and the key machine and components had been selected. This review provided a formal look at the methodology and results from the HAZOP study of the ACCP performed by LKT.

This review examined, through several documents (e.g., failure, severity and risk analysis for human and machine protection and operation safety, mitigation, and follow-up action items), the design choices that have a potential impact on humans, equipment, and availability during operation. The review committee acknowledges the extensive efforts produced by LKT to deliver and conduct a detailed and complete HAZOP study, which allowed us to clearly identify the main failure events that could have a potential impact on safety. The list of control measures to be implemented and proposed by Linde is globally satisfying and meets the ESS's safety objectives. No technical and safety showstoppers were identified, although additional safety studies had to be carried out and delivered by both Linde and ESS. Although the safety aspects were mainly covered during the HAZOP review, details of failure events affecting the operability of the ACCP and the LINAC required further investigated in a quantitative study.

- The second part of the preliminary design review (PDR-2) took place in November 2015. ESS received copies of order confirmation from contract subsuppliers for large vessels (CBx shell, adsorbers, etc), turbines, and control valves, and the contractor was allowed to proceed with the order. This meeting reviewed a more detailed version of the preliminary design of the ACCP, including warm helium compressors and their related oil and oil removal systems, CBx components (turbines and CCs), and interconnections between the components. Significant time was spent on developing and discussing acceptance test procedures for various operating modes, particularly for the CCs. Quality assurance plans, safety issues, and P&IDs were reviewed. The status of previous recommendations was also discussed. More detailed discussions regarding controls, operations, test plans, and recovery from trips were required, which were planned in future meetings on controls, a failure model analysis (FMEA) review, and in upcoming CDRs.
- Upon ESS's request, a FMEA review took place in January, 2016. The FMEA focused on the ACCP's RAMI. The FMEA matrix includes the longitudinal coordinates consisting of possible failures and trips from various subsystems or components (e.g., warm compressor system, CCs, turbines, adsorbers, and utilities) and the horizontal coordinates consisting of root cause, consequences to the ACCP and the LINAC, time to recovery, occurrences, and corresponding actions or comments.
- The first part of the critical or detail design review (CDR-1) took place in March 2016. The purpose of this CDR-1 was to confirm that the design for the ACCP was likely to meet all requirements and was specified in sufficient detail for production (including material purchase for manufacture, procurement of manufacture services, and in-house manufacture) and assembly. ESS received copies of receipts of main materials when they arrived at the contractor's or their subcontractor's fabrication premises, together with copies of the respective work orders, and the CBx fabrication was allowed to start.

- Due to the complexity, upon ESS's request, an intermediate design review took place in June, 2016 with a focus on the compressor building and cold-box hall layout, as well as the compressor skid foundation plan.
- The logic design review took place in August, 2016. The plant control narrative, cause and effect matrix, alarm and trips, the control loop and step chains, and the CC control logic were discussed and reviewed in detail.
- The second part of the critical or detail design review (CDR-2) took place in September 2016. The outputs of the design, such as CAD models, supporting calculations and analysis reports, and procurement and manufacturing specification, were compared and reviewed against the inputs of design, including technical and interface requirements or input specifications. ESS received copies of the receipts of the main materials when they arrived at the contractor's or their subcontractor's fabrication premises, together with copies of the respective work orders, and the rest of the fabrications were allowed to start, except for the CBx (in CDR-1).
- After the main components had been delivered on-site and before the installation started, the installation readiness review (IRR) took place in April 2017 on the ESS site. The IRR was meant to be the final technical review of the system prior to the start of installation. As such, it examined the final technical design of the integrated system with an emphasis on interfaces between components and subsystems, controls integration, and included a detailed look at the plans, staff, and tooling required for the installation work itself.
- After the on-site installation had finished and before starting the compressor and the CBx, the compressor/CBx start-up review was conducted to review and check the mechanical completion of the warm compressor system and the CBx system, electrical installation and completion, the certificates for the pressure vessel, loop/signal check report/protocol, personal safety, environment, and master documentation.
- The site acceptance test review (SAT) took place in 2020. The ACCP passed all acceptance tests as specified, ESS received all final documentation, and a punch list of minor items that still needed to be fixed was issued and signed by both the contractor and ESS. Afterward, the Provisional Acceptance Certificate was issued.

3.11 Intermediate inspection

After the manufacturing started, the ACCP was closely followed up by intermediate inspections. They were held for critical production steps and were defined by the contractor in a schedule, normally in the inspection and test plan (ITP). Acknowledged inspection procedures were defined and followed for these inspections. These inspection procedures were prepared by the contractor and approved by ESS. These inspections useful in understanding the project status. The inspections also allowed ESS staff to understand the interior construction of equipment that would later be contained within the cryostats. This is valuable for future operations and maintenance.

The following major inspections were performed during the ACCP execution:

- Heat exchanger factory acceptance tests in April 2016, at Linde Schalchen, Germany.
- Motor factory acceptance tests in June 2016, at ABB, Helsinki, Finland.
- Bare compressor performance tests in July 2016, at Aerzen, Germany.
- CBx internals fabrication check before superinsulation is mounted in February 2017, BERO, Switzerland.
- Compressor skid mechanical completion in February 2017, at Aerzen, Germany.
- Compressor skid overall leak test (LT) and pressure test (PT) in March 2017, at Aerzen, Germany.
- Liquid helium storage tank factory acceptance tests in April 2017, at Statebourne, UK.
- GMP and final oil removal fabrication check(s) in May 2017, at KASAG, Switzerland.
- CBx internal piping pressure and LT in June 2017, at BERO, Switzerland.
- CBx externals and valve and instruments panel(s) fabrication check in June 2017, at BERO, Switzerland.

Some photos from these inspections are given in figures 3.7–3.12.

Figure 3.7. Compressor skid mechanical completion in February 2017 at Aerzen, Germany. Courtesy of ESS

Figure 3.8. CBx internal piping pressure and LT in June 2017, at BERO, Switzerland. Courtesy of ESS

Figure 3.9. Liquid helium storage tank factory acceptance tests in April 2017, at Statebourne, UK. Courtesy of ESS

3.12 Delivery

The buildings were finished and ready for goods reception (figures 3.13 and 3.14) in May 2017. Some key components of the ACCP started to arrive in July 2017. Significant planning and preparation were required for these deliveries, which included determination of access routes both to ESS and on the site itself

Figure 3.10. Bare compressor performance tests in July 2016, at Aerzen, Germany. Courtesy of ESS

Figure 3.11. Motor factory acceptance tests in June 2016, at ABB, Helsinki, Finland. Courtesy of ESS

(figures 3.15 and 3.16), scheduling the delivery, and creating specific lifting plans for on-site lifting and positioning prior to any delivery.

Deliveries had to be scheduled to ensure that they did not interfere with other construction activities and safe handling of the materials was of highest priority. The ESS logistics, rigging, and alignment departments were responsible for taking delivery and positioning the equipment properly within the buildings.

Despite the number and size of the ACCP components, delivery and installation generally went quite well and no significant interferences between the ACCP components and the buildings were found. Figures 3.17–3.20 give some examples of component deliveries.

Figure 3.12. Heat exchanger factory acceptance tests in April 2016, at Linde Schalchen, Germany. Courtesy of ESS

Figure 3.13. Cold-box building—ready for goods reception. Courtesy of ESS

Figure 3.14. Compressor hall—ready for goods reception. Courtesy of ESS

Figure 3.15. Access routes to ESS premises in Lund area, Sweden. Courtesy of ESS

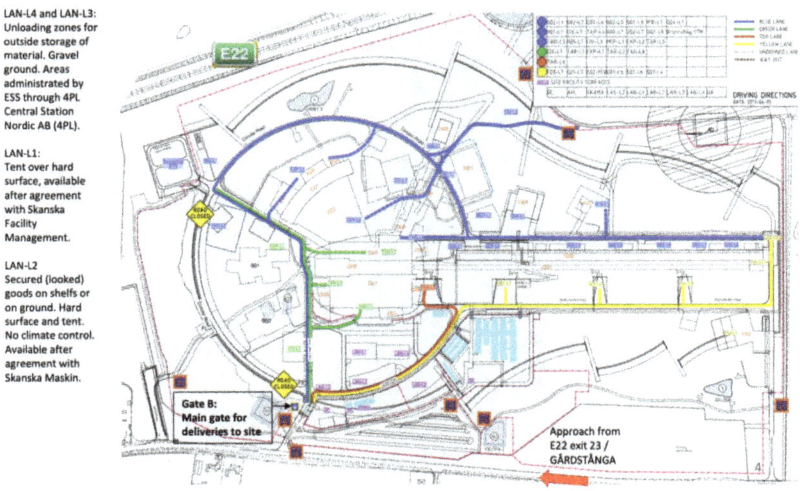

Figure 3.16. ESS building site access and uploading areas. Courtesy of ESS

3.13 Installation

The scope and sequence of ESS on-site installation are summarized below:

- The interconnecting piping between the compressor hall and the cold-box building.
- Concrete block and grouting the empty space between the compressor skid and the concrete block.

Figure 3.17. CBx delivery. Courtesy of ESS

Figure 3.18. Compressor skid delivery. Courtesy of ESS

- Equipment placement and installation.
- CBx platform installation.
- The interconnecting piping inside the cold-box building and the compressor hall from the vendor provided equipment to the wall interface.
- Mechanics verification tests, consisting of PT, LT, and radiographic test.
- Electrical installation, including cabling, wiring, earthing, etc.

Figure 3.19. Warm helium storage tank delivery. Courtesy of ESS

Figure 3.20. 20,000 l liquid helium storage tank delivery. Courtesy of ESS

- Electrical verification tests, such as loop check and test protocol.
- Final documentation.

Some pictures after the installation was finished are shown in figures 3.21–3.27.

An area supervisor is dedicated to an area or building and takes responsible for coordination of all installation work on-site. A work and safety coordination plan is a mandatory document required before the start of any work. The main purpose of this document is to identify the list of preparatory and organizational measures required prior to the start-up of the installation activities to be carried out by the contractor, as well as listing of associated hazards and safety control measures to be implemented. A work order had to be registered in the booking system and approved by the area supervisor before work can start. In order to perform specific tasks, the workers or technicians required special training and certification, e.g., for

Figure 3.21. Cold-box building installation-1. Courtesy of ESS

Figure 3.22. Cold-box building installation-2. Courtesy of ESS

hot work, fall protection, forklift, and crane operation, first aid training, and cryogenic safety training.

Installation, however, took much longer than expected. While six months were allotted for installation, it took about one year to finish. Some lessons learned are listed below.

- Parallel work and conflictions should be very planned, managed, and coordinated. While the building is ready, the contractor, subcontractor and ESS would like to perform the work including electrical, mechanics, and civil engineering at almost the same time. Managing this parallel work and solving the conflicts are the big issues at the beginning. An experienced area supervisor is essential and a framework with one company at the early stage will help to speed up the procurement procedure.
- The civil engineering work should not be underestimated. It takes a lot of time to get permission to build the compressor skid foundation, grouting the

Figure 3.23. Compressor hall installation. Courtesy of ESS

Figure 3.24. Cabling and wiring on the wall of the cold-box building. Courtesy of ESS

space between the skid and the foundation, or even drilling some holes to fix the warm storage tank.

- Failure or damage of the key equipment seems unavoidable, and it takes time to repair and replace (see figure 3.27). After assembling into the skid and compressor test run at Aerzen workshop, a leak between oil and water in the HP oil cooler was detected. After being delivered to ESS site, surface corrosion in the SP/LP oil cooler was found by an endoscope inspection. The replacement of these two heat exchangers by a new manufacturer, and the fabrication and installation caused at least five months delay.

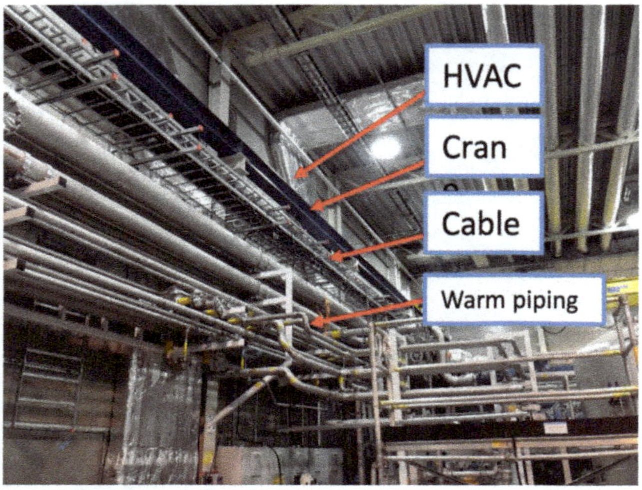

Figure 3.25. Piping and cabling on the wall of the compressor hall. Courtesy of ESS

Figure 3.26. Cooling water station installation (next to the compressor hall). Courtesy of ESS

Figure 3.27. Oil cooler corrosion and replacement. Courtesy of ESS

- The escape paths and space requirement in the case of fire were introduced into the equipment layout in the compressor building towards the end of the design phase, which caused some extra efforts to place the components and redesign the piping routing.
- Design drawbacks or mistakes are inevitable, and they must be corrected on-site. For example, the contractor overlooked the requirement to reserve a piping port connection for the redundant compressor and the installed piping had to be cut and reworked, while some piping and platform conflicts could only be found on-site. Even if the design seems perfect in the 3D model, an in-house piping design is a good solution and is strongly recommended.
- Searching and fixing leaks is always a headache for the cryogenic team. Even though a LT may have been performed and the system found to be perfect in the workshop, leaks are still possible due to transportation and on-site handling. Unfortunately, a tiny leak was detected inside the CBx during the PT on-site, which took a lot of time and effort to find and fix. It turned out that the welds of the connection of a pressure transmitter capillary were leaky.
- In summary, our recommendations focus on the installation: prepare your staff as early as possible, make the interfaces as simple as possible, develop a framework with one well-experienced installation company, the piping should be designed in-house, the helium recovery system should be brought online early, make your utilities available as early as possible, start the commissioning as early as possible, and, all in all, reserve enough margin in your schedule!

3.14 Warm compressor commissioning[2]

Following successful installation and commissioning in 2018, the final 100 h test run of the warm compressor system at maximum nominal design condition and the part load tests under various ACCP operation modes were carried out at the beginning of 2019. The key parameters, including mass flow, power consumption, and isothermal and volumetric efficiencies, are well fulfilled with the design data. This section describes the system's features, the project's challenges, the lessons learned, and the acceptance test results.

The ACCP WCS consists of three oil injected helium screw compressors and related bulk oil separators (BOSs), oil and helium coolers manufactured by Aerzen, the FOR, a dryer to adsorb moisture, and the GMP to control the pressure levels in the plant. The FOR, dryer, and GMP were separately ordered by LKT from their subcontractors. Schematics of the ACCP WCS and the overview of assemblies are shown in figures 3.28 and 3.29.

The SP compressor generates the compressor station's lowest pressure on the intake side, which directly comes back from the CCs in the CBx. The helium gas is compressed to MP. The SP compressor can be operated with the same process data

[2] This section is reproduced from Wang X L *et al* 2020 *IOP Conf. Ser.: Mater. Sci. Eng.* **755** 012087 used with permission.

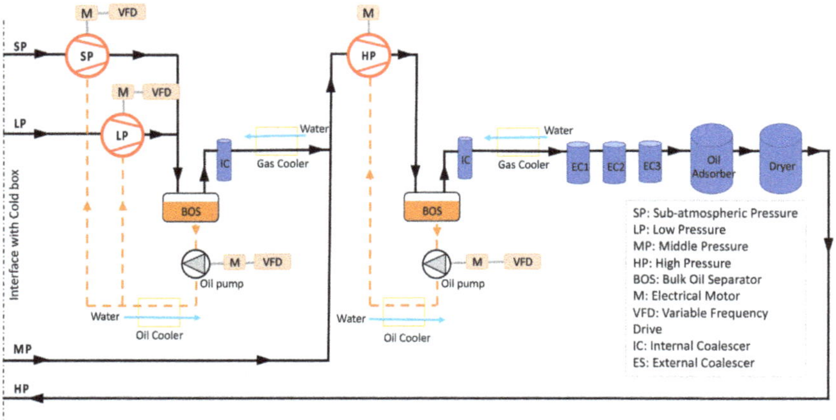

Figure 3.28. Schematic of the ESS ACCP warm compressor system. Reproduced from [16]. © IOP Publishing Ltd. CC BY 3.0.

Figure 3.29. Overview of the warm compressor system assemblies. Reproduced from [16]. © IOP Publishing Ltd. CC BY 3.0.

as the LP compressor. The LP compressor generates the pressure on the intake side, which is slightly above the atmospheric pressure and compresses the gas to the same MP as the SP compressor. Both the SP and LP compressors are supplied with oil by a common oil and gas skid, which also separates the gas–oil mixture coming from the compressors and cools the process gas by means of a gas cooler. The process gas from the oil and gas skid of the SP and LP compressors flows through a line to the intake side of the HP compressor. Additional helium coming back from the CBx is fed into this line. The HP compressor then compresses the entire gas volume to a high-pressure level and then releases it into the system on the discharge side. The MP is consistently regulated by the operation conditions. The adaption of the MP ensures operation of the turbines and screw compressors at highest efficiencies during all operation modes. Both the SP and LP compressors are equipped with a VFD to adapt the capacity of the compressor to the load case. The HP compressor runs with fixed speed because load adaption by adjustment of the MP makes a VFD unnecessary.

A guard system has been installed to prevent the entry of air into areas of the SP compressor that are constantly under subatmospheric condition during normal operations. This system consists of double seals whose intermediate space is pressurized to a level above the ambient pressure. This means that the only leaks that could occur are those resulting from intermediate pressure or guard pressure in the direction of the environment or the process system and the entry of air is prevented, even in case of leaks. All flanges and instruments for the SP compressor under subatmospheric conditions are guarded.

The oil and helium mixed after the compressor have to be separated completely in an oil removal system located downstream of the compressor. In the first step, the oil is removed from the helium in the first coalescer located in each BOS skid (internal coalescer). Downstream of the BOS the oil content of the helium is below 100 ppm. Three external coalescers are located downstream of the HP BOS and the oil content after the first of these coalescers is lower than 0.5 ppm. The first and second coalescers are equipped with a level switch to control drains for automatic return of the accumulated oil back to the compressor suction, while the third coalesce has only a manual valve because no aerosol oil collection is expected at this stage. The final removal of the oil vapor takes place in the charcoal adsorber vessel, which is designed for an operation time of more than 18,000 h. Within the adsorber, both the vapor phase and the remaining liquid impurities are removed from the helium to a final concentration of less than 10 ppb by weight. A full flow dryer is located downstream of the ORS to control the moisture level lower than 0.1 ppm by volume.

All valves and related piping terminals for the process control in the compressor station are grouped together as much as possible and are installed into one common panel, called a GMP, which is used to control the pressure levels in the plant, and for loading and unloading of the helium to/from the pure helium buffer vessels.

The design parameters for the ACCP WCS at the nominal design conditions are shown in table 3.12.

3.14.1 General test procedures

The WCS was ordered by LKT in September 2015 and delivered to the ESS site in July 2017, almost exactly on schedule. It took about 10 months to finish the installation in May 2018, and an additional nine months to complete the commissioning and the final performance tests in January 2019, which is a big deviation from the original project schedule. This delay was caused by several unexpected issues, which have been discussed in the previous section.

Each compressor was function tested, as well as pressure and leak checked, before shipment to ESS took place. The functional tests were performed at the workshop with ambient air in the Aerzen test bench for a minimum of 2 h. A thermodynamic equivalent operation of the compressor is not possible due to restrictions of the test-bench setup and of test motor capacities. These tests proved that the compressors operated within their design limits.

After the installation on-site, a series of inspections and reviews were performed together by LKT, Aerzen, and ESS's staff, including the completion of the

Table 3.12. ACCP WCS parameters at the nominal design conditions. Reproduced from [16].

Stage	SP	LP	HP
Number of units	1	1	1
Compressor type	Aerzener Screw Compressor VMY 536 H		
Internal volume ratio, Vi	4.0	2.6	2.6
Intake pressure, bar a	0.607	1.05	4.05
Intake temperature, °C	33	33	40
Intake volume flow, $m^3 h^{-1}$	4550	6345	4272
Mass flow, $g s^{-1}$	117	287	735
Discharge pressure, bara	4.25	4.25	20.5
Discharge temperature, °C	71	83	83
Motor speed, rpm	2670	3600	2950
Compressor capacity at motor speed, %	100	100	85
Shaft power consumption, kW	303	516	1383
Equipped with VFD	Yes	Yes	No
Impurity removal	Oil	BOS (1000 ppm)—Coalescer in skid (100 ppm)—FOR (0.5 ppm aerosols and 10 ppb vapor)	
	Water	One-bed online dryer (0.1 ppmV)	

mechanical and electrical installations, control loop checks, and documentation checks. A continuous action list of open points was established to keep track of all the issues found and to implement a plan to work on these issues. The compressor system was allowed to start up when all of the showstoppers were removed.

The acceptance tests at ESS for the WCS included functional tests and capacity tests with the helium. During these tests, the WCS remained isolated from the CBx. The functional tests measured the pressures and temperatures of the helium, oil, and cooling water, checked the performance of the oil removal system, tested the control software and interlocks for all modes of operation and simulated failures, and measured the main characteristics, such as helium flow rates, pressures, temperatures, and power consumption. The capacity tests were undertaken when all of the partial tests described above were successfully completed. For each compressor, a 100-hours, full load, steady-state run at the conditions necessary to run the ACCP in the nominal design conditions defined in table 3.12 was performed and an additional 2-hours run at various plant operation modes, as described in [9], was tested.

3.14.2 Final acceptance test results

The commissioning took much longer than the planned two months due to some issues, abnormal conditions, and troubleshooting. The commissioning and test runs indicated by the compressor mass flows are shown in the figure 3.30. The compressors experienced six test runs in total, two or three weeks for each time, and had to be stopped in-between to solve the problems. However, the 100-hours test run carried out from January 10 to 15, 2019 went smoothly and successfully. All of the parameters are according to the design data, and the expectations of mass flow and power consumption of all three stages are well fulfilled.

Volumetric and isothermal efficiency are the fundamental performance parameters commonly used for performance evaluation and comparisons of oil-flooded

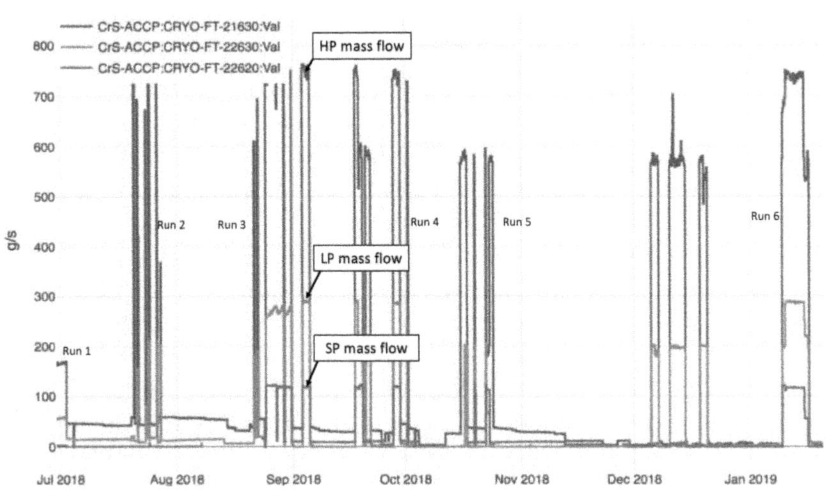

Figure 3.30. Commissioning and test run of the ACCP WCS. Reproduced from [16]. © IOP Publishing Ltd. CC BY 3.0.

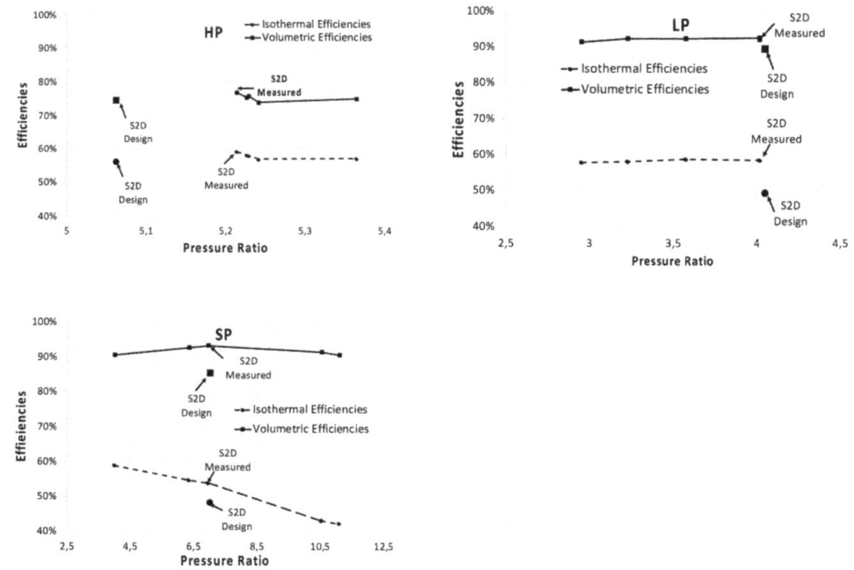

Figure 3.31. Isothermal and volumetric efficiencies of three compressors (the efficiencies at the normal design condition are shown for comparison with the actual measured results and S2D condition experienced 100 h, and the rest have about two hours test run). Reproduced from [16].

screw compressors [17–20]. Volumetric efficiency is the ratio of the actual (measured) mass flow rate to the theoretical mass flow rate calculated using the swept volume (displacement) at the measured inlet process conditions. Isothermal efficiency is the ratio of the theoretical input power to isothermally compress the actual mass flow rate to the 'measured' shaft power being delivered to the compressor.

Figure 3.31 shows the behavior of the volumetric and isothermal efficiencies with respect to the pressure ratios (various operation conditions). The shaft power is given by the measured output power from the switchgear with accuracy of 1%, and is corrected by the VFD and motor efficiencies, which is specified by the manufacturers. The mass flow rates are measured by the orifice flow meters with an expected error of 1.3%, which are installed in the GMP and calibrated before use. The temperatures, with an accuracy of 1%, and the pressures, with an accuracy of 0.2%, are measured directly at the suction and discharge side of each compressor. Besides the stage two design conditions (S2D) shown in figure 3.4, the other points include various actual operation conditions, such as stage two turn down, stage one design, stage one turn down, and stage two maximum liquefaction. Under the S2D conditions, the isothermal efficiencies show the HP of 59.1%, the LP of 58.4%, and the SP of 53.9%, and the volumetric efficiencies indicate the HP of 76.9%, the LP of 92.5, and the SP of 93.1%, which all reach the state-of-the-art compressor performance and are in advance of the guaranteed values. It should be noted that the HP compressor has a rather low volumetric efficiency because only 81% of HP capacity is used to perform the work and its slide valves stay at the fixed position of 95%. For LP and SP compressors, the VFDs are used to drive the motors. The slide valves stay open at 100% during nominal operation, resulting in much better volumetric efficiencies. In addition, at various operation modes, all compressors show rather good performance. As a result of the 100 h acceptance tests, the overall performance of each of the compressors and the overall compressor system, including the GMP and process control, met all of the goals set for ESS ACCP operations and were handed over to LKT for future commissioning together with the CBx.

3.14.3 Issues and improvements during commissioning

During commissioning several major issues have been encountered and improvements were incorporated rapidly, which are listed below:

- During the compressor start-up the oil is still cold and under certain circumstances the discharge pressure of the oil pump is high enough to trigger the external safety valve in the oil discharge line, causing the safety valves to become leaky. To prevent this, a mechanical overflow valve with lower set pressure than that of the safety valve was internally implemented into each of the oil pumps.
- It was found that about 1 l of oil collected at the bottom of the gas cooler after several days of continuous run. To minimize the helium bypass, a sight glass with manual isolating valves and the related piping were installed at the lowest point of each gas cooler to allow a local visual check and send the collected oil back to the compressor if necessary.
- For a long while (test run 1–4 in figure 3.3), the LP compressor showed ~20% less mass flow rate with ~10% less power consumption. It was eventually found that the LP mass flow was bypassed in the GMP. After closing the HP–LP bypass in the GMP, cross-checking the mass flow with a venturi flow meter through the CBx and oil injection optimization, the design mass flow was obtained and confirmed with a 10% less power consumption. This means that the actual LP performance is much better than the design.
- Oil leakage across the LP compressor shaft seal of ~12–15 ml h^{-1} is much higher than the ESS requirement of 5 ml h^{-1}. The shaft seal has already been replaced without effect. Further investigation is in progress.
- Wrong sized non-return valve and resulting extreme vibration. A non-return valve is installed at the suction side of the HP compressor to prevent the oil-rich helium flowing back to the pipeline in case the compressor is switched off. It was found during the commissioning that this valve is oversized and flapping continuously in certain operation cases, causing severe vibration to the surrounding equipment and piping. A correctly sized valve has been ordered, is being delivered, and the installation has been scheduled.
- To keep the CBx LP and LINAC SP pressures smooth and below acceptable limits, even during compressor malfunction or during transition cases, check valves used as backpressure regulator valves for each LP and SP compressor are installed to prevent the overpressure in these lines and release the helium to the gas bag. It turns out that the flow capacity of both check

valves combined is almost doubled than the gas bag safety relief valve capacity (~100 g s^{-1}), imposing a substantial risk for the gas bag. Implementations to restrict the check valve flow were carried out as soon as this was known.

- It is difficult to disassemble some parts of the skids, such as oil pumps or suction strainers, due to obstructed access, resulting in higher efforts for further maintenance. Special procedures must be developed and plenty of spare gaskets should be held in stock.
- The cooling ducts for the VFDs were pointing horizontally towards to the stairway, which is risky to the personal in case of an incident blowing up the VFDs. The ducts have now been modified to point vertically up to the roof.
- The helium guard box for the SP compressor has not been pressure tested in the workshop. The vessel had to be tested and certified by the notified body on-site.
- The compressor document package is, more than two years after compressor delivery, not fulfilling minimum requirements under the consideration of the complexity, completeness, and clearness. It is, however, being worked on by the supplier.
- There was an incident resulting in compressor damage, caused mostly by missing interlocks. This happened because when the LKT and Aezen PLC communication was lost, both the suction and discharge line valves upstream and downstream the HP compressor unit were closed but the compressor stayed in operating conditions. The suction pressure dropped to nearly 0.05 bar abs and the oil supply increased due to the low suction pressure. The oil between the female rotor and discharge side plate was pressurized and pushed the pressed-in radial bearing to the outer side of the side plate. The movement of the bearing sheared off the bearing temperature sensor, causing the compressor to finally trip because the suction pressure and vibration signals were not foreseen as trip initiations. After the compressor shutdown, a high oil flow flushed backward to the suction strainer, causing damage to the suction strainer. The dust and particles collected in the suction strainer dropped into the compressor during the equalization process, finally causing the scratches at the rotors. The damaged compressor had been disassembled at the workshop and the root causes have been identified. New trip signals have been implemented as a result, such as low suction pressure, both vibration sensors over-range, and PLC communication loss. An investigation into the possibility to reuse the damaged compressor with necessary repair or to replace it with a completely new one is ongoing.

3.14.4 Conclusion

The ESS ACCP compressor system was commissioned and successfully passed the final 100 h acceptance tests at the beginning of 2019. To reach the design mass flow, both LP and SP only consume 90% of the electrical power. The compressors' performances are beyond the guaranteed levels. This has allowed a very wide range of operation and enabled us to keep the reasonably good isothermal and volumetric efficiencies at various modes. After about 2000 h of operation for each compressor, they have proven to be efficient, reliable, and can be maintained.

3.15 ACCP commissioning experience

3.15.1 ACCP SAT

The ACCP will supply supercritical helium to the CMs in the LINAC through the CDS. According to the design specification [9], the ACCP has a cooling capacity of ~2.5 kW at 2 K, 4.5 K liquefaction rate of 6.8 g s^{-1}, and 8.5 kW at 40–50 K in stage one, and ~3.0 kW at 2 K, 4.5 K liquefaction rate of 9 g s^{-1}, and 11.4 kW at 40–50 K in stage two. In both stages, there are five operation modes, in which the stage one nominal design mode (S1ND) will be the most important operation mode in the coming years. In order to maintain the 2 K operation for the CMs, three stages of CCs and one SP warm compressor are adopted. In a collaboration between the supplier, Linde Kryotechnik (LKT), and the ESS, the ACCP has been successfully commissioned and tested in different operating modes. A simplified ACCP process flow diagram with a test vessel and heaters is shown in figure 3.32.

After the successful completion of the acceptance test and agreeing on the residual punch points, ACCP was handed over to ESS in October 2020. The SATs were carried out for more than six months at ESS with a joint effort of the engineers from Linde and the ESS Cryogenics group. Heaters built in the CBx were used to imitate the heat load from 2 K isothermal heat load, 2 K non-isothermal heat load, and TS heat load, and an external ambient heater is used to imitate coupler cooling flow. During the CBx acceptance test, seven working modes were performed and met the guaranteed performance requirements, i.e., stage two nominal design mode, stage two turn-down mode, stage two 4.5 K standby mode, stage two TS standby mode, stage 2 maximum liquefaction mode, stage one nominal design mode, and stage one nominal turn-down mode.

In the first year's operation, ESS will operate in the working mode of S1ND. A comparison of the parameters and the measurements, ESS requirements, and the

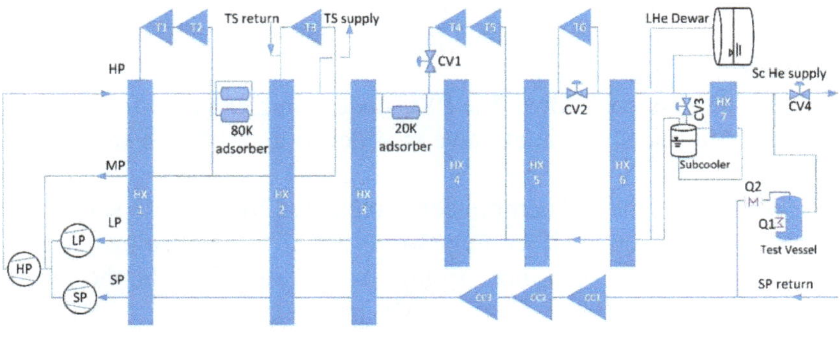

Figure 3.32. The simplified ACCP process flow diagram with a test vessel (heat exchangers: HX1–HX7; turbines: T1–T6; different pressure stages: HP, MP, LP and SP; heaters: Q1, Q2; TS [21]). Reprinted/adapted by permission from Zhejiang University Press: [23]

Table 3.13. A comparison between the measurement and design parameters in S1ND. Reprinted/adapted by permission from Zhejiang University Press: [23]

	Measurements	Requirements	Design*
2 K heat load, W	2519	2478	2603
4.5 K coupler cooling, g s^{-1}	6.9	6.8	7.14
TS heat load, W	8962	8551	8979
2 K mass flow, g s^{-1}	95.3	92.5	97.0
CC1 suction pressure, mbar	25.93	$\leqslant$ 27.00	26.00
4.5 K equivalent heat load, W	7585	7292	7670
WCS input power, kW	1900	—	2034
WCS exergy efficiency, %	48.5	—	45.1
CBx exergy efficiency, %	54.0	—	55.0
System exergy efficiency, %	26.2	—	24.8

*The ACCP design parameters have a 5% margin compared with the requirements of ESS.

design of ACCP in S1ND is listed in table 3.13. All the measurements exceed the ESS requirements. The design parameters have a 5% margin with the requirements. Using the exergy efficiency with the same definition as in [21], the measured WCS and the CBx exergy efficiencies are 48.5% and 54.0%, respectively. The measured system exergy efficiency is 26.2%, which is 1.4% higher than the design. The performance of ACCP is better than the expectations.

3.15.2 ACCP commissioning and performance tests

After taking over ACCP, the Cryogenics group had around two years to perform the ACCP commissioning and different tests in order to characterize the operation envelope, the tolerance for isothermal heat load, non-isothermal heat load and TS heat load deviations on CDS and CMs, as well as the plant flexibility.

With the operation experience, ACCP is operated in a very efficient and stable condition. The major commissioning experience includes:

- Before cool down, the impurities of the system are removed by a small flow circulation among ACCP, pure helium tanks, gas bag, and purifier until the impurity condensation is below 10 ppm. In this way, we could save a lot of helium compared with pumping and purging. For ACCP only, it took 4 days to make the system clean.
- During the CBx cool down, the PLC code is implemented for the opening of the third turbine string inlet valve in order to avoid Mach shock of turbines T4 and T5.
- The capacity of the HP compressor is regulated by the slide valve, which will affect the efficiency of HP compressor. Normally, the opening of the HP slide valve should make sure the opening around 20%–30% of the two HP-MP bypass valves in order to keep the MP pressure stable.

In future operation, the heat loads might be less or more than the design, and therefore several different scenarios have been tested [22], the test results include:

- Based on the measured CM's heat load, the measured 2 K static heat load and 2 K dynamic heat load are less than the design, while the TS heat load is ~10% more than the design. After testing, the ACCP operated steadily with the corresponding measured heat loads.

- For 2 K non-isothermal heat load, with the design 2 K static heat load, i.e., 850 W, the maximum allowable 2 K non-isothermal heat load is 40% more than the design (627 W). The maximum allowable 2 K non-isothermal heat load increases by increasing the CC1 suction pressure or adding the extra 2 K isothermal heat loads by heaters in the 2 K test vessel.

- For 2 K isothermal heat load, with the CC1 suction pressure at 26 mbar, the maximum allowable 2 K isothermal heat load is 2000 W, which is 7.5% higher than the design. The maximum allowable 2 K isothermal heat load could be increased by increasing HP pressure or increasing CC1 suction pressure.
- For TS heat load, the maximum allowable TS heat load is at least twice as much as the design and the test is limited by the capacity of the heater.

Based on the test results in different operating conditions, the HP pressure in GMP and the system efficiency are shown in figure 3.33, in which the HP pressure has a linear correlation with the 4.5 K equivalent heat load [23]. Generally, the system efficiency increases with the 4.5 K equivalent heat load increase. When ACCP operates in floating mode, i.e., the GMP load/unload valves are closed, a 20,000 l dewar is used to balance the system capacity. The slide valve in the HP compressor could control the bypass flow from HP to MP, which affects the WCS efficiency. The HP pressure could be set according to the 4.5 K heat load to fine tune the system's stability before operating in the floating mode.

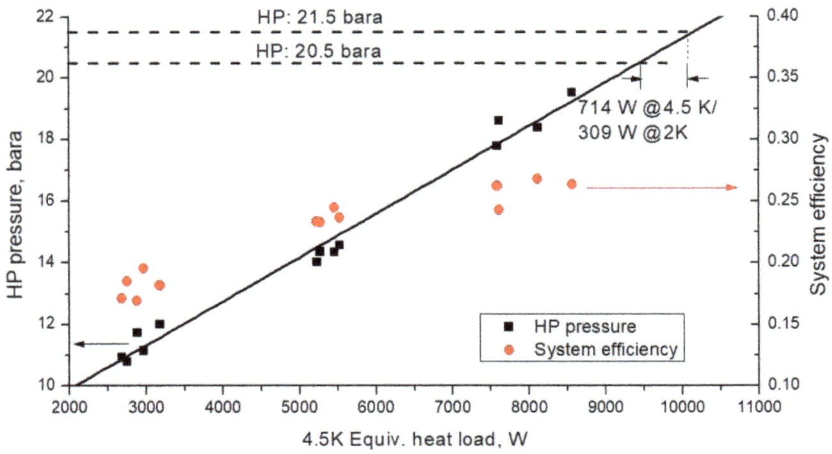

Figure 3.33. HP pressure in GMP and system efficiency versus 4.5 K equiv. heat load for ACCP. Reprinted/adapted by permission from Zhejiang University Press: [23]

3.15.3 The cold compressor operation

In ACCP, the three CC string combined with a SP compressor are used to keep all CMs in the LINAC at 2 K. The three CCs were installed in the CBx and can be replaced easily with the CBx keeping cold. The suction pressure of SP compressor could range from 0.3 bara to 1.05 bara has a large margin to compensate the flow variation for CCs, especially during the pump down or in off-design conditions. The flow scheme for the 2 K part is presented in figure 3.34. There are three controllers to control the CC1 suction pressure, CC1 suction temperature, and the CCs' mass flow. The bypass valves CV34998 and CV33801 are used to regulate the CCs mass flow (FC34905) and CC1 suction temperature (TC34650). The set point of FC34905 is based upon the SP suction pressure, which means the mass flow set point is regulated by setting the SP suction pressure. The CC1 suction pressure can be set in auto mode. If the CCs have not arrived at the maximum working capacity, then the CC1 suction pressure will reach the set point.

We tested the CCs with the 2 K test vessel and heaters in the CBx, and also tested the CCs with the 374 meters long 2 K return line and two CMs connected. The CCs met the expectations of the design and work stably.

With 2 K test vessel and heaters, the CCs were tested with mass flow from 55 g s^{-1} to 95 g s^{-1}, the measured parameters of CCs working in S1ND mode are shown in table 3.14. With the CC1 suction pressure of 25.93 mbar and the suction temperature

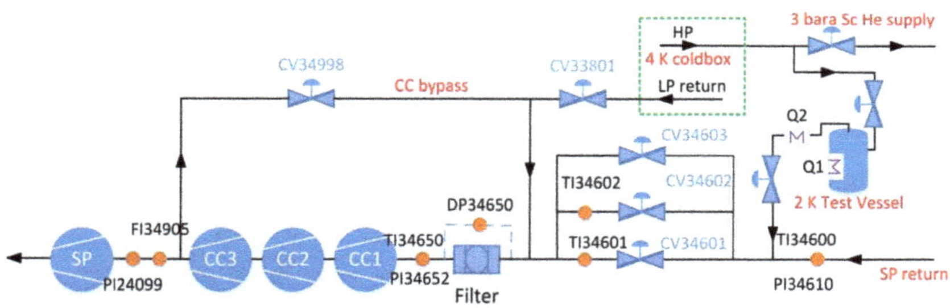

Figure 3.34. The flow scheme of CCs and SP compressor. Reprinted/adapted by permission from Zhejiang University Press: [23]

Table 3.14. The measured parameters of CCs in S1ND mode with 2 K test vessel. Reprinted/adapted by permission from Zhejiang University Press: [23]

	CC1	CC2	CC3
T_{inlet}, K	4.5	8.9	15.7
P_{inlet}, mbar/bara	25.93 mbar	0.10 bar	0.30 bar
T_{outlet}, K	8.9	15.7	22.7
P_{outlet}, bara	0.10	0.30	0.62
Mass flow, g s^{-1}	95.3	95.3	95.3
Polytropic efficiency, %	79.1%	75.7%	81.5%
Isentropic efficiency, %	73.1%	72.1%	75.5%

of 4.5 K, the mass flow of the CCs is stabilized at 95.3 g s^{-1} and the isentropic efficiencies of all the CCs are >72%. When the CCs work in off-design conditions, such as higher suction temperature up to 7 K, the CCs might work with lower efficiency.

With 2 K test vessel and heaters, the capacity, flexibility and stability of the CCs and SP compressor were tested, the results include:

- The SP compressor has a margin to deal with the extra 2 K heat load. When the SP suction pressure is 0.31 bara, the 2 K isothermal heat load can be increased to 1700 W, which is twice the design 2 K static heat load. Therefore, we can operate the CCs with higher flow than the design with lower SP suction pressure because the SP compressor can speed up to deal with the extra flow.
- The lower the CC1 suction pressure is, the higher capacity of CCs will be required. With the lowest CCs' mass flow of 55 g s^{-1}, the minimum CC1 suction pressure is 20.5 mbar. With the design S1ND mass flow 95 g s^{-1}, the minimum CC1 suction pressure is 25.6 mbar.
- The CC mass flow controller FC34905 and CC1 inlet temperature controller TC34650 are used to regulate the CCs' mass flow and CC1 suction temperature by CC bypass valves, so as to keep the operation of CCs stable when the 2 K heat loads change. The default minimum opening of bypass valves is optimized by several tests to make sure there is no trip when the bypass valves start to open.
- In different working conditions, the fluctuation of CC1 suction pressure is within ±0.1 mbar.

In June 2023, the CCs had 2 K commissioning with the 374 meters long 2 K return line and two CMs connected. First, the CC line was cooled automatically by setting SP suction pressure to 0.8 bara, and the return helium from VLP line went back to CC line instead of the bypass line. Then the VLP line, two CMs, Endbox, and 2 K test vessel were pumped down from 0.8 bara to 0.3 bara by SP compressor within 1 h. When the SP suction pressure was 0.3 bara, the CCs started to pump down to 30 mbar within 2 h, as shown in figures 3.35 and 3.36. With the set pressure of SP compressor at 0.3 bara, the corresponding CCs' mass flow set point is 55 g s^{-1}. At the beginning of CCs' pump down, the CCs' mass flow was between 60 g s^{-1} and 90 g s^{-1}. As a result, CC2 and CC3 operated in a choked flow with a very high speed to handle the high flow. In the end, all CCs operated in the correct stable working area. The stable operation OPI of CCs in S1ND with two CMs connected is shown in figure 3.37.

The CCs were tripped several times during commissioning. There are two main causes of a CC trip: one is SP compressor trip and the other is CC1 inlet temperature too high caused by the warm connection pipes of the bypass valves. When the mass flow has a large change, especially during CCs' pump down, the speed of the SP compressor cannot quickly react, so that the SP compressor suction pressure drops to 0.1 bara and then the SP compressor has a trip. In order to cope with a SP compressor trip, we implemented a new control logic to make the SP suction

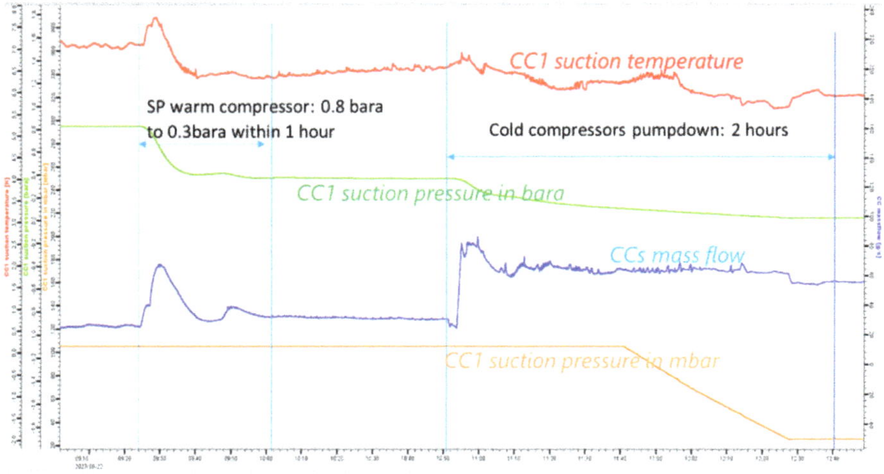

Figure 3.35. The trends of CC1 suction temperature, CC1 suction pressure, and CCs' mass flow during the CCs' pump down. Courtesy of ESS

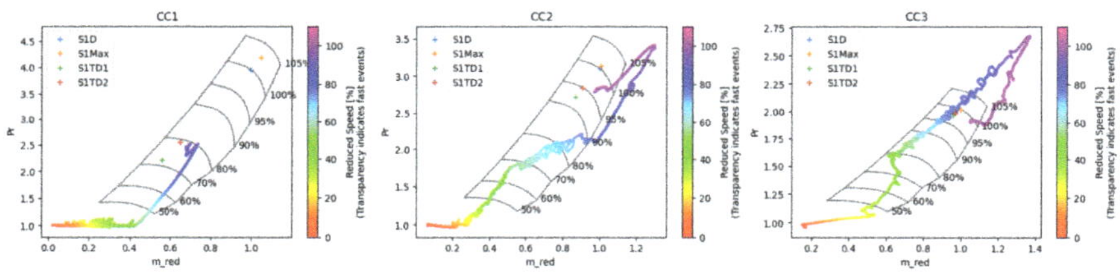

Figure 3.36. The CCs' map during CCs' pump down. Courtesy of ESS

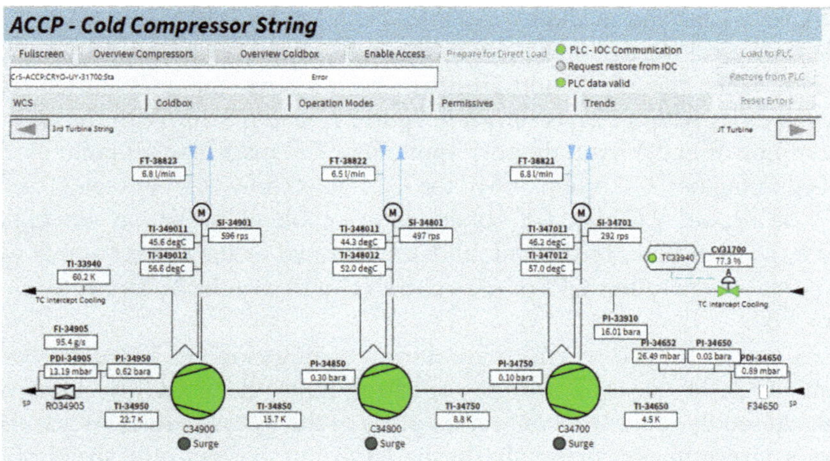

Figure 3.37. The stable operation OPI of CCs in S1ND with 2K return line and two CMs connected. Courtesy of ESS

pressure maintain 0.3 bara by sending flow from MP line to SP suction side when the SP suction pressure $<$ 0.3 bara. When there is a 2 K heat load drop or CC1 suction temperature increase, the CC bypass valves will open to maintain the mass flow and CC1 suction temperature. However, if the connection pipe between the bypass valve and the CC1 suction side is warm, then it will also cause a CC trip. The default opening of the CCs' bypass valves was optimized to keep the connection pipe cold. The CCs now operate stably during pump down and heat load change.

3.15.4 ACCP-CDS heat-load test preparation

After the installation, the CDS includes a 374 meters long cryogenic multi-transfer lines, 43 valve boxes, and one Endbox (see also chapter 4). In order to perform the CDS heat-load test, vapor helium with a supply temperature at 8 K was used for the ACCP and CDS integrated commissioning [24]. Therefore, ACCP was implemented with two new working modes for the ACCP-CDS integrated tests by the supplier, Linde.

Compared with the normal operation modes during ACCP acceptance tests, the ACCP-CDS tests are more challenging and need new controllers to supply stable gas helium to CDS. A comparison between ACCP acceptance tests and ACCP-CDS tests is given in table 3.15.

The two new working modes for ACCP-CDS tests were implemented in PLC, i.e., CDS heat-load test and CCs' test. There are two new controllers, shown in figure 3.38: one new temperature controller (TC1) is added, which is deployed on Turbine 4 (T4) inlet valve CV1; one new pressure controller (PC1_2) is added, which is deployed on JT turbine (T6) bypass valve CV2 in CDS heat-load test; and one already existing pressure controller (PC1_1) is deployed on subcooler inlet valve CV3 in the CCs' test.

In September 2021, the new control logic was tested with ACCP and the heaters. All the new controllers worked well [25]. For the CDS heat-load test, the ACCP could supply stable mass flow to CDS, which ranged from 11 g s^{-1} to 55 g s^{-1} at 3 bara and 8 K. With a 4.5 K supply temperature of 6.2 K, the CCs' test was carried out successfully and the CCs' mass flow was 94.5 g s^{-1}.

Table 3.15. A comparison between ACCP acceptance tests and ACCP-CDS tests. Reprinted/adapted by permission from Zhejiang University Press: [24]

No.	ACCP acceptance tests	ACCP-CDS tests
1#	Helium in liquid and vapor phase	Only gas phase
2#	Temperature stable with 4.5 K subcooler (600 l) and 20,000 l LHe dewar	New temperature controller with different principle
3#	JT turbine is in operation	JT turbine does not work in CDS heat-load test
4#	Load calibration and stabilization with LHe dewar	New pressure controllers
5#	Mass flow is decided by heaters	Different mass flows are set in CDS heat-load tests by control valves

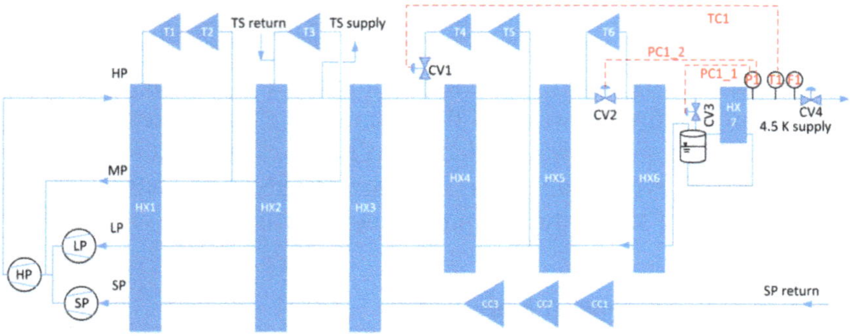

Figure 3.38. The simplified ACCP flow diagram with new controllers (heat exchangers: HX1–HX7; turbines: T1–T6; different pressure stages: HP, MP, LP, and SP). Reprinted/adapted by permission from Zhejiang University Press: [24]

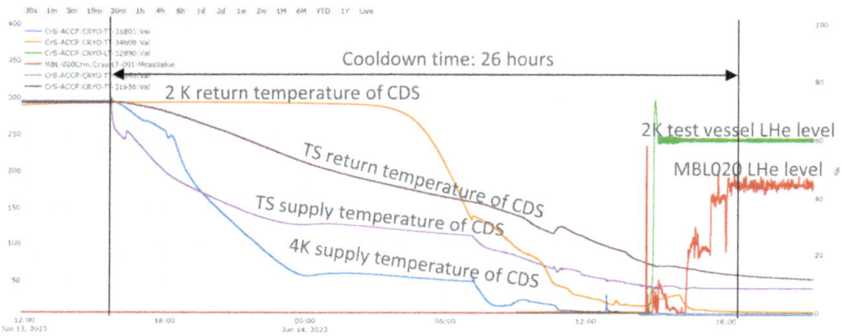

Figure 3.39. Trends of the second CDS cool down. Courtesy of ESS

3.15.5 ACCP-CDS integrated commissioning

In December 2022, CDS and ACCP were first cooled down together to 8 K. The CDS heat-load tests were performed with different mass flows. Unfortunately, thermoacoustic oscillation (TAO) occurred in most of the 2 K return valve CV04 in CDS-EL and the Endbox safety valve SV91 inlet pipe. TAO could induce extra heat load to the system and make the system unstable, e.g., pressure fluctuation. More information about CDS commissioning is given in chapter 4.

Before the second ACCP-CDS commissioning, most of the issues that occurred in the first cool down were fixed and two pilot CMs (one Spoke cryomodule SPK110 and one elliptical cryomodule MBL020) were installed in the tunnel. It took 26 h to cool the ACCP, CDS main lines, 2 K test vessel, two CMs, and Endbox to 4.5 K (see figure 3.39). The CCs then started to pump down the system to 2 K. The 2 K stable operation state of MBL020 and SPK110 are shown in figure 3.40. At 2 K, the automatic control and interlock were tested for the CMs, the highlights are:

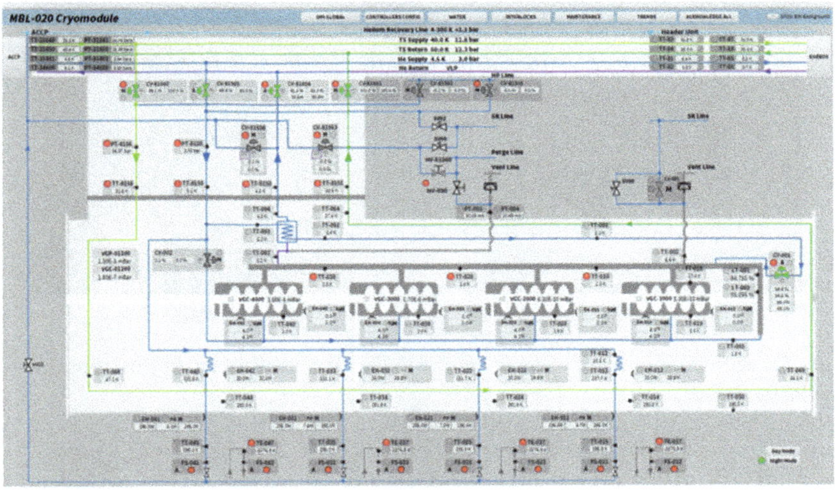

(a) 2 K stable operation of MBL020 with helium level of 85% at 2 K

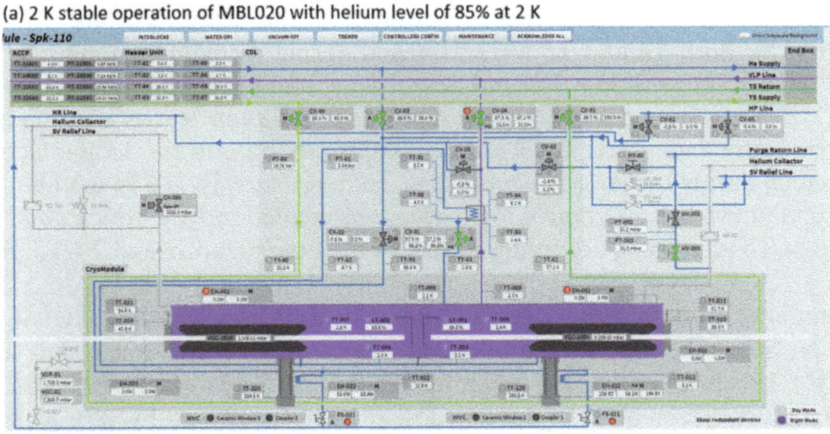

(b) 2 K stable operation of SPK110 with helium level of 89% at 2 K

Figure 3.40. The 2 K operation of CMs MBL020 (a) and SPK110 (b). Courtesy of ESS

- The helium level is automatically controlled by the inlet valve CV01.
- The helium pressure is automatically controlled by the outlet valve CV04 with a pressure fluctuation of ±0.1 mbar.
- During the night, the night mode was activated to avoid helium overfill, i.e., if one of the level gauges is higher than the set point, then the 4 K inlet valve located in the valve box CV03 will be closed immediately.

3.15.6 Issues

During the ACCP commissioning and tests, there were some issues and challenges. The outstanding issues include turbine damage, CC2 speed sensor failure, CCs' filter blockage, and 20,000 l dewar leak to air.

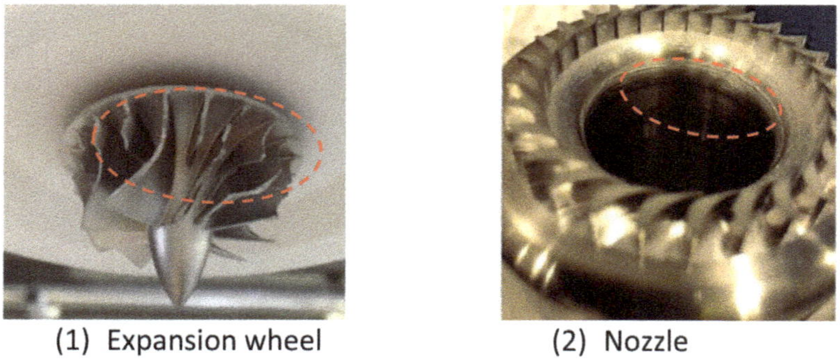

Figure 3.41. The damaged expansion wheel and nozzle in T5. Reprinted/adapted by permission from Zhejiang University Press: [23]

3.15.6.1 Turbine damage

There are two serial turbines in the CBx third turbine string, Turbine 4 (T4) and Turbine (T5). Both T4 and T5 were damaged during the commissioning. On June 14, 2021, the T5 expansion wheel and nozzle were found to have been damaged after the ACCP had suffered from a performance degradation for several months, as shown in figure 3.41. After LKT's investigation, the reason for T5 damage is the Mach shock during the cool-down process. To make sure T4 /T5 Mach number <1 during cool down, a new control logic was implemented in PLC which controls the opening of T4 inlet valve according to the T4 inlet temperature and T4/T5 pressure ratio.

T4 suffered from bearing damage and inlet filter deformation after a power outage on September 30, 2021. Following LKT's inspection it was found that the reason for T4's bearing damage might have been that T4 was incorrectly assembled in the LKT workshop and the deformed filter might be caused by very high debris load. The spare T4 was then installed.

The repair was still under warranty and the damaged turbines were swiftly carried out by LKT. Both T4 and T5 have nominal performance.

3.15.6.2 CC2 speed sensor failure

On October 22, 2021, the CC2 speed sensor was broken and sent back to LKT for replacement. During the future operation of the LINAC, it may happen that we need to replace one CC and keep the rest of the cryogenic system and CMs working at 4.5 K. Therefore, we arranged the test to warm the CC string under the CBx cold condition. The process includes CC string warm up, CC spare replacement, CC string cool down, and CC string pump down, as shown in figure 3.42. It took 24 h to warm the CCs, 1.2 h to cool the CC string, and 0.6 h to pump down the CC string. In the CCs' pump down with VLP line and two CMs, 3 h are needed for the CCs' pump down. According to our experience, around 6 h are required to replace the spare CC. Therefore, 35 h in total are needed to replace one CC with CBx cold.

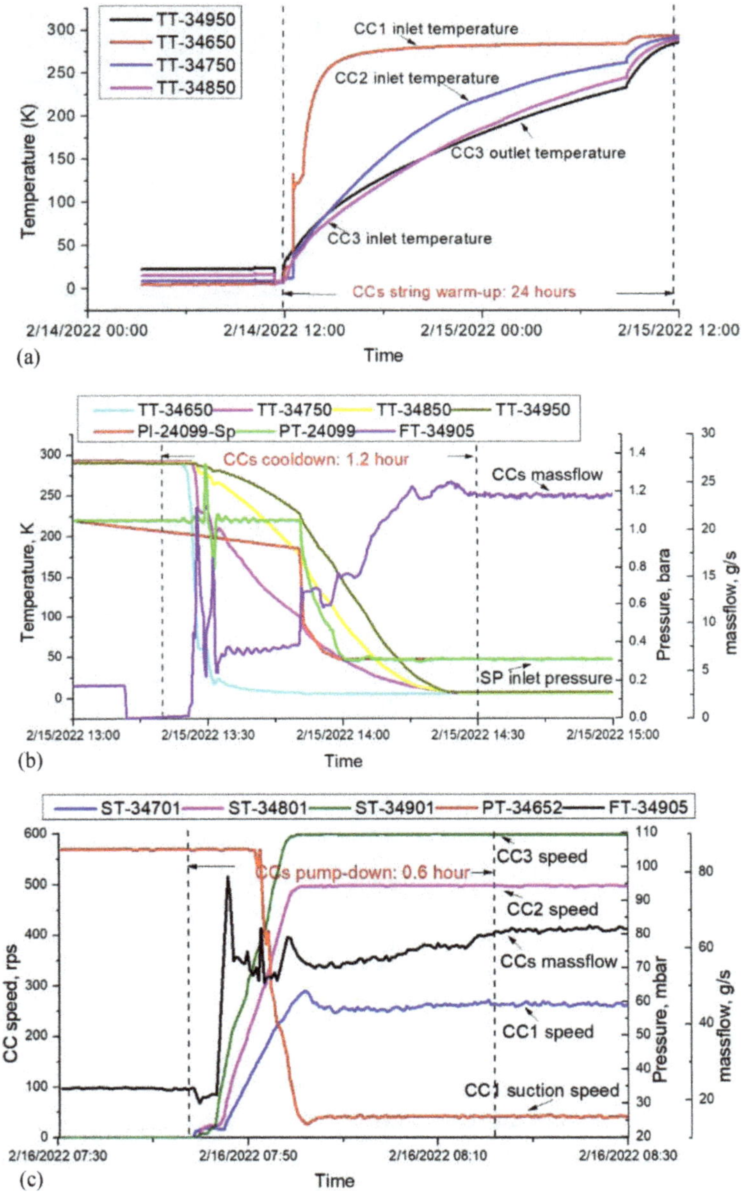

Figure 3.42. (a) The trends for CC string warm up, (b) the trends for CCs' cool down, and (c) the trends for CCs' pump down. Reprinted/adapted by permission from Zhejiang University Press: [23]

3.15.6.3 CC filter blockage

The CC upstream filter F34650, shown in figure 3.37, is used to protect the CCs from damage caused by debris and small particles. In the CCs' normal operation with a mass flow of 55–95 g s^{-1}, the pressure drop of filter F34650 is within 1 mbar. However, when impurities accumulate in the filter, the pressure drop across the CC

(1) Dewar leak to air at 1.9 bara

(2) No leak after tightening the bolts on top flanges at 1.9 bara

Figure 3.43. Dewar leak and fix. Courtesy of ESS

filter will rise. A CC filter blockage occurred twice in the previous commissioning: the first time was when there was an air leak in the CBx cold end, and the second time was when the Endbox helium vessel cooled after warm up to 200 K during CDS commissioning. The solution for CCs filter blockage is to warm the CC filter to 300 K to remove the impurities and then cool the CC line again.

3.15.6.4 20,000 l dewar leak to air

During the second CDS cold commissioning there was a large helium leak to air on the top flanges of the 20,000 l dewar and ice formation on the top of the dewar. The estimated helium loss was 10 kg day^{-1}. After depressurizing the dewar pressure to 1.25 bara and tightening the bolts on the dewar flanges, there was no leak when the dewar pressure was 1.9 bara (see figure 3.43).

3.16 Lessons learned

1. The ACCP is a large cryogenic system working at 2 K that has three warm compressors. If one of the warm compressors stops working, then the operation of the CMs in the LINAC will be affected. In order to compensate for a possible malfunction of one of the warm compressors during 2 K operation, ESS will procure a new helium compressor system that can back up any stage of the ACCP warm compressor. This will increase the availability of the cryogenic system.
2. In order to have more flexibility for the 2 K operation, ESS chose a CC string with three serial CCs and one SP warm compressor to pump down helium to the saturation pressure at 2 K. In the CCs' commissioning with a large volume 2 K return line, the SP compressor could easily pump down the 2K return line from 1 bara to 0.3 bara thanks to its large capacity, and then the CCs were able to pump down the 2K return line steadily from 0.3 bara to 26 mbar within 2 h.

The CCs are able to handle mass flow rates from 55 to 95 g s^{-1} and keep the suction pressure stable, even though there is heat load change in the CMs, thanks to the mass flow, suction pressure, and suction temperature controllers.

3. Due to the project delay in the cryomodule installation and other key components, ESS had more than 2 years to operate the CBx of the ACCP. This gave us the opportunity to:
    - Train our staff to be familiar with the operation procedures and operate the system in different work conditions.
    - Optimize the control logic of the ACCP to have better automatic and stable operation.
    - Implement new control logic for safety relief after ACCP shutdown.
    - Find the design deficiency of two turbines during CBx commissioning, which were damaged during cool down, and then optimize the cool-down procedures in PLC code in order to avoid Mach shock.
    - Get precious experience of the CCs' operation, including replacing one CC while the remaining system was working cold, removing impurities for the upstream filter of the CCs, and optimize the default settings of the bypass valves to avoid CC trip, etc.
4. In the design phase of the ACCP, the working mode of testing the CDS without CMs should also be considered. In ESS, we added two new working modes to test the CDS heat load at 8 K and test the CCs at 6 K with helium gas only. One new temperature controller and one new pressure controller were implemented and tested successfully. The CDS heat-load test mode was very helpful for CDS testing (see also chapter 4). The CC at 6 K testing was not successful due to dysfunctional CC bypass controllers at this elevated temperature.
5. Before the CDS commissioning, the implementation of CDS control and CDS safety interlock started a little late due to the lack of skilled cryogenic control system engineers. With the CDS commissioning of first cool down and second cool down, a number of instrumentation and control system related issues were found and solved.
6. Commissioning of the CDS without connecting to CMs made it easier to find issues and to have more space in the tunnel to do the repair work on the valve boxes. For example, there are 8 or 10 control valves in each valve box; however, the space between the cryomodule and valves box is very narrow. Consequently, it would be very hard for operators to dismount valves from the valve box if the cryomodule is connected. After the first cool down, the CDS repair work related to valve boxes included:
    - Seat seals were replaced in more than 50 control valves to fix seat leaks.
    - Twelve 4 K circuit safety valves were dismounted and certified for working pressure due to opening at a pressure lower than the set point.
    - Convection inhibiters were added to the stems of all 30 2 K return valves (CV04) in CDS-EL to eliminate TAOs.

- TAOs in the 4 K supply line of the endbox at the inlet to the safety valves were detected. Mitigating measures included moving the safety valve as close as possible to the vacuum vessel and adding a hydraulic RLC system consisting of throttle valve, flexible hose, and damper vessel (see also chapter 4).

Appendix A: ACCP technical specification

Table of contents

(*Continued*)

(*Continued*)

Appendix B: ACCP bid evaluation criteria

1. PRINCIPLES OF EVALUATION

The contract award shall be decided on the most economically advantageous tender presented based on the criteria below:

(1) CAPEX criteria;
(2) OPEX criteria;
(3) Qualitative criteria.

The most economically advantageous tender is based strictly on a combination of capital costs, operational costs, and qualitative impact.

CAPEX	Capital cost for the entire baseline scope of supply requested in this tender without options and offered by the tenderer in EUR
OPEX	Operating cost for the offered cryoplant, taking into account the electrical power consumption for the warm compression system in the operation scenario sequence as assumed in section 3 of this document
SC	Score for CAPEX, maximum score is 30
SO	Score for OPEX, maximum score is 30
SQ	Score for quality, maximum score is 40
MIN	MIN is the figure characterizing the minimum value given by one tenderer out of all tenderers
OT	Operation time [h], total operation time $OT_{\text{Total}} = 66\ 000$ h
OC	Operation cost per time [€/h]
EC	Electricity cost 50 €/MWh
P_{el}	Electricity consumption for warm compressor station [MW]

Definitions/overview of process

The score will be calculated with an accuracy of one decimal, the maximum score being 100.

Total Score = SC+SO+SQ

2. CAPEX EVALUATION

SC = Score of CAPEX	Maximum Score: 30
$SC = 30 - \frac{CAPEX - CAPEX_{MIN}}{100\ 000\ EUR}$	

3. OPEX EVALUATION

The score for operating cost is similarly defined to the score for capital cost.

SO = Score of OPEX	Maximum Score: 30
$SO = 30 - \frac{OPEX - OPEX_{MIN}}{100\,000\ EUR}$	

For the operation cost (OC) of the ACCP, only the cost for electric power of the warm compression system is considered. The electric power of the warm compression system will be measured during the final acceptance tests at ESS with ESS AB provided measurement equipment. Only the electric power consumptions measured during the final acceptance test will be taken into account, not the actual operating costs during plant operation.

The OC of the ACCP (OPEX in EUR) is calculated by multiplying the specific OC in EUR/hour for three specified operation modes with their respective estimated fraction of operation time and the total operation hours over the course of 10 years, regardless of the actual future operation of the system. The specific OC of the relevant operation modes are defined as

Stage one—Nominal design: $OC_{St1,ND}$
Stage two—Nominal design: $OC_{St2,ND}$
Stage two—Nominal turn down: $OC_{St2,TD}$
Operation during ten years is considered (OT_{Total}).

The specific OC of each operation mode is calculated by multiplying the assumed energy cost (EC in EUR/MWh) with the electric power consumption of the warm compressor system as measured during the respective final acceptance tests ($\boldsymbol{P_{el}}$ in MW)

$$OC = EC \times P_{el}$$

The OC of the ACCP is then defined as

$$OPEX = (0.25 \times OC_{St1,\ ND} + 0.5 \times OC_{St2,\ ND} + 0.25 \times OC_{St2,\ TD}) \times OT_{Total}$$

To calculate the score of the OC SO, the respective electricity consumptions as given in the tender will be used. These values will be verified during the performance tests at ESS.

In case the OPEX as defined here are higher than the OPEX calculated by using the electrical power consumptions as stated in the Contractor's tender, the Contractor shall pay ESS AB this difference. The total sum due will be deducted from the last milestone payment in Clause 5.1 above.

In case the OPEX as defined here are lower than the OPEX calculated by using the electrical power consumptions as stated in the Contractor's tender, the ESS AB shall pay the contractor 50% of this difference. The payment is due at the end of the warranty period.

4. QUALITATIVE EVALUATION

The maximum score for qualitative impact SQ is 40.

(a) **Completeness** of proposal with regard to requested documentation in appendix 3 *Instruction to Bidders*; please note that poor documentation may lead to score reduction due to lacking means of comparison and proposal evaluation.	Maximum Score: 3
(b) **Conformity** to commercial terms and conditions as set forth in ESS draft contract.	Maximum Score: 2
(c) **Understanding** and compliance with requirements of the technical specification, i.e., no or only few exemptions from the technical requirements.	Maximum Score: 4
(d) **Simplicity** and feasibility of plant layout, design, and interface concept to limit cost and effort for site preparation and installation work at ESS.	Maximum Score: 4
(e) **ESS Option 1**: On-site piping, cabling, and installation as described in appendix 4 *Statement of Work*—technical and organizational approach.	Maximum Score: 3
(f) **ESS Option 2**: LHe storage tank and transfer lines as described in appendix 4 S*tatement of Work* and specified in appendix 5 *Technical Specification*—technical solution.	Maximum Score: 2
(g) **Additional features** implemented in the proposed design approach that are technically and/or commercially advantageous for ESS.	Maximum Score: 4
(h) **Availability** of the cryogenic system, taking into account the submitted data for MTBF and MTTR for all equipment, the concept and number of cold and hot spare components, and the ability to keep operating at lower performance upon equipment failure.	Maximum Score: 4
(i) **Soundness** of the control strategy as described in the tender, considering feasibility, stabilization after load adaptions, and simplicity.	Maximum Score: 5
(j) **Recovery time** from a trip of rotating machinery without damage, taking into account submitted descriptions of failure scenarios, mitigation strategies, and references that can be contacted for verification.	Maximum Score: 4
(k) **Liquefaction** capacity at constant level in the liquid helium storage tank and full shield performance as specified in appendix 5, Technical Specification, section 5.2.5 (increased shield temperatures up to 60 K are permitted); the liquefaction capacity will be tested during the plant acceptance test at ESS.	Maximum Score: 3
(l) **Response time** during warranty period upon notification of relevant mechanical failures.	Maximum Score: 2

References

[1] Peggs S *et al* 2013 Technical Design Report, ESS-2013-001 https://europeanspallationsource.se/sites/default/files/downloads/2017/09/TDR_online_ver_all.pdf
[2] Roland G *et al* 2018 The European Spallation Source design *Phys. Scr.* **93** 014001
[3] Wang X, Weisend J, Koettig T, Wolfgang H and Darve C 2014 ESS accelerator cryogenic plant *HVAC&R Research* **20** 296–301
[4] Duthil P 2014 Heat loads evaluation for the spoke cryomodule *IPNO Technical Note, IPNO-DA-ESS-NT-20140126*
[5] Olivier G 2014 Cryomodule for elliptical cavities: Global heat balance *IPNO Technical Note*
[6] Molloy S, Darve C and Wang X L 2014 Heat load estimates for the ESS superconducting cryomodules *ESS Technical Note, ESS-0008356*
[7] Jarosz M 2014 Preliminary calculations on heat loads due to beam losses in spokes and elliptical cavities *ESS Technical Note, ESS-0011645*
[8] Fydrych J 2013 Preliminary heat load estimations for the cryogenic distribution system of the ESS accelerator *ESS Technical Note*
[9] Wang X, Arnold P, Fydrych J, Hees W, Jurns J M, Piso D and Weisend J 2015 Specification of the ESS accelerator cryoplant *Phys. Proc.* **67** 89–94
[10] Arnold P, Boros M and Nilsson P 2021 ESS cryogenic controls design *EPJ Techn. Instrum.* **8** 8
[11] Ganni V and Knudsen P 2010 Advances in cryogenic engineering *AIP Conf. Proc. (Tucson, AZ, USA) vol 1218 pp 1057–71*
[12] Petersen B 2010 Presentaion, *Project X Collabration Meeting* (FNAL, USA)
[13] Wilhelm H 2015 Presentation, *HEPTech Academia Meets Industry on Cryogenics* (Grenoble, France)
[14] White M and Martinez A 2014 *Advances in cryogenic engineering AIP Conf. Proc.* (Alaska, USA: Anchorage) 1573 pp 179–86
[15] Arenius D, Ganni V, Creel J, Dixon K, Knudsen P and Wilson J 2006 *Cryogenic Operation Workshop 2006, Presentation* (Stanford, CA: Stanford Linear Accelerator Center)
[16] Wang X L *et al* 2020 Commissioning of the Warm Compressor System for the ESS Accelerator Cryoplant *IOP Conf. Ser.: Mater. Sci. Eng.* **755** 012087
[17] Martinez A and Pallaver C B 1994 Test of an improved oil injected helium screw compressor at Fermilab' 1994 *Adv. Cryog. Eng.* **39A** 887–92
[18] Ganni V, Knudsen P, Creel J, Arenius D, Casagrande F and Howell M 2008 Screw compressor characteristics for helium refrigeration systems *AIP Conf. Proc.* **985** 309
[19] Knudsen P, Ganni V, Dixon K, Norton R, Creel J and Arenius D 2014 Commissioning of helium compression system for the 12 GeV refrigerator *AIP Conf. Proc.* **1573** 962
[20] Knudsen P, Ganni V, Dixon K, Norton R and Creel J 2015 Commissioning and operational results of the 12 GeV helium compression system at JLab *IOP Conf. Ser.: Mater. Sci. Eng.* **101** 012126
[21] Wang X L, Arnold P, Hees W, Hildenbeutel J and Weisend J G 2015 ESS accelerator cryoplant process design *IOP Conf. Series: Materials Science and Engineering* **101** 012012
[22] Zhang J 2022 ACCP new test results in 2022 *ESS Internal Document (controlled): ESS-4001140*
[23] Zhang J, Arnold P, Kolev N and Rueegge A 2023 Commissioning of the large-scale 2 K helium refrigeration system at ESS *Proc. 28th Int. Cryogenic Engineering Conf. and Int.*

Cryogenic Materials Conf. 2022. ICEC28-ICMC 2022 (Advanced Topics in Science and Technology in China **vol 70**; L Qiu, K Wang and Y Ma (Cham: Springer) pp 101–8

[24] Zhang J, Arnold P, Fydrych J and Weisend J G 2023 Preparation of the integrated test of the accelerator cryoplant and the cryogenic distribution system at ESS *Proc. 28th Int. Cryogenic Engineering Conf. and Int. Cryogenic Materials Conf. 2022. ICEC28-ICMC 2022 (Advanced Topics in Science and Technology in China* **vol 70**; L Qiu, K Wang and Y Ma (Cham: Springer) pp 172–8

[25] Zhang J 2021 Test results of ACCP-CDS tests with ACCP only *ESS Internal Document (Controlled): ESS-3727633*

Chapter 4

The cryogenic distribution system

J Fydrych

This chapter presents the cryogenic distribution system (CDS) of the European Spallation Source (ESS) linear accelerator and the course of the CDS project from the conceptual design to the site acceptance test. The system is composed of the CDS for the elliptical linear accelerator (LINAC) and the CDS for the spoke LINAC. The two subsystems were provided as two separate in-kind contributions from Wroclaw University of Science and Technology from Poland and Laboratory of the Physics of the Two Infinities Irène Joliot-Curie from France, respectively. The project started in 2013 with the preparation of technical specifications and negotiations between ESS and in-kind partners. The first successful commissioning campaign was carried out by ESS from October 2022 to February 2023. It demonstrated the mechanical stability of the system; however, several technical issues were revealed, mainly with the tightness of valve seats and thermoacoustic oscillations (TAOs). Both ESS and the in-kind partners analyzed all the observed issues and caried out the required repairs. The second commissioning campaign was carried out in June and July 2023 with successful verification of the applied modifications. The measured heat load to the cold circuit was 4% below the specified value.

4.1 Introduction

The ESS CDS is one of the main parts of the ESS LINAC cryogenic system. It transfers and distributes the cooling power produced in the accelerator cryogenic plant to the cryomodules by means of the constant flows of cold helium. A general layout of the ESS cryogenic system is shown in figure 4.1. The CDS consists of a cryogenic transfer line (CTL), cryogenic distribution line (CDL), and a number of auxiliary process lines. The CTL includes a splitting box that is equipped with a plugged cryoline terminal. The terminal allows for an easy and relatively quick connection of another CDL. This additional line will be dedicated to distributing cold helium to the contingency and upgrade cryomodules in the case of a future extension of the LINAC lattice [1]. The CDL comprises 43 valve boxes and an end box. The

doi:10.1088/978-0-7503-3223-1ch4

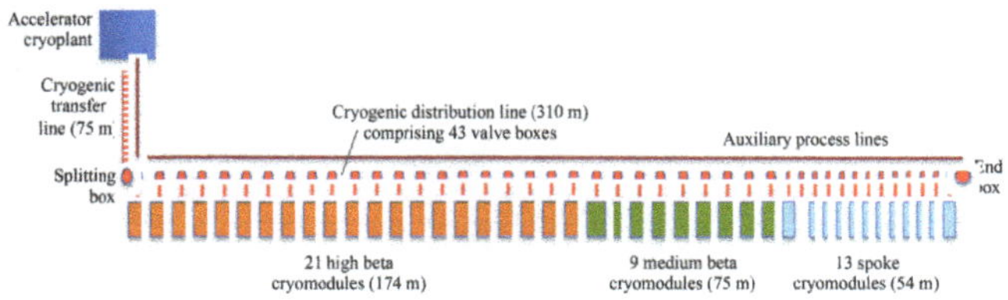

Figure 4.1. Layout of the cryogenic system for the ESS LINAC. Reprinted from [15], Copyright (2015), with permission from Elsevier.

main function of the valve boxes is to allow for warming up and cooling down one or more cryomodules without affecting the others. The end box is for returning the remaining flows of helium in the main circuits back to the cryoplant. These flows ensure the proper temperature of the cold helium in the inlet to the last cryomodules. The auxiliary lines are mainly used for conditioning the cryomodule helium circuits before cool down, for supporting the warm-up and cool down of a single cryomodule, and for connecting the cryomodule and valve box safety valves to the cryoplant.

In 2014 the CDS was split into two subsystems: the CDS for the elliptical LINAC and the CDS for the spoke LINAC. The two subsystems were provided as two separate in-kind contributions from two in-kind partners, namely Wroclaw University of Science and Technology (WUST) from Poland and Laboratory of the Physics of the two Infinities Irène Joliot-Curie (IJCLab) from France. WUST delivered the CDS for elliptical LINAC, which includes the CTL and a part of CDL comprising valve boxes for high-beta and medium beta cryomodules. IJCLab provided the CDS for Spoke LINAC, which includes 13 valve boxes for spoke cryomodules and the end box. The two in-kind partners were responsible for the design, procurement, manufacturing, transportation, installation, and all testing related to fabrication and installation activities. ESS provided controls, front-end electronics for the CDS for the spoke LINAC, and vacuum pumps. ESS was also responsible for connecting the CDS to the cryoplant, leak tests of control valves, and the final commissioning at low temperature.

4.2 Execution of the CDS project

The CDS project started in 2013 and was completed in 2023. The project was executed in seven main phases: specification and negotiations with in-kind partners, preliminary design, final design, procurement, production, installation, and commissioning. The time lines of the CDS-EL and CDS-SPK subprojects are shown in figure 4.2. The installation phases lasted significantly longer in respect to the other phases due to several technical issues combined with the impact of the COVID-19 pandemic (table 4.1).

ESS and partners held a number of project meetings and reviews. The specification and negotiation phases were concluded in kick-off meetings, while the design phases were completed with preliminary and critical design reviews. The approved critical design reviews allowed the teams to proceed to procurement. The next phases were preceded by readiness reviews to verify the preparation of the team for carrying

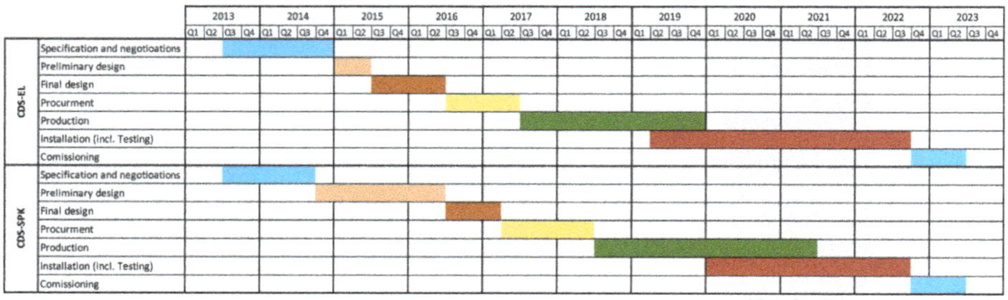

Figure 4.2. Main phases of the CDS project. (Courtesy of ESS)

Table 4.1. CDS project reviews. (Courtesy of ESS)

	KoM	PDR	CDR	TRR1	IRR	TRR2	TRR3	SAR
CDS-EL	February 2015	May 2015	April 2016	June 2017	March 2019		September 2022	Scheduled for August 2023
CDS-SPK	October 2014	June 2016	April 2017 Feb. 2021	September 2018	Dec. 2019 March 2021			
Controls	May 2019	-	September 2019	-	-	-	April 2022 September 2022	

KoM – Kick off Meeting
PDR – Preliminary Design Review
CDR – Critical Design Review
TRR1 – Test Readiness Review (Production)
TRR2 – Test Readiness Review (Installation)
TRR3 – Test Readiness Review (Commissioning)
IRR – Installation Readiness Review
SAR – System Acceptance Review

out all the activities related to tests and verifications at production, installation, and commissioning phases. The CDS project is planned to be completed with the system acceptance review initially scheduled for August 2023.

4.3 Functional and technical requirements

At the specification and negotiation phases, ESS and partners discussed and agreed on all the functional and technical requirements for the CDS subsystems and their interfaces to the adjacent systems. A number of main functional and technical requirements for the CDS resulted from the top-level requirement of 95% availability of the ESS LINAC itself [2]. It imposed that the LINAC equipment, and the CDS as well, must be extremely reliable throughout the expected lifetime of 40 years and must require a minimum of maintenance. The LINAC shall be operated remotely and continuously, 24 h per day, for about 250 days per year, so the design of the equipment had to facilitate easy repairs or even easy replacements of faulty components. Therefore, the design and construction of the CDS had to meet the following functional requirements:

- Suitable for smooth continuous operation with limited scheduled interruptions only.

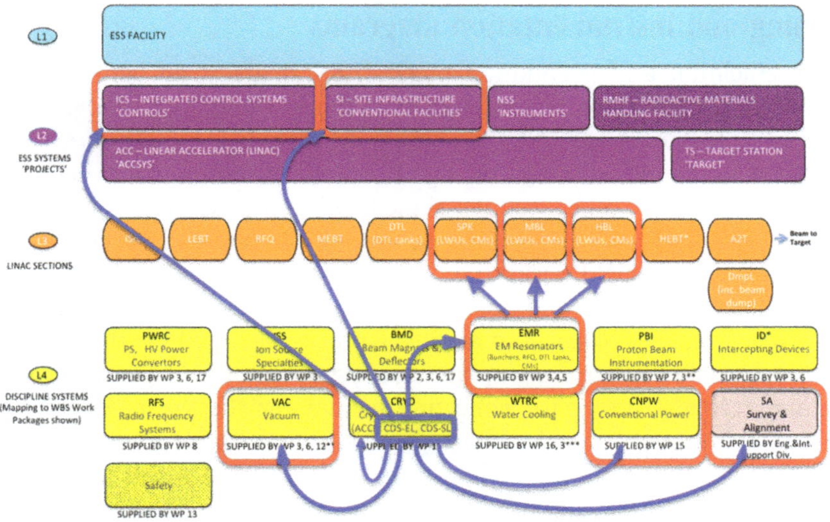

Figure 4.3. Map of the CDS interfaces. (Courtesy of ESS)

- Ensure no deterioration of thermal and mechanical properties within the operation lifetime.
- Tolerate all possible pressurization and cool-down rates.
- Allow for warm-up and cool-down of a single cryomodule, while keeping the rest of the system at cryogenic temperatures.

All the CDS requirements are collected in [3], as well as in the technical specifications for CDS-EL [4] and CDS-SPK [5]. Some of the main high-level technical requirements for the CDS are as follows:

- Heat loads not higher than 420 W and 3.66 kW to the cold helium circuit and thermal shield, respectively.
- Supercritical helium temperature in the interfaces to the cryomodules below 5.2 K at nominal operation conditions.
- Vacuum insulation below 10^{-6} mbar at nominal working condition (below 5×10^{-3} mbar at ambient temperature with active vacuum pumping).
- Integral helium leak rate into the insulation vacuum below $5 \times \bullet 10^{-7}$ mbar$\bullet$l s^{-1}.
- Tightness of the valve seats $\leqslant 1 \times 10^{-4}$ mbar$\bullet$l s^{-1}.
- All materials and components located in the LINAC tunnel resistant to the radiation dose of 5×10^5 Gy.

Apart of the ACCP and cryomodules, the CDS is integrated with some other components of the ESS facility, such as vacuum system, control system, and buildings. To allow smooth integration and cofunctioning, the design of the CDS had to also meet a number of technical requirements for its interfaces. Figure 4.3 shows a map of all CDS interfaces. All the technical requirements for the CDS interfaces are described in detail in [6–8], as well as in the appendixes to the CDS specifications.

4.4 Piping and instrumentation diagrams

In the specification and negotiation phase, ESS created and communicated the piping and instrumentation diagrams of the elliptical and spoke valve boxes, end box, and CTL to the partners.

Figure 4.4 shows the simplified piping and instrumentation diagram of the CTL. The CTL, as well as the other sections of the CDS, comprises four cold process lines that form the cold helium circuit (i.e., helium supply line and vapor low pressure line) and the thermal shield circuit (i.e., TS supply and return lines). These lines are used at nominal operation conditions and for the cool downs and warm-ups of the entire ESS LINAC. The other four auxiliary process lines, which run alongside the CDL, are necessary for warming up and cooling down of a single cryomodule (HP and helium recovery lines), for collecting helium for the power coupler cooling circuits (helium recovery line), for purging and flashing a single cryomodule (HP and purge lines), and for recovering helium from pressure relief valves (SV relief line).

The flow schemes of the valve boxes were developed to ensure that each cryomodule can be operated independently from the others. Figure 4.5 shows a simplified piping and instrumentation diagram for the elliptical valve box. It is equipped with six cryogenic control valves (i.e., CV03, CV04, CV06, CV60, CV61, and CV63) and some other process control elements, such as check valves (i.e.,

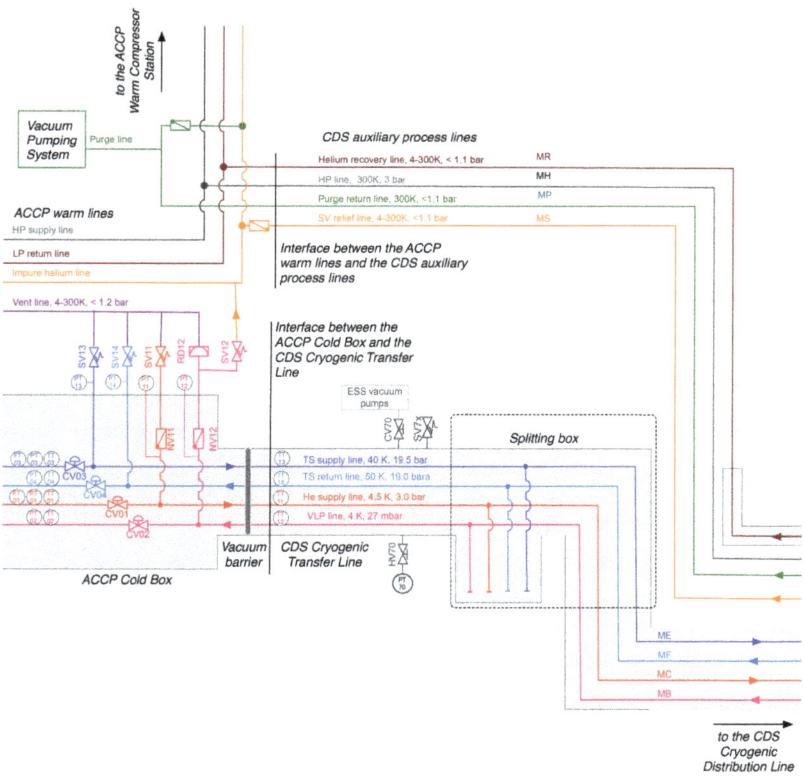

Figure 4.4. Piping and instrumentation diagram of the CTL. (Courtesy of ESS)

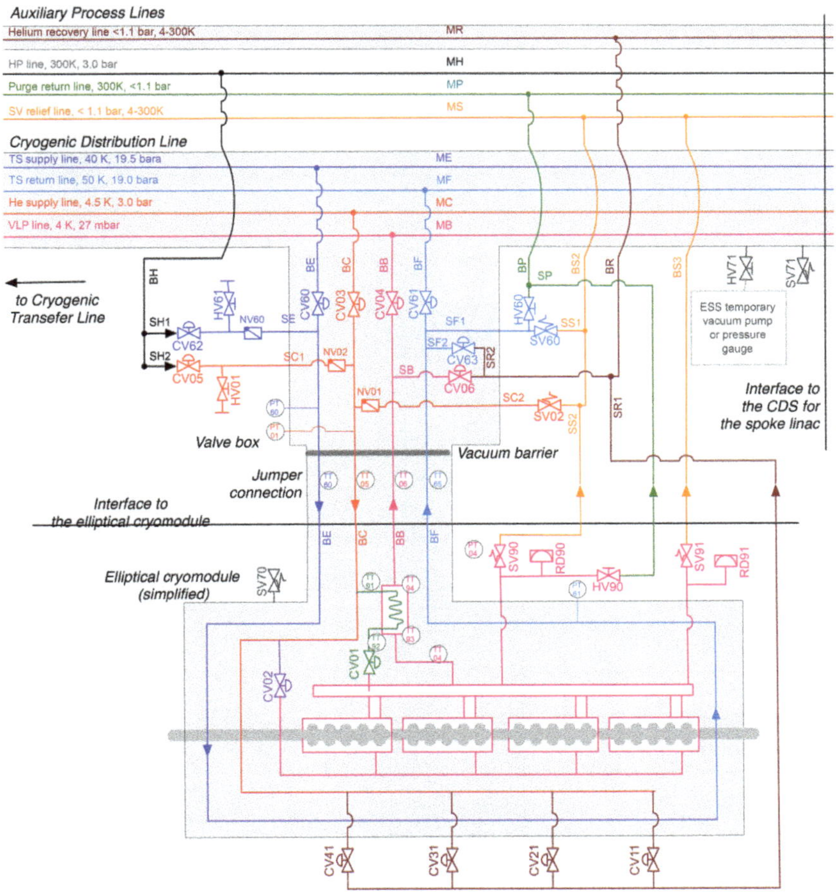

Figure 4.5. Simplified piping and instrumentation diagram of the CDS-EL valve box. Reprinted from [15], Copyright (2015), with permission from Elsevier.

VN01 and VN02), warm control valves (i.e., CV05 and CV62), safety valves (i.e., SV02 and SV60), as well as pressure transmitters (i.e., PT01 and PT60) and temperature sensors (i.e., TT05, TT06, TT60, and TT65). The cryomodule is connected to the valve box via a branch cryoline, a so-called jumper connection. Since the cryomodule insulation vacuum has to be separated from that of the CDL, the jumper connection includes a vacuum barrier. During nominal operation conditions, as well as during the cool-down and warm-up of the whole LINAC cryogenic system, only CV03, CV04, CV60, and CV61 will be kept open in all the valve boxes. However, to bring a single cryomodule to the ambient temperature, only CV05, CV06, CV62, and CV63 will be open. Then warm helium from the HP line will flow into the cryomodule circuits and back to the ACCP via the helium recovery line.

The piping and instrumentation diagram of the CDS for the spoke LINAC differs only slightly from this for the elliptical LINAC. Since the spoke cavities are larger in diameter, there is less space for the other equipment in the spoke cryomodules.

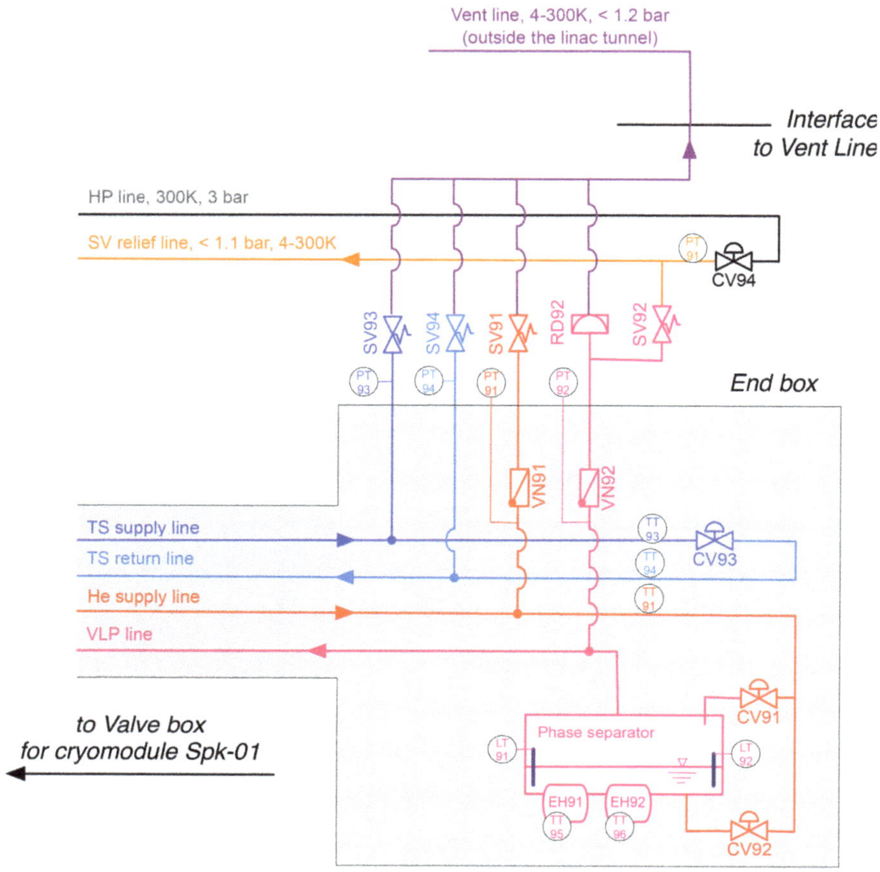

Figure 4.6. Simplified piping and instrumentation diagram of the CDS-SPK end box. (Courtesy of ESS)

Therefore, to enable cryomodule assembly, the heat exchanger (HX01), JT valve (CV01), filling valve (CV02), and some related temperature sensors were moved to the CDS-SPK valve box.

The CDS-SPK is terminated with the end box. Its simplified piping and instrumentation diagram is shown in figure 4.6. The main function of the end box is to reverse excess flows in the cold helium and thermal shield circuits. The helium that is reversed in the cold helium circuit, i.e., from the helium supply main line to the VLP main line, expands isenthalpically in the JT valve (CV91) to a pressure around 30 mbar and its temperature falls to 2 K. In this throttling process, around 55% of the supplied helium remains in a liquid phase. Therefore, the end box is equipped with a phase separator with electrical heaters and level gauges for collecting and evaporating the liquid phase in a controllable way. Pressure transmitters PT91–PT94 and temperature sensors TT91–TT93 monitor the parameters of the helium flowing in the process pipes.

The end box is also equipped with a set of safety devices protecting the LINAC CDS circuits against excessive pressure. In a failure mode of losing insulation vacuum in the CDS vacuum jacket resulting in excessive heat loads to the helium in the process pipes, safety valves SV91, SV92, SV93, and rupture disk RD92 will open. Then, the helium from the process pipes will be vented to the atmosphere via a dedicated vent line.

The VLP line is additionally protected by safety valve SV92, which is intended to recovering cold helium from the VLP line and returning it to the cryoplant via the SV relief line. The flow capacity and opening pressure of this valve are lower than those of RD92, so SV92 only protects against opening the rupture disk and venting significant amounts of helium during some minor failures.

The pressure safety devices that are installed on the end box are designed for protecting only one half of the CDS process pipes. The other halves are protected by another set of safety devices that are installed on the cryoplant cold box. This set consists of SV11, RD12, SV13, SV14, and SV12, and is schematically shown in the CTL flow diagram (see figure 4.4).

The final general scheme and detailed piping and instrumentation diagrams for CDS valve boxes and end box are in documents [9–12].

4.5 Conceptual design

In the specification and negotiation phase, ESS developed a conceptual design of the system in order to check the feasibility of the CDS construction and installation in the LINAC tunnel with respect to the functional and technical requirements. Figure 4.7 presents the conceptual design of the CDL section with the valve box for the elliptical cavity cryomodules.

The proposed conceptual design is based on a modular structure. Each module is 8.26 m in length and is connected to the adjacent modules with special interconnections. The module includes a valve box with two main cryoline sections and a jumper connection. The vacuum jacket of the jumper connection is equipped with two lateral compensators, while the branch process lines have flexible hoses. These components allow some adjustments for connecting the cryomodules to the CDS valve boxes, as well as thermal-shrinkage compensation required at nominal conditions and some failure modes [13].

Due to limited space in the LINAC tunnel, the valve box and cryoline sizes were limited to DN1000 and DN600, respectively. The design and location of the external supports were strongly affected by the routing of the cryomodule waveguides [14]. There are four and two waveguides under every CDL module of the CDS-EL and CDS-SPK, respectively.

Both the vacuum jacket and thermal shield of the valve box have demountable bottom plates that provide access to the piping and cryogenic valve bodies when repairs are needed. The cryogenic control valves will be fitted with significantly extended spindles equipped with heat sinks thermally connected to the TS return pipes.

Neutron radiation in the LINAC tunnel has a strong impact on the CDL design, especially on material choices and on the architecture of the control and instrumentation system. Materials sensitive to this radiation are not allowed. Therefore, the control valves have pneumatic actuators and electro-pneumatic positioners with

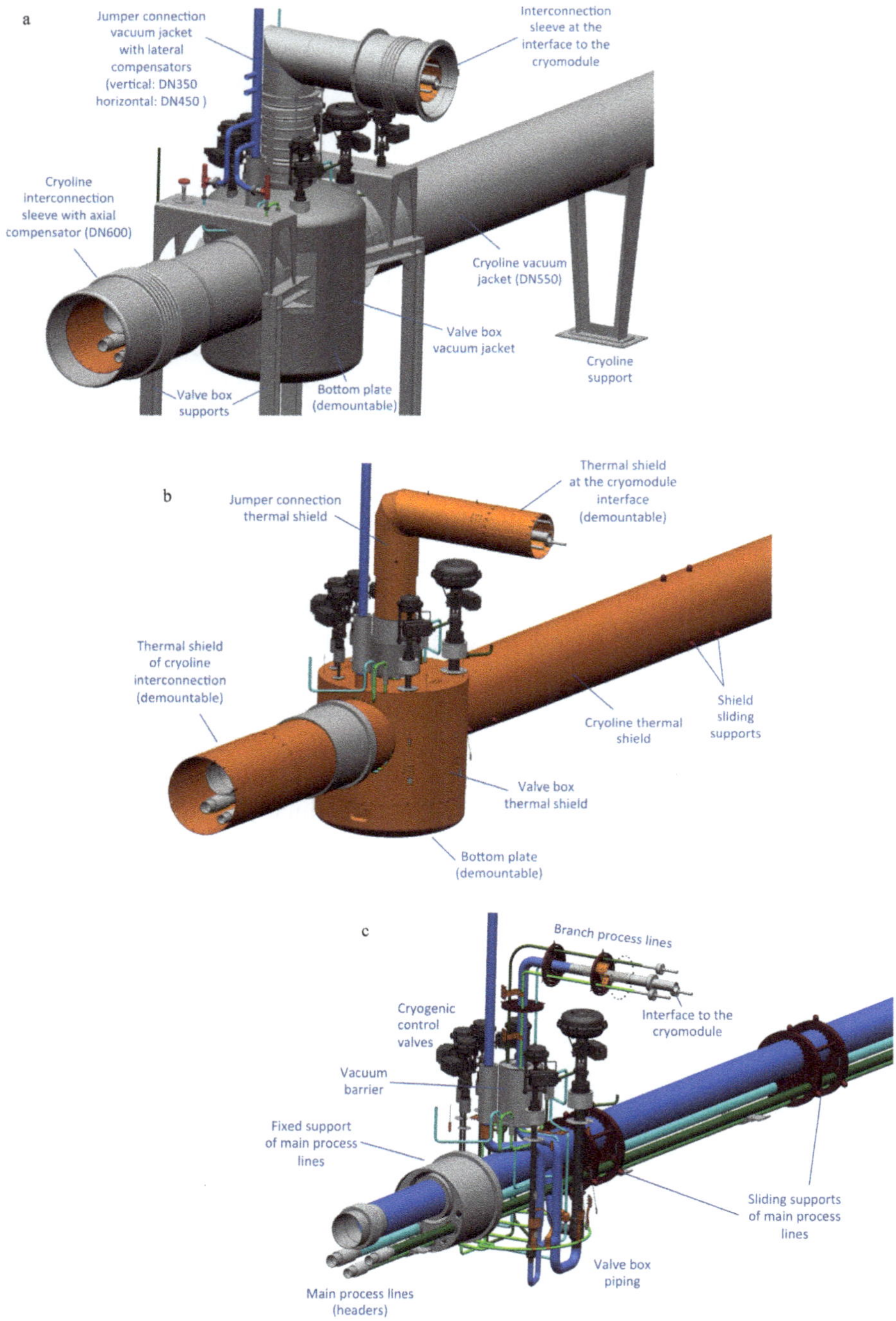

Figure 4.7. Conceptual design of the CDS-EL valve box: (a) vacuum jacket, (b) thermal shields, and (c) process pipes and supporting system. Reprinted from [15], Copyright (2015), with permission from Elsevier.

remote electronic parts located in a radiation free area. The electronic racks for the CDS valve boxes are placed in the klystron gallery, in a distance of 50–100 m away from the valve boxes.

4.6 Detailed design

In the preliminary and detail design phases, the WUST and IJCLab teams modified and developed the designs of the CDS-EL and CDS-SPK, respectively. The progress of the design work was demonstrated and verified first in the preliminary design review and finally in the critical design reviews. The development of the detailed design apart from 3D modeling included a massive amount of engineering work. The major bulk of engineering work done by WUST and IJCLab teams included:

- Selecting materials.
- Sizing process pipes and external envelope components.
- Sizing and selecting cryogenic control valves.
- Routing process pipes inside valve boxes and jumper connections.
- Selecting metal hoses and expansion joints for compensating thermal contractions.
- Designing process pipe supports and vacuum barriers.
- Designing thermal shield components including their thermal bridges to the TS return line.
- Designing vacuum jacket components.
- Designing external supports and their anchoring to the tunnel floor and ceiling.
- Positioning methods and adjustment system calculations.
- Thermomechanical calculations of the process pipes, thermal shields, vacuum jackets for nominal operation, pressure test, and failure modes.
- Heat load estimations.
- Pressure drop calculations.
- Hazard analysis.
- Safety analyzes focused on estimating potential flow rates of helium to be potentially discharged from the process pipes.
- Sizing and selecting safety valves and rupture disks.
- Analyzes of thermoacoustic oscillations in the side process pipes and pressure transmitter tubes.
- Instrumentation lists.
- Electronic block diagrams.
- Specification of main maintenance tasks and spare parts.

Figure 4.8 shows the final design of the CDS-EL developed by WUST. The modular structure remained as it was proposed in the ESS conceptual design. The main modifications made by WUST designers in the CDS-EL design were made in the designs of supports of cold and auxiliary process pipes, supports of the valve boxes and cryolines, and the jumper connection. The two lateral compensators in the conceptual design (shown in figure 4.7(a)) were replaced by one axial compensator in

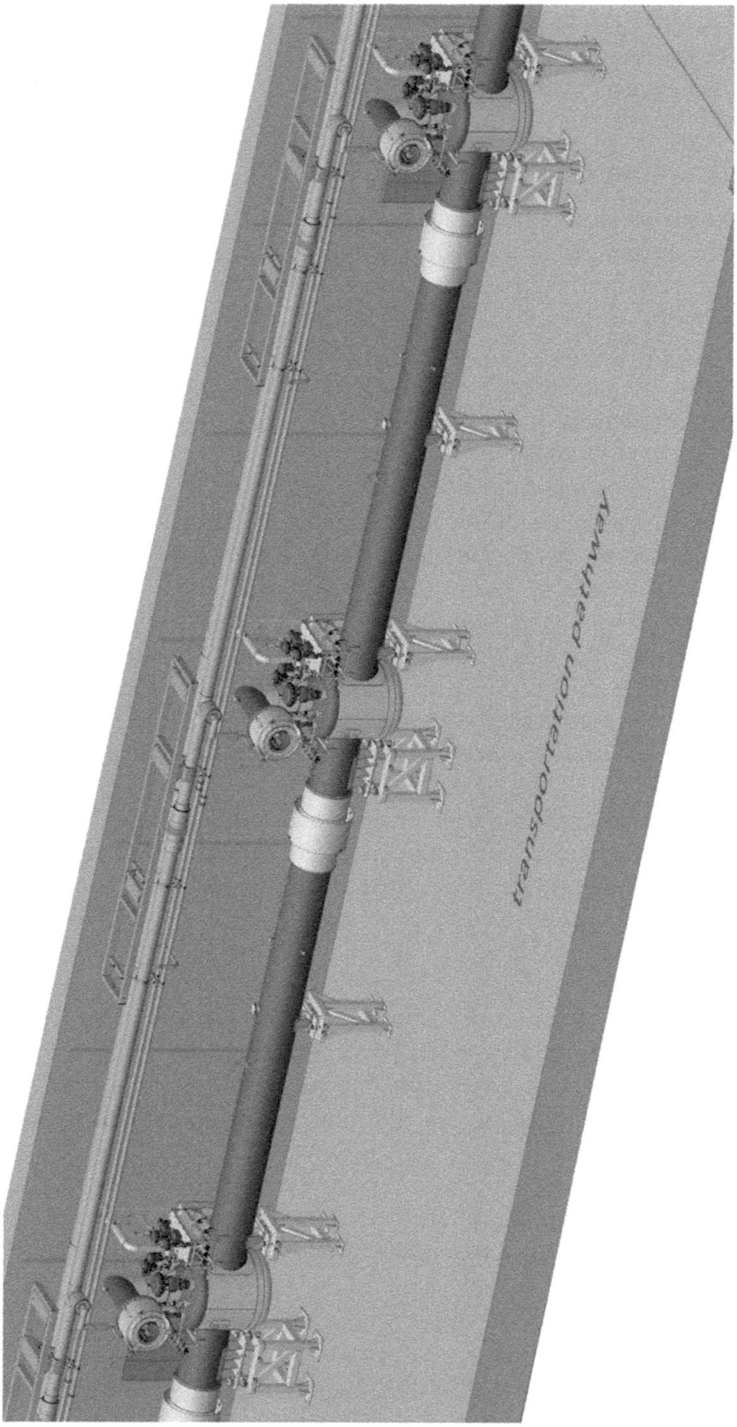

Figure 4.8. CDS-EL detailed design developed by WUST in the scope of Polish in-kind contribution to the ESS accelerator project. (Courtesy of ESS)

the interconnection sleeve at the interface to the cryomodule. This interconnection sleeve is fixed to the valve box jumper via a special DN450 flange with enlarged bolt holes, allowing for adjustments of ±10 mm in radial directions. WUST'S designers vastly modified the external supports and equipped them with positioning systems and fiducials. The positioning system and fiducials were added for allowing adjustments of the valve box and CTL module positions in a range of ±35 mm horizontally and ±45 mm vertically to fulfill requirements specified by the ESS Survey and Alignment Group.

The final design of the CDS-SPK developed by IJCLab is shown in figure 4.9. The CDS-SPK valve box has more components on its top plate because IJCLab designers had to integrate inside the valve box two additional control vales (JT valve and filling valve) and a heat exchanger with eight temperature sensors without increasing the valve box external diameter. They decided to place the jumper eccentrically to the valve box top plate, which resulted in shortening the jumper connection and better optimization of the routing of inner process pipes. In this design, the valve box jumper has rigid process pipes and vacuum jacket. The thermal-shrinkage compensation components were moved into the cryomodule jumper. IJCLab also decided to design the cryoline linking two adjacent valve boxes as separate components. These cryoline sections were named 'header units' and consisted only main process lines (headers), their internal supports, and expansion joints (thermal-shrinkage compensators). This solution simplified the manufacturing and transportation of these components but doubled the number of interconnections and increased the scope of factory acceptance tests and installation works.

The IJCLab team added required fiducials and proposed their own design of the valve box, header unit and end box supports. The supports were also equipped with positioning system for providing the required adjustability.

The design phase also consisted of design works on the CDS instrumentation, cabling, and front-end electronics. WUST designed the architecture of the CDS-EL front-end electronics and cabling, selected valve box instrumentation (temperature and pressure sensors), control valves, and their positioners. IJCLab selected instrumentation and control valves for the CDS-SPK valve boxes, end box, and special header unit (at the interconnection to the CDS-EL). Finally, the ESS Integrated Control System Division designed the architecture of the CDS-SPK front-end electronics and cabling, selected positioners for the CDS-SPK control valves and designed the entire CDS controls.

4.7 Procurement

As soon as the critical design reviews were approved, the in-kind partners proceeded to open tendering of the manufacturing and installation of the CDS components in the ESS tunnel. For the CDS-EL, WUST was awarded the contract for production, transportation, and installation of the CTL modules, valve boxes, and auxiliary lines to KrioSystem from Poland. In case of the CDS-SPK, IJCLab organized several individual open tenders. This resulted in awarding the following separate contracts:

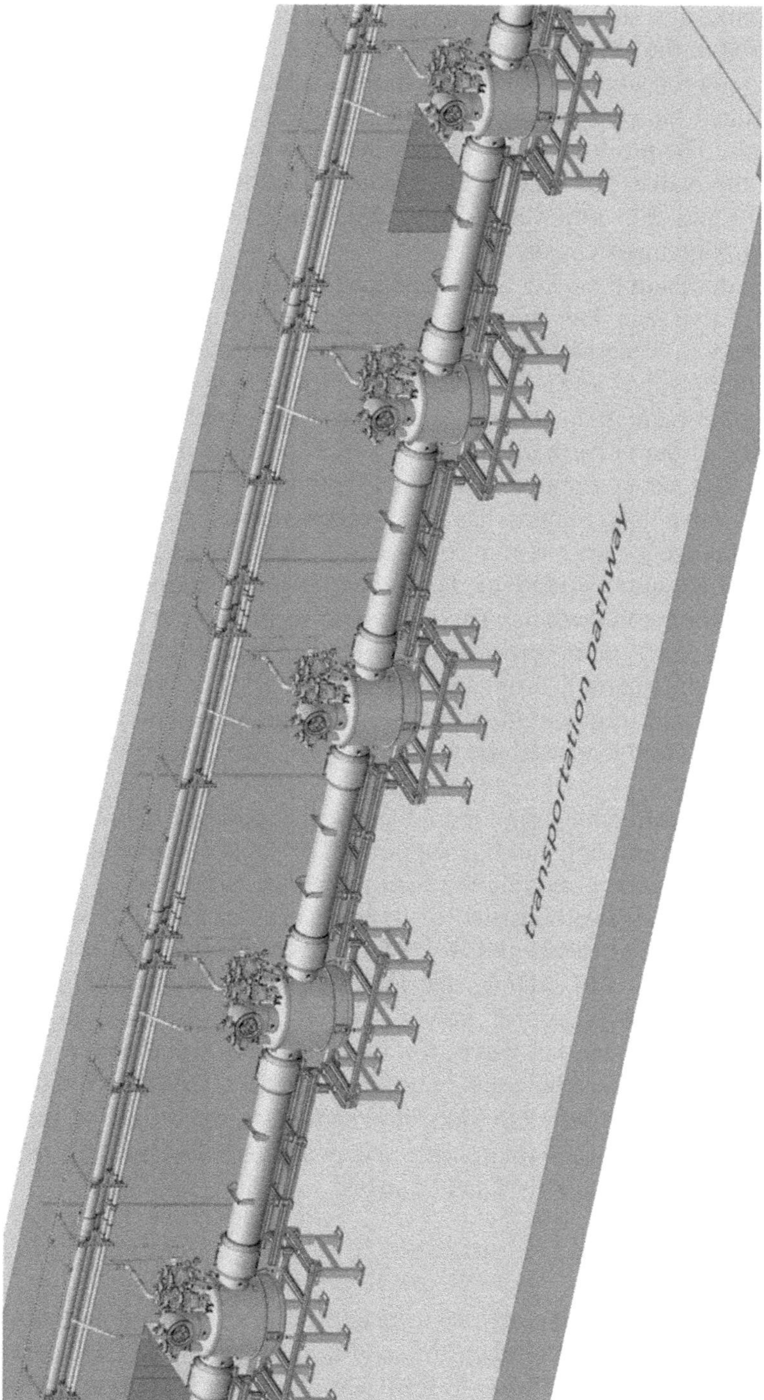

Figure 4.9. CDS-SPK detailed design developed by IJCLab in the scope of French in-kind contribution to the ESS accelerator project. (Courtesy of ESS)

—For production of the valve boxes to CryoDiffusion Ltd from France.
—For production of the header units to CryoDiffusion Ltd from France.
—For production of the end box to SDMS from France.
—For installation of the valve boxes, header units, and end box to KrioSystem from Poland.
—For design, manufacturing, and installation of the auxiliary process lines to KrioSystem from Poland.

The in-kind partners procured separately all cryogenic and warm control valves and provided them to the valve box manufacturers (KrioSystem and CryoDiffusion). WUST awarded WEKA from Switzerland with the contract for the CDS-EL control valves, while IJCLab contracted manufacturing of the CDS-SPK control valves to VELAN from France.

4.8 Production and transportation

The production of the CDS components started in Q2 2017 and Q3 2018 for the CDS-EL and CDS-SPK, respectively. Just at the beginning of the production phases, ESS, partners, and manufacturers held two separate test readiness reviews for production of the valve boxes for CDS-EL and CDS-SPK, in June 2017 and September 2018, respectively. The reviews aimed to evaluate if the procedures of the manufacturing tests, test equipment, and support personnel were ready for testing the CDS elements, components, and assemblies to be manufactured during the production phase. These reviews were also focused on confirming that all the changes that were introduced after the critical design reviews supported the CDSs' designs to meet all the requirements and were specified in sufficient detail for production and assembly works.

All process pipes, fittings, expansion joints, metal hoses, and valves which were classified as pressure equipment had to be manufactured and tested to fulfill the requirements of Pressure Equipment Directive 97/23/EC (later replaced with PED 2014/68/EU). The fabrication and installation of the piping systems, including supports, had to meet the requirements specified in EN 13 480–4, while the inspections and tests of the piping to be performed in accordance to EN 13 480–5. The process pipes had to be seamless pipes if their nominal diameter was lower or equal to DN300.

At the production phase, the partners' contractors carried out visual examinations and radiographic tests of the process pipe and vacuum jacket welds. Visual examinations were performed according to ISO 17 637. The quality level criteria of weld imperfections were level B of ISO 5817, while the acceptance criteria for the weld imperfections had to comply with EN 12 517. Later, all subassemblies were subjected to pressure tests, helium leak tightness tests, and thermal shocks with liquid nitrogen.

During the production phases, the partners continuously monitored the progress of all manufacturing works and reported it to ESS. The ESS delegates, including representatives from the ESS Cryogenic Group, Quality Group, Survey and

Figure 4.10. ESS delegation witnessing the assembly of the CDS-EL valve boxes in KrioSystem's production hall in August 2018. (Courtesy of ESS)

Figure 4.11. ESS quality engineer inspecting welds and process pipe cleanliness in the CDS-SPK valve boxes in CryoDifussion's production hall in January 2021. (Courtesy of ESS)

Alignment Group, Vacuum Group as well as Integrated Control System Division, regularly witnessed production and assembly works carried out in the manufacturer's premises. Figures 4.10 and 4.11 shows photos taken at the ESS witnessing of the manufacturing of CDS-EL and CDS-SPK valve boxes, respectively.

Figure 4.12. CDS-EL valve boxes transported on a semitrailer. (Courtesy of ESS)

Figure 4.13. CDS-SPK valve box in a transport wooden box with an anti-shock metal frame. (Courtesy of ESS)

As soon as the CDS components passed factory acceptance tests, they were transported to the ESS site. All the components had to be protected to withstand all impact forces that could appear during transportation. Figure 4.12 shows a semi-trailer with three CDS-EL valve boxes just before a receiving inspection. For transportation of the CDS-SPK valve boxes, IJCLab designed and fabricated special wooden boxes equipped with an anti-shock metal frame (see figure 4.13).

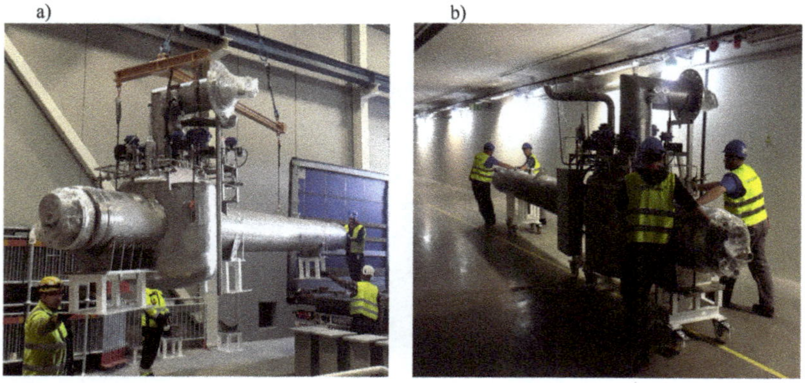

Figure 4.14. Unloading of CDS-EL valve box in the HEBT loading bay (a) and transporting in the LINAC tunnel (b). (Courtesy of ESS)

In addition, they used accelerometers to monitor that the valve boxes were not subjected to transport loads that were too high.

All the CDS components were unloaded in the HEBT loading bay by ESS riggers. Then, the KrioSystem team carried out a receiving inspection and transported the components to their installation place in the LINAC tunnel. Figure 4.14 shows unloading of the valve boxes and their transportation in the tunnel. The CDS-EL valve boxes were transported on specially designed trolleys with a support for the valve box jumper connection. For the CDS-SPK valve box, ESS riggers carried them to the installation area on an electric forklift.

4.9 Installation

The CDS installation in the ESS LINAC tunnel started in Q2 2018 and Q1 2019 for the CDS-EL and CDS-SPK, respectively. Before these phases, ESS, partners, and their contractor for installation activities organized installation readiness reviews (IRRs) combined with test readiness reviews (TRRs) for the CDS installation. The IRR for CDS-EL was held in March 2019, while for the CDS-SPK first in December 2019. The main objectives and purposes of these reviews were to evaluate if the system installation procedures, installation equipment, and supporting personnel were ready for installing the system components and assemblies in their final positions. The reviews also aimed at assessing the initial reports of fabrication and factory acceptance test, inspection reports, and material certificates. Whereas, the TRRs evaluated if the procedures of the installation tests and test equipment were ready for testing the CDS components and assemblies during the installation phases.

The CDS installation phase included the following activities:

- Installation and anchoring of the CDS supports to the tunnel, CTL gallery, and cold box hall floors.
- Installation of the valve boxes, end box, and CTL modules on their supports.

- Positioning of the valve boxes in respect to the interface points to the cryomodules.
- Installation of header units (only for CDS-SPK).
- Welding process pipe expansion joints in the CDS interconnections.
- Pressure test the process pipe connections.
- Rough leak test of the process pipe welds in the interconnections.
- Wrapping MLI onto the process pipe connections.
- Installation of the thermal shield plates in the CDS interconnections.
- Wrapping MLI blankets onto the thermal shield plates.
- Closing and welding the interconnection sleeves of the CDS vacuum jacket.
- Final leak test of the process pipe and vacuum jacket interconnection sleeves.

At the installation of first three CDS-SPK valve boxes, some additional quality inspections revealed severe weld defects in the inner process pipes of the CDS-SPK valve boxes and header units. As a result of these observations, ESS and IJCLab agreed on shipping the delivered components back to their manufacturer and requesting them to repair the faulty welds. The weld repair campaign was finished in Q1 2021 and the second IRR for the CDS-SPK valve boxes was held in March 2021.

The CDS-EL valve boxes were one of the first components installed in the LINAC tunnel. The valve boxes were lifted and moved from transport trolleys and placed on their final supports by using two connected gantry cranes, as shown in figure 4.15. The installation of the CDS-SPK valve boxes was done in a much more confined space, which required the use of an electric forklift, as shown in figure 4.16.

The positioning of all valve boxes had to be done in respect to the virtual interface points specified in the center of the interface plane between CM and Valve box

Figure 4.15. Installation of CDS-EL valve box on its supports in the tunnel. (Courtesy of ESS)

Figure 4.16. Installation of CDS-SPK valve box on its support. (Courtesy of ESS)

Figure 4.17. Positioning of CDS-EL valve boxes. (Courtesy of ESS)

jumper flanges. The required accuracy was 1 mm. The valve box positioning systems provided the course alignment ranges in the vertical direction of ±45 mm and the horizontal plane of ±35 mm, with respect to the theoretical position. The systems allowed for fine adjustment in the horizontal and vertical directions with precision below 1 mm/turn. The valve boxes were equipped with a set of fiducials and the positioning was monitored with use of a laser tracker. Figure 4.17 shows the positioning of CDS-EL valve box.

The positioning of valve boxes was followed by the activities of closing interconnections. These activities included welding process pipe connections, installation of thermal shield plates, wrapping MLI blankets, and closing and welding the external sleeve of the vacuum jacket envelope. Figures 4.18 shows welding of process pipe connections in the interconnections of CDS-EL and CDS-SPK valve boxes.

Figure 4.18. Welding process pipes in CDS interconnections. (Courtesy of ESS)

Figure 4.19. CDS-EL ready for the final leak test. (Courtesy of ESS)

Figures 4.19 and 4.20 show the elliptical and spoke sections ready for the final leak test after closing all interconnections. The final leak test of the CDS-EL was carried out before connecting it to the CDS-SPK and ACCP cold box. For the process pipes, this activity included pumping vacuum in the vacuum jacket to a vacuum pressure of below 5×10^{-3} mbar and a residual helium background below 10^{-8} mbar•l s^{-1}, filling the process lines with a 50–50 helium-air mixture up to their design pressures, and measurements of the total leak rates. For testing the tightness of the vacuum jacket interconnections, the tested welds were sprayed around with helium. The leak test revealed several leaks on the vacuum jacket sleeve welds and one leak at a welding ring of the VLP process line (10^{-4} mbar•l s^{-1}). After repairing the leaking welds and welding ring, the following leak test showed the total leak rate below 10^{-7} mbar•l s^{-1}, which confirmed the appropriate tightness of the system.

The last total leak test was carried out after the installation of the special header unit between the two CDS sections and closing the CDS-EL interconnection to the

Figure 4.20. CDS-SPK prepared for the final leak test. (Courtesy of ESS)

ACCP cold box. This test revealed one leak in the CDS-SPK vacuum jacket weld (10^{-5} mbar•l s^{-1}), one leak in the process pipe weld inside the CDS-SPK valve box jumper (>10^{-3} mbar•l s^{-1}), and one leak in the welding ring of the ACCP-CTL interconnection. All these leaks were repaired and followed by another leak test for verifying the specified tightness requirement.

In the CDS installation phase, the partners' contractor, KrioSystem, installed 43 valve boxes, 1 end box, 13 CTL modules, and 4 process auxiliary lines of a total length of 1500 m. They also closed 70 CDS interconnections, executed all required pressure and leak tests, and installed the CDS-EL front-end electronics in electronic racks in the klystron gallery. In addition, ESS carried out a certain fraction of the installation works, and provided technical gases and vacuum pumping stations for pressure and leak tests. The ESS Survey, Alignment and Measurement Group positioned the valve boxes and end box. ESS Cryogenic Group carried out the calibration of pressure transducers and leak tests of the control valves. The ESS Vacuum Group strongly supported KrioSystem in pumping vacuum in the CDS vacuum jacket as well as in executing the leak tests. The ESS Integrated Control System Division installed the entire front-end electronics for CDS-SPK and CTL. The ESS Infrastructure Group installed the cables connecting the valve box instrumentation and control valves with the CDS front-end electronics. Finally, ESS contracted to Power Heat the installation of 13 CDS-SPK header units, welding process pipes in 26 CDS-SPK interconnections and connecting the CTL to the ACCP cold box.

4.10 Commissioning

The CDS commissioning stared in October 2022 and was carried out by ESS Cryogenic Group according to procedures described in [15]. This phase was proceeded by a test readiness review held in September 2022. This TRR aimed in evaluating if the CDS components, commissioning procedures, test equipment, and support personnel were ready for the final commissioning at cryogenic temperatures. ESS Integrated Control System Division also held two other TRRs, in April and September 2022, which aimed in assessing the readiness of the CDS controls for the system commissioning.

The CDS commissioning campaign included the following activities:

- Cleaning and conditioning of the CDS helium circuits.
- Leak tests of the CDS control valve seats.
- Cool down.
- Instrumentation tests at cryogenic conditions.
- Heat load measurements.
- ACCP cold compressor tests and 2 K mode verification.
- Warm-up.

The tightness tests of the control valve seats were verified with pressure change tests similar to technique D3 of EN 1779:1999. Each tested valve seat was pressurized from one side and the pressure increase in the enclosed volume in other side was continuously monitored. The specified volume and measured pressure increase rate allowed for calculating the leak rate though the valve seat. The course and results of the leak tests were reported in [16]. The initial measurements revealed that a number of valves did not meet the specified leak rate of 10^{-4} mbar•l s^{-1} and inspections of several valves revealed contaminations in the valve seals and seats, scratches on the valve plugs, and in the case of CDS-EL CV-04 valves some significant misalignments between the valve bodies and their inserts. The cleaning of valve seats and replacement of seals partly improved the tightness of the valve seats. Table 4.2 shows the measured leak rates.

Although 91 of 344 tested valves were not sufficiently tight, it was decided to proceed to the cool down due to schedule constrains and postpone the repair of leaky valve after tests in cryogenic conditions. The leakiest valves were the TS return valves in the CDS-EL valve boxes (located in the MBL and HBL sections of the LINAC tunnel). For 14 of 30 CV60 valves, the measured leak rate was higher than 10^{-1} mbar•l s^{-1} and for 26 of 30 CV61 valves the leak rate exceeded 30 mbar•l s^{-1}. Since there was no need to close the TS circuits inside valve boxes at the CDS cold test, it was agreed to keep all the CV61 valves open and regulate the helium flow in the valve box thermal shield loops with valves CV60.

The CDS cool down started on December 5, 2022. On December 6, as soon as the temperatures dropped below 70 K, the ESS CRYO team walked down the CDS to inspect its external envelope. Temperature measurements were done with a thermal imaging camera and a contact thermometer. In many points, the measured temperatures were between 16.5 °C to 18 °C. However, the temperature of the jumper connections at their vacuum barriers varied between 11.9 °C and 13.5 °C. The air temperature and relative humidity in the tunnel were around 18.7 °C and 35%, so the dew point was around 7 °C. Therefore, there was no water condensation visible on these surfaces. Figure 4.21 shows example infrared images and pictures of the temperature measurements.

Table 4.2. Measured leak rates of the CDS control valve seats after cleaning and installing new seals in 25 CVs (outlined in yellow) [16]. (Courtesy of ESS)

Valvebox	CV-03	CV-60	CV-04	CV-61	CV-05	CV-62	CV-06	CV-63
Spk-010	6.31E-05	2.77E-01	5.94E-02	0.00E+00	3.18E-07	0.00E+00	0.00E+00	6.19E-04
Spk-020	0.00E+00	3.95E-05	0.00E+00	0.00E+00	0.00E+00	0.00E+00	0.00E+00	0.00E+00
Spk-030	4.00E-08	1.59E-01	1.26E-07	2.63E-06	0.00E+00	0.00E+00	7.89E-06	0.00E+00
Spk-040	6.31E-04	0.00E+00	0.00E+00	0.00E+00	0.00E+00	4.10E-04	0.00E+00	4.40E-06
Spk-050	0.00E+00	0.00E+00	0.00E+00	0.00E+00	0.00E+00	0.00E+00	0.00E+00	1.13E-05
Spk-060	0.00E+00	2.05E-05	0.00E+00	1.11E-06	0.00E+00	0.00E+00	0.00E+00	0.00E+00
Spk-070	0.00E+00	1.12E-04	2.36E-04	1.33E-04	0.00E+00	2.01E-05	0.00E+00	0.00E+00
Spk-080	0.00E+00	4.10E-04	0.00E+00	0.00E+00	5.48E-08	0.00E+00	3.11E-07	1.12E-05
Spk-090	1.19E-05	0.00E+00	0.00E+00	0.00E+00	0.00E+00	0.00E+00	0.00E+00	0.00E+00
Spk-100	0.00E+00	5.34E-04	0.00E+00	2.12E-02	1.58E-05	0.00E+00	0.00E+00	1.98E+01
Spk-110	1.58E-04	2.05E-04	0.00E+00	1.84E-07	0.00E+00	0.00E+00	0.00E+00	4.10E-04
Spk-120	3.18E+00	0.00E+00	3.73E-06	0.00E+00	0.00E+00	0.00E+00	0.00E+00	0.00E+00
Spk-130	0.00E+00	0.00E+00	0.00E+00	6.13E-04	0.00E+00	0.00E+00	0.00E+00	6.11E-04
MBL-010	7.59E-01	1.57E+01	7.50E-04	6.26E+01	3.22E-03	0.00E+00	1.09E-03	0.00E+00
MBL-020	8.07E-04	2.75E-01	0.00E+00	8.63E+01	0.00E+00	0.00E+00	2.25E-03	0.00E+00
MBL-030	5.74E-03	2.56E-01	0.00E+00	6.26E+01	2.51E-03	1.23E+00	1.45E-03	0.00E+00
MBL-040	4.20E-04	1.28E-02	0.00E+00	8.97E+01	0.00E+00	0.00E+00	0.00E+00	4.77E-04
MBL-050	0.00E+00	1.66E-03	0.00E+00	7.49E+01	0.00E+00	0.00E+00	4.49E-02	1.33E-02
MBL-060	0.00E+00	0.00E+00	0.00E+00	7.67E+01	0.00E+00	0.00E+00	0.00E+00	4.38E-05
MBL-070	0.00E+00	6.25E-02	0.00E+00	9.26E+01	0.00E+00	0.00E+00	0.00E+00	0.00E+00
MBL-080	0.00E+00	1.47E-01	5.01E-04	1.72E-03	0.00E+00	0.00E+00	0.00E+00	0.00E+00
MBL-090	1.74E-06	2.27E-01	2.79E-06	6.44E+01	0.00E+00	4.07E-04	7.41E-04	0.00E+00
HBL-010	0.00E+00	2.69E-01	1.87E-05	9.29E+01	0.00E+00	0.00E+00	0.00E+00	0.00E+00
HBL-020	4.95E-04	2.32E-01	0.00E+00	6.57E+01	0.00E+00	0.00E+00	1.24E-03	0.00E+00
HBL-030	0.00E+00	2.79E-01	0.00E+00	9.34E+01	0.00E+00	0.00E+00	0.00E+00	0.00E+00
HBL-040	3.96E-04	2.63E-01	7.72E-02	9.13E+01	4.83E-05	0.00E+00	1.06E-03	0.00E+00
HBL-050	0.00E+00	5.95E-03	7.17E-03	9.52E+01	0.00E+00	0.00E+00	0.00E+00	0.00E+00
HBL-060	7.25E-04	5.09E-02	5.22E-02	9.08E+01	0.00E+00	1.10E-06	4.98E-04	0.00E+00
HBL-070	0.00E+00	6.06E+01	2.21E-06	3.01E+01	0.00E+00	6.91E-05	0.00E+00	4.38E-02
HBL-080	5.14E-01	1.85E-03	0.00E+00	3.36E+01	0.00E+00	2.69E-04	0.00E+00	0.00E+00
HBL-090	7.49E-04	1.64E-03	0.00E+00	6.25E+01	0.00E+00	3.73E-03	1.24E-03	3.90E-05
HBL-100	0.00E+00	4.14E-04	0.00E+00	9.14E+01	0.00E+00	1.19E-06	2.53E-03	0.00E+00
HBL-110	0.00E+00	2.74E-01	0.00E+00	9.33E+01	0.00E+00	0.00E+00	1.04E-03	2.55E-06
HBL-120	0.00E+00	1.43E-07	0.00E+00	9.08E+01	0.00E+00	0.00E+00	0.00E+00	0.00E+00
HBL-130	0.00E+00	7.87E-04	5.95E-03	3.63E+01	0.00E+00	0.00E+00	1.93E-06	2.78E-06
HBL-140	8.41E-03	0.00E+00	6.64E-02	1.12E-02	8.24E-05	1.46E-01	1.92E-03	3.31E-03
HBL-150	4.98E-04	1.74E-02	0.00E+00	8.89E+01	9.89E-05	0.00E+00	1.24E-03	1.95E-04
HBL-160	9.66E-04	8.16E-05	1.73E-05	5.31E+01	0.00E+00	0.00E+00	3.34E-04	1.39E-04
HBL-170	0.00E+00	2.38E-01	9.90E-05	8.39E-01	0.00E+00	5.11E-06	1.23E-03	1.82E-08
HBL-180	0.00E+00	2.66E-01	2.15E-03	3.08E-02	0.00E+00	0.00E+00	3.08E-03	0.00E+00
HBL-190	0.00E+00	1.04E-03	0.00E+00	9.41E+01	0.00E+00	4.28E-02	0.00E+00	0.00E+00
HBL-200	0.00E+00	2.24E+00	5.29E-04	9.49E+01	0.00E+00	0.00E+00	0.00E+00	3.41E-04
HBL-210	1.41E-06	6.86E+01	0.00E+00	3.69E+01	0.00E+00	0.00E+00	4.95E-04	2.76E-07

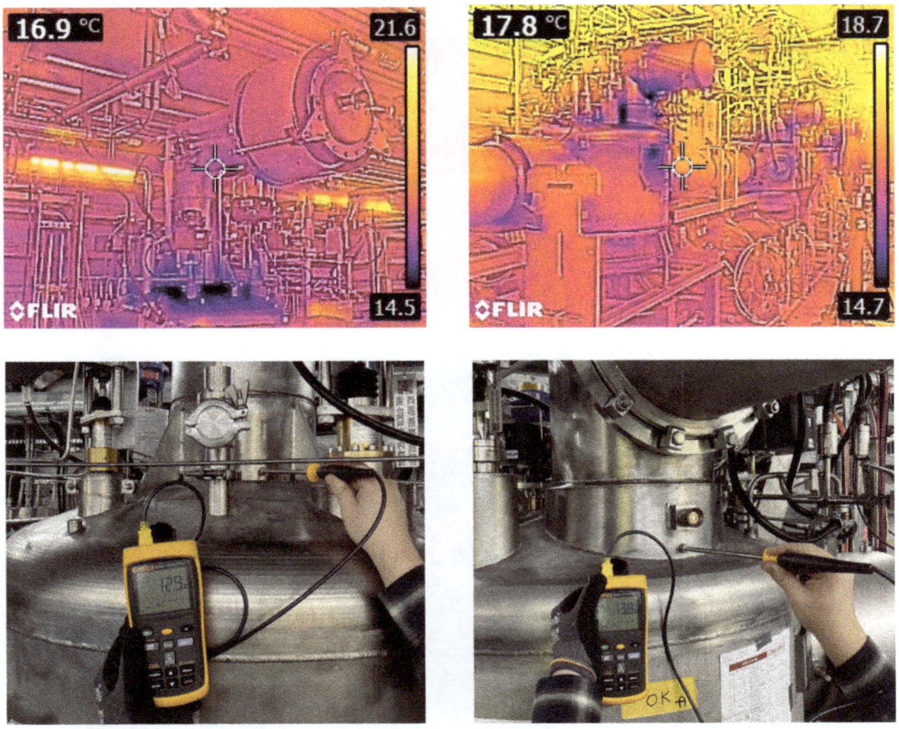

Figure 4.21. Infrared imaging and temperature measurements of the valve box vacuum jackets after CDS cool down below 70 K. (Courtesy of ESS)

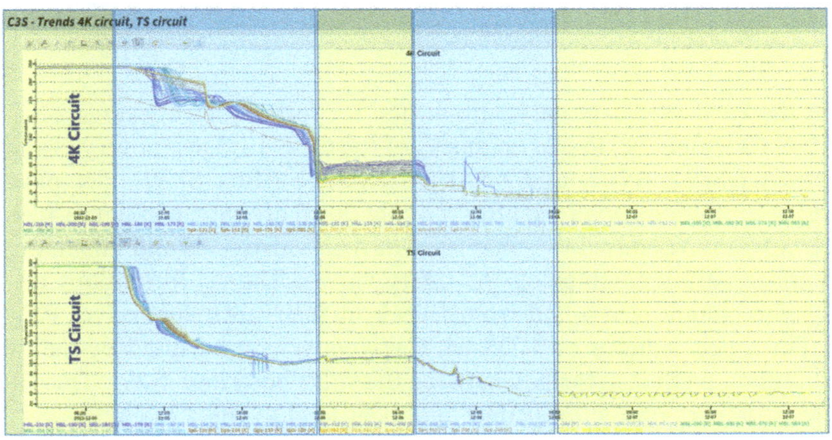

Figure 4.22. Evolution of temperatures in the process lines during the CDS cool down. (Courtesy of ESS)

After two days, the temperatures of the 4 K and thermal shield circuits were stabilized around 8–10 K and 35–40 K, respectively. The temperature evolutions in the process lines measured at the cool down phase are shown in figure 4.22. The lowering of temperature in the He supply and VLP lines revealed a number of cold

Figure 4.23. Ice formations at colds spots on the CDS components, (a–c) CDS-EL CV04 valve bonnets, (d) chimney of the He recovery branch line downstream CV06 and CV63 of CDS-EL valve boxes, (e) CDS-EL SV02, and (f) CDS-SPK EBox SV91 inlet pipe. (Courtesy of ESS)

spots on the valve box external components. Ice formations appeared on 22 of 30 CDS-EL CV04 bonnets, on two chimneys of the CDS-EL He recovery branch lines, on several CDS-EL SV02, and on the CDS-SPK EBox SV91 inlet pipe. Some example ice formations observed after the cool down are shown in figure 4.23. Those on SV02 valves were caused by some leaks at their valve seats and inlet flanges, and they disappeared after tightening the safety valves. The ice formations on the He recovery line chimneys resulted from a leak in their upstream valves (CV06 and CV63). The ice formations on CDS-EL CV04 valves were caused by the thermoacoustic oscillation effect in the gap between the valve body and insert. This effect had a significant impact on the measured total heat load to the 4 K circuit.

The heat load measurements were done for 19 different configurations of helium mass flow rates and valve openings. The measurement results are described in detail in [17]. Figure 4.24 shows the summary of measurement results. When ice was observed on 22 of 30 CDS-EL CV04 valves, the total heat load was around 900 W. Closing the CV03 valves to 20% reduced the flow in the 4 K branch circuits in the valve boxes and decreased the number of CV04 valves affected by TAO effect to 15.

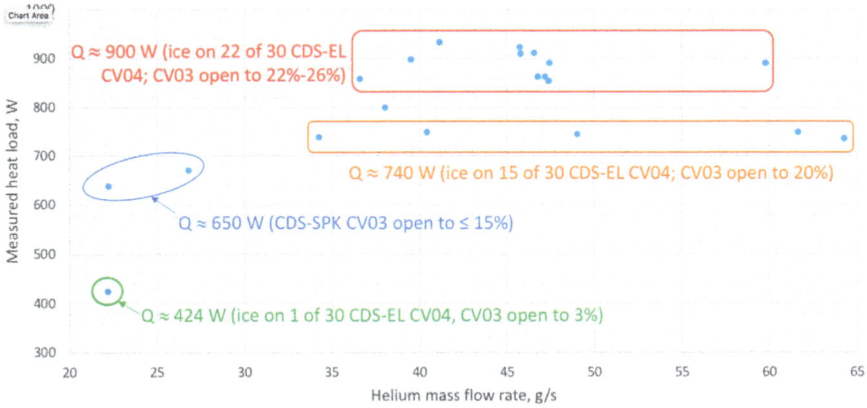

Figure 4.24. Measurement results of the heat load in the CDS 4 K circuit. (Courtesy of ESS)

Then, the total heat load dropped to around 740 W. For CV03 valves open to 3%, there was still sufficient flow of cold helium in the 4 K branch circuits to keep them in the required temperature ranges and simultaneously the TAO effect disappeared in next 14 valves. In this configuration the total heat load dropped to 424 W, which is only 1.2% higher than the specified value (419 W). This result was very promising and showed that eliminating the TAO effects in the CV04 valves would reduce the total heat load below the acceptable value.

The activities on the ACCP cold compressors were mainly focused on testing the behavior of the cryoplant at the subatmospheric mode. These tests also helped to verify the performance of the VLP line. Lowering the pressure in the VLP line to 31 mbar(a) allowed 2 K helium to be produced in the end box phase separator and the performance of its electrical heaters, level probes, and control loop to be verified at their nominal operating conditions.

During these tests, when the temperature in the He supply line was lowered to 4.5 K, another ice formation appeared on SV91 of the CDS-SPK End Box. Simultaneously, the pressure measured by PT91 on the end box and PT11 on the ACCP cold box fluctuated between 2.6 bara and 3.4 bara with frequency around 0.4 Hz. These measurements revealed that TAO effect appeared in the SV91 inlet pipe and caused the pressure oscillation in the entire He supply line.

4.11 Lessons learned and corrective actions

Both ESS and the in-kind partners have gained a lot of knowledge during the executing the CDS projects. All observed issues were analyzed and used to improve the course of the project. For example, after revealing weld defects in the inner process pipes of the CDS-SPK valve boxes and header units, ESS supported the partners in quality inspections of components still under repair. ESS quality engineers regularly inspected the production of CDS components in the partners' contractors' premises, as well as at the installation activities in the tunnel.

One of significant lessons learned was related to the cleanliness of process pipes. Some endoscopic inspections done at the repair of the CDS-SPK valve boxes revealed a lot of metal chips inside the process pipes. After this observation, IJCLab and ESS carried out many regular inspections of the repaired CDS-SPK valve boxes and header units, both in the production workshop and in the tunnel just before closing their process lines.

Another endoscopic inspection of randomly selected 12 cryogenic control valves of the CDS-EL valve boxes already installed in the tunnel revealed some contamination, including liquid in several branch process pipes. WUST and their contractor immediately responded to this observation and carried out a full-scale inspection and cleaning campaign for all the CDS-EL valves.

The most substantial technical issues observed at the CDS commissioning were leaks in the control valve seats and TAO effects in the CDS-EL CV04 valves and in the CDS-SPK End Box SV91 inlet pipe. Since the TAO effects strongly affected the heat load measurement results, it was decided to repair defective components and carry out another commissioning test, initially scheduled for June 2023.

In order to eliminate the TAO effect in the CDS-EL, ESS removed all the valve inserts and installed 6 convection breaks on each insert, as shown in figure 4.25. The convection break is made of a PE-UHMW ring, which closes the gap between the valve and body and insert almost along its entire circumference. The ring is held by two metal rings tack welded to the valve insert. The selected solution is designed to strongly minimize convection movements of helium but still allow for evacuating air from the entire volume of the gap during pumping and purging of the process pipes.

Detailed investigations of the TAO effect that was observed at SV91 valve were done by IJCLab. As a solution they proposed to decrease the volume of the warm section of the SV91 inlet pipe by moving the valve from its initial position on the CDS vent line onto the end box, as shown in figure 4.26. Since there is a risk that this change will not fully eliminate the TAO effect, it has been decided to use an additional countermeasure of a buffer volume with a flow restriction element. This countermeasure is designed to suppress potential remaining oscillations. It includes a

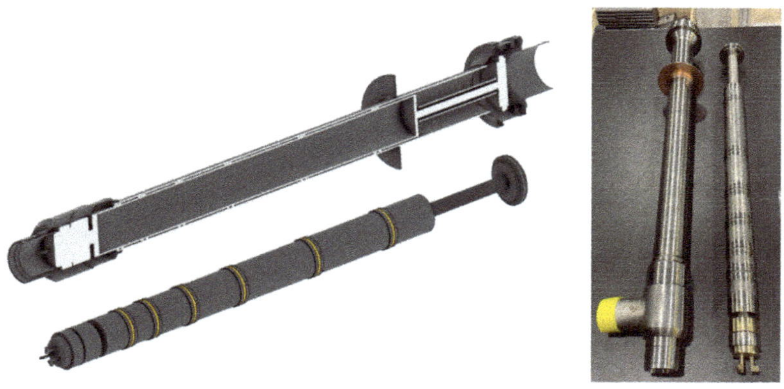

Figure 4.25. Convection breaks on the CV04 valve insert. (Courtesy of ESS)

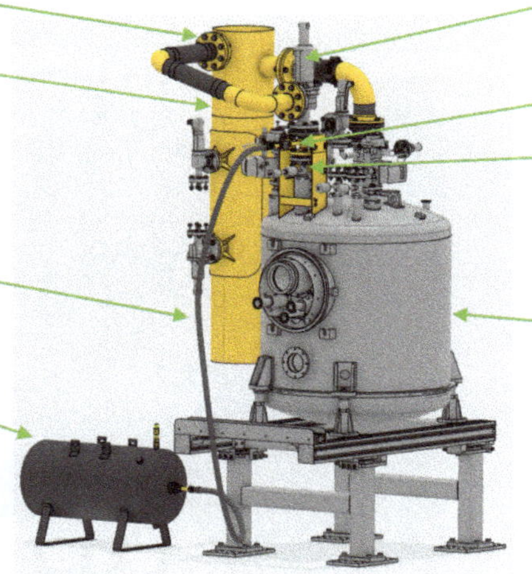

Figure 4.26. Simplified 3D model of the CDS-SPK end box and CDS vent line with relocated SV91 valve and a buffer tank for dumping TAO effect in the SV91 valve inlet pipe. (Courtesy of ESS)

100 dm^3 tank connected to the warm end of the SV91 inlet pipe via a flexible hose and a needle valve, as shown in figure 4.26.

For leaks in the control valve seats, ESS have discussed the issue with the manufacturers of the valves and planned to improve the tightness by increasing preloading of the valve inserts, installing new seals, and replacing plugs with digital ones (on-off) for valves that do not need provide flow control function. Due to time constrains and relatively long delivery time, these activities are planned to be carried out simultaneously with the installation of other cryomodules after the second commissioning.

Regarding the issue of some misalignments in the CDS-EL CV-04 valves, during the first commissioning campaign ESS had already carried out numerical analyzes to find the root cause of these misalignments. The analyzes did not reveal any issue the CDS-EL valve box design. ESS also constructed a test rig to investigate a potential correlation between the valve body bend and the seat tightness. The test rig comprises a DN10 WEKA valve of the same design as CV04 installed in the CDS-EL valve boxes and allows the valve to be bent in a controlled way up to 15 mm. Figure 4.27 shows the design of the test rig and its valve under verification of its seat leak at different valve body deflections.

These tests showed that the tested valve seat already had a leak of 10^{-2} mbar•l s^{-1} when its body was straight and the leak did not increase when the valve was bent, even to 15 mm. Further discussions with the valve manufacture revealed that the given preload on the valve steam was too small. Increasing this preload from 1 mm to 6 mm improved the seat tightness to 10^{-6} mbar•l s^{-1} for both the straight and bent configurations. However, an initial leak test done on one CV61 valve in the tunnel with the increased preload and showed a decrease of the seat leak from almost 10^2 mbar•l s^{-1} to 10^{-3} mbar•l s^{-1}. This is a very significant improvement of the valve tightness, but the present leak rate is still below the specified value (10^{-4} mbar•l s^{-1}).

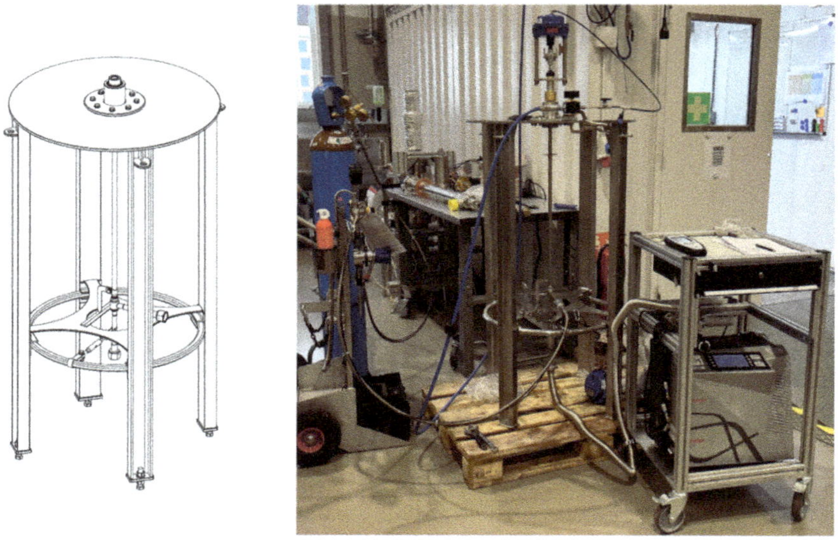

Figure 4.27. Test rig for experimental investigations of the impact of CV61 valve misalignments on the valve seat tightness. (Courtesy of ESS)

Therefore, more detailed leak tests on the valves in the tunnel are planned after the second commissioning campaign, together with investigations of other potential root causes of insufficient tightness of valve seats as well as potential solutions.

4.12 Second commissioning

The second commissioning campaign was carried out in June and July 2023. It was done just after completing the pilot installations of one spoke and one elliptical cryomodules in the tunnel and connecting them to the CDS valve boxes. The campaign included testing the installed cryomodules at 2 K and measuring heat loads to the cold circuit at 8 K. At the heat load measurement stage, the circuit was composed of the He supply and VLP lines with their branch loops in all valve boxes and end box, as well as in the two cryomodules. During cool down and heat load measurement phases there was no ice formation on any of the CDS-EL CV04 valves and their bonnets were at a temperature around 10 °C. This showed that the applied convection breaks described in section 4.11 efficiently eliminated TAO effect inside the valves. For the TAO effect in the end box SV91 inlet pipe, the applied solution significantly minimized this effect. There was only very tiny ice formation on the SV91 inlet pipe and the pressure oscillations in the He supply line were only around ± 2–4 mbar around the average value of 3 bara.

The heat load measurement was done for four mass flow rates of helium at 8 K and 3 bara in the inlet to the CDS He supply line, namely 38.3 g s^{-1}, 43.7 g s^{-1}, 48.0 g s^{-1}, and 54.3 g s^{-1}. The temperature in the outlet of the CDS VLP line was 9.48 K, 9.22 K, 9.05 K, and 8.85 K, respectively, while the pressure was constantly maintained at 1.08 bara. In order to eliminate an impact of the systematic errors of temperature

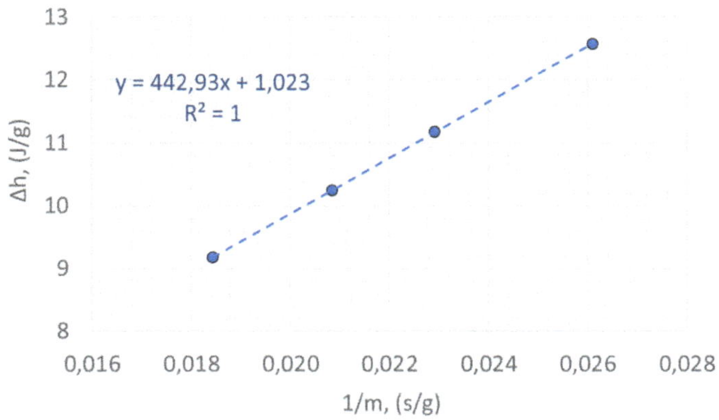

Figure 4.28. Heat load measurement results plotted in Δh-m^{-1}. (Courtesy of ESS)

measurements on the heat load calculation, the obtained results were analyzed with regression method. Figure 4.28 shows the measurement results plotted in a Δh-m^{-1} diagram, where Δh is enthalpy difference and m is mass flow rate.

It must be noted that the measurement points perfectly fit the line $\Delta h = 442\ 93\ \mathrm{m}^{-1} + 1.023$, with regression coefficient $R^2 = 1$. This means that the measured heat load to the tested cold circuit, including the two connected cryomodules, is equal to 443 W with very high accuracy, and the systematic error of enthalpy difference is 1023 J g^{-1}. This systematic error can be converted to the systematic error of the measured temperature difference between the cold circuit inlet and outlet, measured by TT11 and TT12, as shown in the piping and instrumentation diagram in figure 4.4. The product of the Δh systematic error (1023 J g^{-1}) and the mean value of helium specific heat at the inlet and outlet of the cold circuit (6.4 J g•K) gives the ΔT systematic error of 0.16 K. This means that the measured temperature differences between TT12 and TT11 are 0.16 K above the real values.

The measured total heat load of 443 W includes heat loads to the cold circuit of the tested cryomodules. Since the total heat load to the cryomodules at 8 K and with closed power coupler cooling loops was estimated around 40 W, the total heat load to the CDS cold circuit only is estimated as equal to 403 W, which is below the specified value (419 W).

References

[1] Lindroos M, Eshraqi M and McGinnis D 2013 The European Spallation source *Proc. PAC2013* (Pasadena, USA) pp 1448–52

[2] Peggs S 2013 *ESS Technical Design Report* (Lund, Sweden: European Spallation Source ESS AB) 650 p

[3] Fydrych J 2016 CDS-EL L4 requirements *ESS Technical Document (Controlled): ESS-0054483*

[4] Fydrych J 2014 Technical specification of the cryogenic distribution system for the elliptical linac *ESS Technical Document (Controlled): ESS-0011735* p 66

[5] Fydrych J 2015 Technical specification of the cryogenic distribution system for the spoke linac *ESS Technical Document (Controlled): ESS-0017178* p 44

[6] Fydrych J 2016 CDS-EL to ICS interface requirements *ESS Technical Document (Controlled): ESS-0054484*

[7] Fydrych J 2016 CDS-EL to ACC interface requirements *ESS Technical Document (Controlled): ESS-0054485*

[8] Fydrych J 2016 CDS-EL to SI(CF) interface requirements *ESS Technical Document (Controlled): ESS-0065708*

[9] Su X T 2022 General scheme for CDS and CM in tunnel *ESS Technical Document (Controlled): Drwg. ESS-3484497*

[10] Su X T 2022 P&ID of CDS and elliptical CM in tunnel *ESS Technical Document (Controlled): Drwg. ESS-3484499*

[11] Su X T 2022 P&ID of CDS and spoke CM in tunnel *ESS Technical Document (Controlled): Drwg. ESS-3484500*

[12] Su X T 2022 P&ID of CDS and end box in tunnel *ESS Technical Document (Controlled): Drwg. ESS-3484501*

[13] Fydrych J, Arnold P, Hees W, Tereszkowski P, Wang X L and Weisend J G 2015 Cryogenic distribution system for the ESS superconducting proton linac *Phys. Proc.* **67** 828–33

[14] Fydrych J and Tereszkowski P 2015 Location of the wave guides under the Cryogenic Distribution Line for the elliptical linac *ESS Technical Document (Controlled): ESS-0012571*

[15] Zhang J 2022 ACCP-CDS integrated test verification *ESS Technical Document (Controlled): ESS-4017250*

[16] Zhang J 2022 CDS control valves seat leak test report *ESS Internal Document (Controlled): ESS-4792471*

[17] Zhang J 2023 CDS head load and thermal acoustic oscillation on valves *ESS Internal Document (Controlled): ESS-4860022*

Chapter 5

The target moderator cryoplant

J Jurns and H Tatsumoto

The target moderator cryoplant (TMCP), which is a 20 K-large scale helium refrigeration system with the cooling capacity of 30.3 kW at 15 K, will cool the cryogenic moderator system (CMS) where liquid hydrogen (17 K and 1.1 MPa) is circulated to cool high-energy neutrons down to cold neutrons in two cryogenic hydrogen moderators (four in the future). Two compressor skids in parallel and a cold box are located close to all the European Spallation Source (ESS) cryoplants. This results in a long cryogenic helium transfer line (CTL) between the TMCP cold box and the CMS in the target building. The first installation of the compressor skids and the cold box was completed in 2019. In the second phase, the CTL and a valve box were installed in 2022. The commissioning of the TMCP without connecting the CMS was completed in December 2022. This chapter describes the requirements, acquisition process, design, construction, and commissioning of the TMCP. Lessons learned from the project are also presented.

5.1 Background and history

The ESS Technical Design Report [1, 2], which was first published in early-2013, defined the preliminary requirements for the TMCP. The TMCP is one of a suite of three helium cryoplants that support ESS operations, the other two being the accelerator cryoplant (ACCP), described in chapter 3, and the test and instruments cryoplant (TICP), described in chapter 7. The 2013 report noted that a key feature of the ESS target is that it has hydrogen moderators, which use hydrogen at 20 K and 1.5 MPa[1] to reduce the energy of the neutrons before they reach the instrument lines. The energy that the neutrons deposit in the hydrogen moderator is removed by the TMCP that supplies a cold helium circuit operating at 16 K to a heat exchanger, which interfaces with the hydrogen moderator circuit and maintains its nominal operating temperature of 20 K.

[1] This was later reduced to 1.0 MPa.

doi:10.1088/978-0-7503-3223-1ch5

As with the ACCP described in chapter 3, specifying and procuring the TMCP in a timely fashion was critical in ensuring that construction, installation, and commissioning was completed consistent with the overall ESS project schedule. Award of the TMCP was made to Linde Kryotechnik (LKT) in April 2016.

John Jurns joined the ESS cryogenics team in 2014, and was appointed the Project Engineer for the TMCP, working under P Arnold, the Work Package Leader for Cryogenics. As with the ACCP, we also had support from senior cryogenic consultants G Gistau (formerly of Air Liquide) and H Quack (formerly of Linde and the Technical University of Dresden).

Conceptual design of the TMCP was largely based on referencing projects from other neutron spallation sources, specifically the Spallation Neutron Source (SNS) at Oak Ridge National Laboratory in Oak Ridge, Tennessee, USA [3], and the J-Parc neutron spallation source at the JAEA Tokai campus in Japan [4]. Both of these facilities have smaller cryogenic cooling capabilities than the ESS TMCP but function in the same way, using a cryogenic helium refrigerator to provide cooling to a supercritical hydrogen moderator.

5.1.1 Why do we need a TMCP?

The ESS target converts the high-energy proton beam produced by the accelerator into neutrons via spallation. Protons are absorbed by the nuclei of the target tungsten causing the emission of a large number of neutrons.

The neutrons produced by the spallation process are reduced in energy as they pass through a moderator that is filled with supercritical hydrogen operating at 20 K. The neutrons scattering through the 20 K hydrogen moderator deposit up to 30 kW of heat, which must be removed by the cryogenic system. This system consists of two parts: the TMCP and the CMS. The CMS [5] is a closed loop of hydrogen that circulates through the moderator absorbing heat from the neutrons (figure 5.1). The

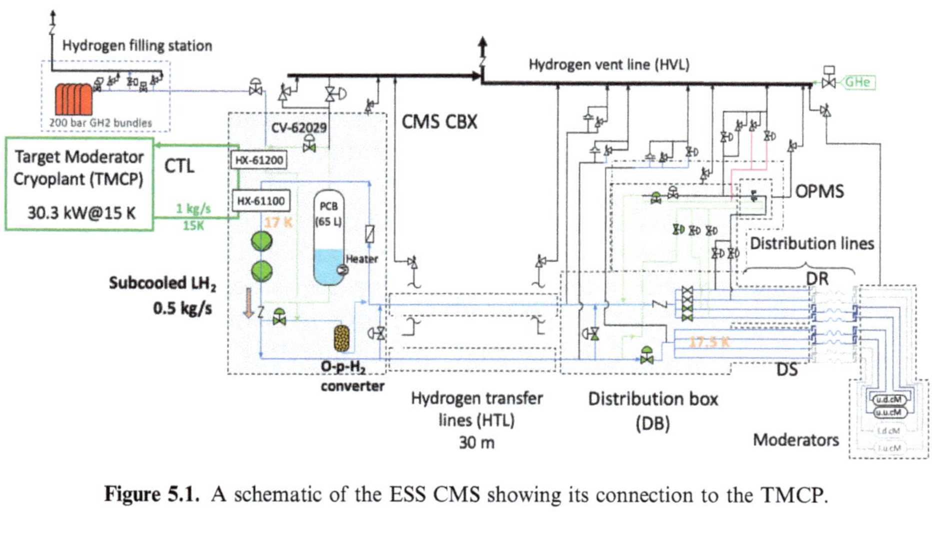

Figure 5.1. A schematic of the ESS CMS showing its connection to the TMCP.

CMS transfers this heat via a heat exchanger to a helium flow at between 16 K and 17 K produced by the TMCP. More details on the CMS can be found in chapter 6.

5.1.2 User discussions

The user of the TMCP cryoplant is the target CMS. At the time of TMCP design and construction, the ESS project was broken into four major divisions—the Accelerator Division, Target Division, Instrument Division, and Conventional Facilities Division. The TMCP project was managed out of the Accelerator Division, but the CMS project was managed out of the Target Division. This required careful coordination and meetings between the two groups to assure that all the system requirements were captured successfully and incorporated into the TMCP specifications. Additionally, the CMS was provided to ESS as an in-kind contribution from the Forschungszentrum Jülich (FZJ) research center in Jülich Germany. Numerous meetings, teleconferences, and design reviews were held between ESS Target Division, FZJ, and TMCP personnel to define requirements and interfaces. Hideki Tatsumoto joined ESS as the ESS Project Engineer for the CMS. He started initially in the Target Division during the CMS procurement and then transferred to the Cryogenics Group within the Accelerator Division for CMS commissioning and operations. This is consistent with the strategy that there would be only one group responsible for cryogenic operations within ESS.

5.1.3 Reference projects—SNS and J-Parc

The design basis of the TMCP fundamentally follows along the same lines as two other neutron spallation source projects—the SNS at Oak Ridge National Laboratory in Oak Ridge, Tennessee, USA [3], and the J-Parc neutron spallation source (JSNS) at the JAEA Tokai campus in Japan [4].

The TMCP, JSNS, and SNS all function in essentially the same way—a helium refrigerator circulates cold helium gas to a heat exchanger situated in an intermediate cryostat which is part of the CMS (see table 5.1). Cryogenic liquid hydrogen

Table 5.1. Comparison of ESS, SNS, and JSNS target cryoplants.

	JSNS[1]	SNS[1]	ESS[1], [2]
He cryoplant capacity	6.45 kW @ 16.5 K	7.5 kW @ 16.5 K [3]	35 kW @ 15 K
H_2 mass flow rate	~180 g s^{-1}	~150 g s^{-1}	~1000 g s^{-1}
He mass flow rate	307 g s^{-1}	~275 g s^{-1}	1125 g s^{-1}
Number of expansion turbines	1	1	2[2]
Number H_2 circulators[1]	2 (parallel)	3 (parallel)	2 (series)
Hydrogen state	Supercritical	Supercritical	Liquid

[1] SNS hydrogen has a separate circulation loop for each moderator, ESS and J-Parc have a single hydrogen circulation loop to the moderators.
[2] ESS TMCP has a third bypass turbine at 90 K level. ESS specified two parallel cold end turbines to provide redundancy and turndown capacity.
[3] Actual capabilities at commissioning ~4 kW at 17 K.

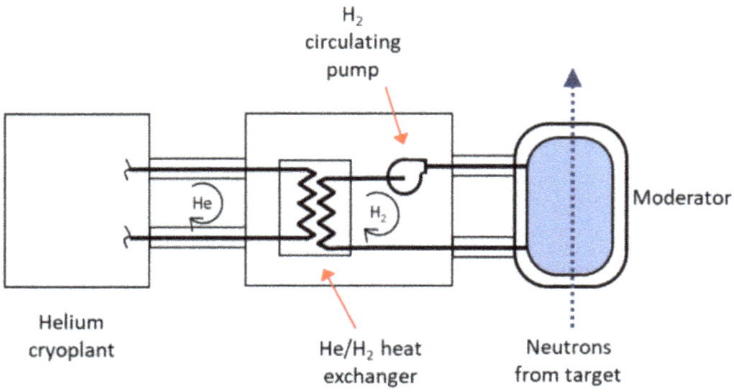

Figure 5.2. Helium cryoplant cooling for a hydrogen moderator.

is circulated via a pump to a moderator, where energy is absorbed from neutrons passing through the moderator. The warmed hydrogen circulates back to the CMS cryostat and is cooled by the helium as is passes through the heat exchanger. A simplified schematic of the process is shown in figure 5.2.

5.1.4 Where the TMCP diverges from previous facility designs

Although functionally similar, there were some significant differences between the design of the TMCP and the other two cryoplants. SNS and JSNS had relatively short distances between the helium cryoplant, an intermediate cryostat, and a target moderator, as shown in figures 5.3 and 5.4. Figure 5.5 shows the overall layout of the TCMP, with the TMCP cold box located approximately 305 meters from the CMS cryostat. The reasoning for locating the TMCP so far from the CMS cryostat was based on a decision early in the project to co-locate all the ESS cryoplant cold boxes in the same building to facilitate their installation, operation, and maintenance.

5.2 Project overview

As mentioned earlier, TMCP is one of a suite of three helium cryoplants that support ESS operations, the other two being the ACCP and the TICP. Chapter 2 noted that all three cryoplants would be collocated to allow them to share common services and permit easier operation, inspection, and maintenance. Although this design decision resulted in certain benefits, it also raised some challenges because the TMCP cold box was separated from the CMS hydrogen system by approximately 305 m, resulting in very long helium transfer piping between the two systems, a large helium inventory of ~336 kg, and greater pressure drop through the helium loop.

5.2.1 Interfaces

5.2.1.1 Organizational interfaces

Before addressing the process and physical interfaces, it is also important to discuss the organizational interfaces. Successful execution of the TMCP project required that TMCP personnel to establish and maintain effective communications with the

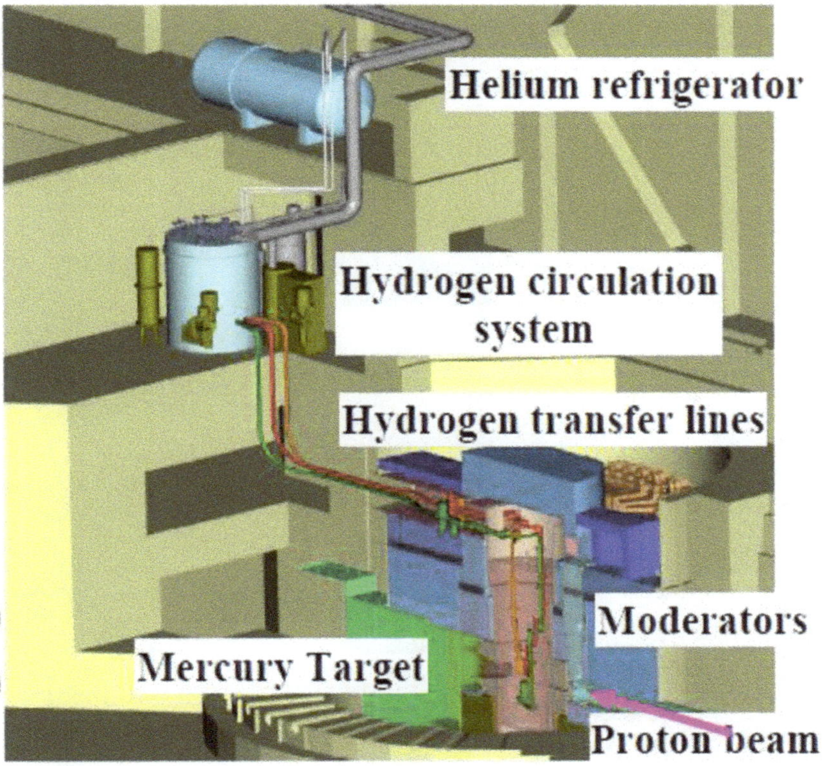

Figure 5.3. JSNS target moderator system showing the cryogenic components. Reprinted from [4].

Accelerator Division, which was responsible for the overall cryogenic infrastructure and how it was integrated into the project budget and schedule; the Target Division, which was responsible for the overall CMS project; in-kind partners responsible for design, fabrication, and installation of the CMS; Conventional Facilities Division, which was responsible for providing the physical space and utilities for the TMCP; the other cryogenic projects (ACCP, TICP, and helium recovery systems) to coordinate physical layouts and avoid interferences; and finally with vendors that supplied the TMCP, CTL piping, and other minor equipment.

Regular project meetings and design reviews were held with all of these groups. In addition, site visits were made to in-kind partner and vendor facilities for meetings and equipment inspections.

5.2.1.2 Process and physical interfaces

TMCP process and physical interfaces were clearly defined early in the project to assure that (1) the TMCP was delivering the required cooling, (2) that the TMCP received the required utilities and physical space, and (3) connected systems clearly understood what the TMCP was delivering.

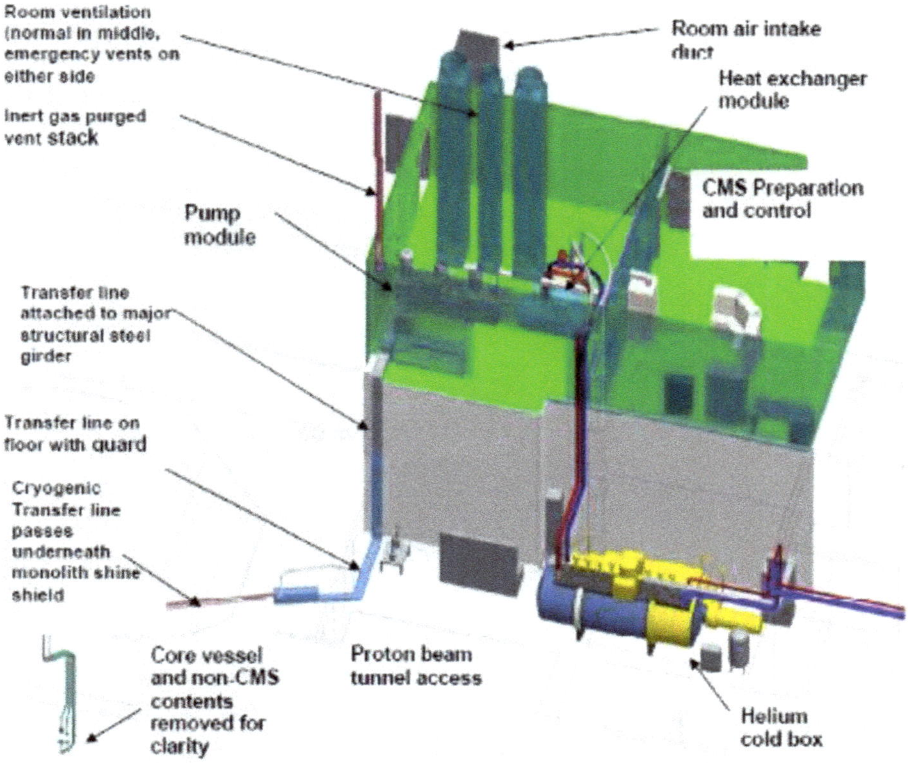

Figure 5.4. SNS target moderator system showing the cryogenic components.

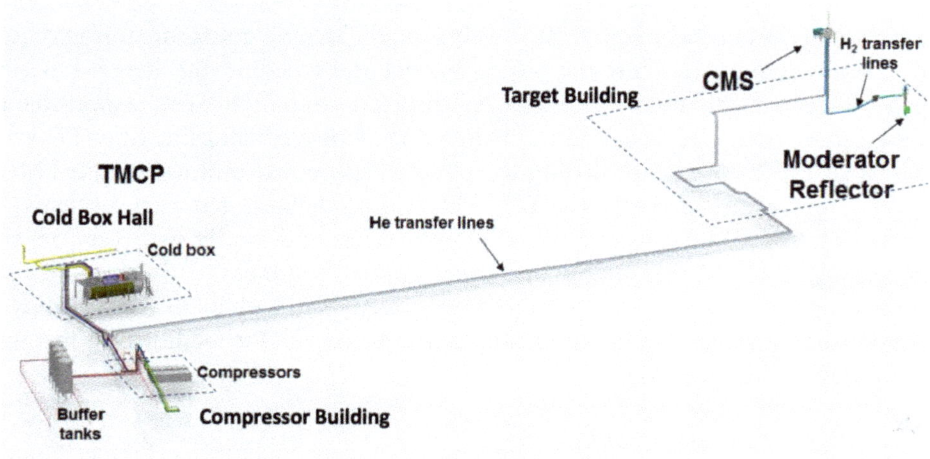

Figure 5.5. TMCP, CMS, and CTL layout.

5.2.1.3 TMCP to CMS process interface

Heat load—The maximum estimated neutronic and static heat loads to the CMS is 32.1 kW. The static heat load is relatively constant for all operating scenarios. However, neutronic heat loads change significantly depending on the proton beam power, which can vary from approximately 500 kW during initial facility commissioning to a maximum of 5 MW at full operation. Additionally, beam trips will result in very large changes in the heat load over short (on the order of seconds) durations. The TMCP was designed to accommodate these heat load variations.

Pressure—The maximum design pressure of the CMS was originally set as 17 bara, based on the maximum allowable design pressure of the moderator. The minimum operating pressure was set at 13 bara[2]. The minimum pressure is based on the requirement to operate at supercritical pressures to avoid the risk of two-phase flow in the hydrogen circuit. The helium supply to the CMS is at approximately 20 bara. This assured that helium pressure was greater than the hydrogen pressure, preventing possible leaks of hydrogen into the helium circuit.

Temperature—The maximum hydrogen operating temperature in the CMS is 20.5 K based on FEA analysis of the moderator temperatures. The moderator design also imposes a maximum temperature difference across the moderator of 3 K. The total CMS allowable temperature difference is 3.5 K, which includes an additional 0.5 K estimated temperature rise through the rest of the CMS. The TMCP supply to the CMS is approximately 15 K. The small allowable temperature difference across the CMS resulted in a large helium mass flow rate to the CMS from the TMCP.

5.2.1.4 TMCP to CMS physical interface

The CTL is the physical interface between the TMCP and CMS, which is a vacuum insulated pipe circuit that delivers 15 K helium to the CMS and returns ~20 K helium from the CMS to the TMCP cold box.

Early in the project, a decision was made to co-locate all of the ESS helium cryoplants. This resulted in a long helium transfer line between TMCP and CMS. This approach had the advantage of consolidating all ESS cryoplants to facilitate maintenance and consolidate utilities. However, the long transfer lines result in a large helium inventory and large pressure drop through the helium loop. Early in the project, the initial heat load estimate for the TMCP was only 20 kW, but changes in the moderator design and additional static heat loads attributed to the TMCP resulted in a heat load increase to 32.1 kW. This increase led to a greater helium mass flow rate in the CTL. Additionally, the CMS cryostat was relocated from its original location to comply with hydrogen safety requirements. This added length to the CTL, resulting in an overall length between CMS and TMCP of ~305 m, and a more significant helium inventory in the CTL of 336 kg.

To accommodate the increase to 32.1 kW refrigeration requirements, the CTL process pipe diameter increased to minimize the pressure drop through the loop. The

[2] Later reduced to 11 bara.

Table 5.2. CTL size and fluid parameters.

DN100 helium process pipe	Unit	Supply	Return
Helium mass flow rate	kg s^{-1}	1.02	1.02
Inlet pressure	bara	19.6	19.2
Temperature	K	15	20
Density	kg m^{-3}	63.81	45.22
Pipe length	m	305	307
Pipe I.D.	mm	108.2	108.2
Pressure drop through CTL	bara	0.055	0.076
CTL helium mass	kg	197	139

pressure drop through the CTL is directly dependent on the helium density, so the TMCP process was set with the expansion turbines on the return side of the helium loop. This allowed higher pressure (and therefore higher density) helium in the entire CTL, thereby minimizing pressure drop through the circuit. Table 5.2 shows details on the CTL size and fluid parameters.

5.2.1.5 TMCP to Cryogenic Systems process interface

Chapter 2 states that each cryoplant has its own warm compressor system, cold box, and controls, and each is operated independently. There are no provisions for sharing loads between the plants. However, the cryoplants do share a common helium gas recovery, purification, and storage system (see chapter 7). Pure helium gas is stored at 20 bar in the pure helium system (PHS) which consists of 19 70 m^3 gas tanks (figure 5.6). Four of these PHS gas tanks are dedicated to storage of TMCP helium.

5.2.1.6 TMCP to Cryogenic Systems physical interface

Although each ESS cryoplant operates independently, the three cryoplants do share utilities. Shared utilities include cooling water for the compressor heat exchangers and cold-box expansion turbine brakes, and instrument air supply for all cryoplant valve operators. Electrical power is distributed to the TMCP from ESS main power supply as follows:

- 6.6 kV 3 Ph 50 Hz compressor motors
- 400 V 3 Ph 50 Hz compressor skid oil pumps and heaters, cold-box vacuum pumps and heaters
- 230 V 1 Ph 50 Hz controls for both compressor skid and cold box

5.2.1.7 Cold-box room

The TMCP shares physical space with both the ACCP and TICP. The TMCP cold box is located in the same building (building G.02) as the ACCP and TICP cold boxes. The TICP helium dewar filling station is also located in the same room.

Figure 5.6. The ESS helium compressor facility with the PHS storage tanks (19 in total). (Courtesy of W Hees and P Arnold ESS.)

5.2.1.8 Compressor building

The cryogenic systems compressor building G.04 is located approximately 100 m from the cold-box building. Warm helium piping connects the cold boxes to their respective compressor stations via underground piping. The building is split by a central wall, with the TMCP compressor skids and oil recovery system (ORS) occupying the western half of the building. The TMCP compressor skids and ORS share this part of the building with the ESS helium recovery system equipment.

5.3 TMCP technical requirements

Design philosophy—The TMCP design was primarily driven by the cooling requirements of the CMS, i.e., it needed to provide adequate cooling to the CMS moderator. Since the CMS heat load varies due to significant variation in the neutronic heat load, the TMCP was configured to accommodate these variations in its design. As noted in paragraph 2.2.3, co-location of all the ESS cryoplants resulted in a long run of vacuum insulated piping between the cold box and CMS, affecting the overall cryoplant heat load.

The CMS cooling requirements, facility constraints, and applicable codes and regulations were used to define the technical specification for the TMCP. TMCP's overall technical requirements were informed by the design philosophy, and defined as follows:

- High operational reliability
- Stable operation for any heat load between minimum and maximum values defined by the CMS
- Power savings by means of capacity adaptation for partial refrigeration loads
- No liquid nitrogen (LN_2) precooling
- Fully automatic operation in all steady-state operation modes
- Capability to handle transient operation modes with minimal operator intervention

Operational reliability—The TMCP is designed for continuous cryogenic supply to the CMS 24 h/day, 7 days/week during normal operations. During normal operations, availability should exceed 99% (not counting utility failures) to contribute to the overall goal of ESS reliability of 95%. During scheduled operational interruptions (planned two times per year, total estimated time 14–16 weeks), the TMCP may not be required to operate. The design and construction of all components of the TMCP were required to have an operational lifetime of at least 25 years.

Stable operation—ESS has defined several steady-state operating modes: nominal design, nominal low power, nominal turndown, and minimum turndown modes. These modes translate into specific ranges of heat loads that the TMCP accommodates.

- *Nominal design mode*
  1. This mode will occur with the beam on and should be the most common mode seen at ESS. This mode features the highest cooling capacity required.
- *Nominal low power mode*
  1. This mode will occur with the beam on but operating at less than full power. During initial operations in particular, beam power could vary from ~2% to 10% of nominal design for extended periods of time. The TMCP shall be capable of holding conditions constant at any point between minimum and nominal design heat loads.
- *Nominal turndown mode*
  1. The mode will occur with the beam off.
  2. For short duration interruptions of the beam, electric heaters in the cryogenic circuits may turn on, or other appropriate means used, to allow the heat load to the cryogenic TMCP to remain constant. For longer duration interruptions, the cryogenic TMCP capacity will be 'turned down' via the control system to allow for this reduced heat load requirement in the most energy efficient manner.
- *Minimum turndown mode*
  1. For the case of no CMS dynamic heat load, the total static heat load from the hydrogen and helium circuits (with no safety or operational margins) may be as low as 2.3 kW.

Table 5.3. TMCP heat loads and cryogenic plant power.

Operation modes		Heat load, W	Mass flow ($g\ s^{-1}$)	Pressure (bar)		Temperature (K)	
				Supply	Return		
Nominal design	Max	31 200	1125	21.5	5.2	$\geqslant$15	$\leqslant$20
	Min	14 900	608	12.1	2.9	$\geqslant$15	$\leqslant$20
Nominal low power	Max	8300	337	12.6	3.1	$\geqslant$15	$\leqslant$20
	Min	5400	251	8.6	2.3	$\geqslant$15	$\leqslant$20
Nominal turndown		4300	207	7.8	1.9	$\geqslant$15	$\leqslant$20
Minimum turndown		2300	126	4.7	1.2	$\geqslant$15	$\leqslant$20

Specific operating heat load requirements—The heat loads and operating temperatures for the various operating modes are given in table 5.3.

Power savings by means of capacity adaptation—As a primary focus for constructing ESS was energy efficiency, the TMCP design had to address this in the design. It was determined that temporary dissipation of excess refrigeration capacity by electrical heating was acceptable, but power saving means of capacity adaptation was important and therefore written into the procurement specification.

No LN_2 precooling—The TMCP schematic shows an 80 K level expansion turbine T1 that eliminates the requirement for LN_2 precooling, results in additional efficiency, and eliminates the need for a liquid nitrogen supply.

Automatic operation in all steady-state operation modes—Automatic operation is accomplished using a refrigeration control system consisting of a main process control PLC that interfaces with local HMIs and the ESS overall EPICS facility control.

Handling transient operation modes with minimal operator intervention—This is accomplished through the automatic operation of PLC controls that will load or unload compressors, switch appropriate valves, and program ramps to the different modes automatically. Diverting helium flow returning from the CMS to a cold-box ambient heat exchanger accommodates rapid changes in neutronic heat load.

5.3.1 Standards

All components and equipment had to meet European and Swedish laws and regulations, as well as standards of good practice commonly accepted in the cryogenic community. A partial list of norms and rules applied for the TMCP is given below:

- ALPEMA: The standards of the brazed Aluminium Plate-fin heat Exchangers Manufacturers' Association
- CGA S-1.3-2008: Compressed Gas Association CGA Pressure relief device standards Part 3—Stationary storage containers for compressed gases
- EN 10204: Metallic products—Types of inspection documents

- EN 12517-1: Non-destructive testing of welds—Part 1: Evaluation of welded joints in steel, nickel, titanium, and their alloys by radiography—Acceptance levels
- EN 12517-2: Non-destructive testing of welds—Part 2: Evaluation of welded joints in aluminium and its alloys by radiography—Acceptance levels
- EN 13445: Unfired pressure vessels
- EN 1435: Non-destructive examination of welds—Radiographic examination of welded joints
- EN 1779: Non-destructive testing—Leak testing—Criteria for method and technique selection
- EN 287–1: Qualification test of welders—Fusion welding—Part 1: Steels
- EN 288–3: Specification and approval of welding procedures for metallic materials—Part 3: Welding procedure tests for the arc welding of steels
- EN 30042: Arc-welded joints in aluminium and its weldable alloys—Guidance on quality levels for imperfection
- EN 439: Welding consumables—Shielding gases for arc welding and cutting
- EN ISO 14731: Welding coordination—Tasks and responsibilities
- EN ISO 15614-1: Specification and qualification of welding procedures for metallic materials—Welding procedure test—Part 1: Arc and gas welding of steels and arc welding of nickel and nickel alloys
- EN ISO 17635: Non-destructive testing of welds—General rules for metallic materials
- EN ISO 3834-2: Quality requirements for fusion welding of metallic materials —Part 2: Comprehensive quality requirements
- EN ISO 5817: Welding—Fusion-welded joints in steel, nickel, titanium, and their alloys (beam welding excluded)—Quality levels for imperfections
- EN ISO 9001: Quality Management Systems—Requirements
- EN13480: Metallic industrial piping
- IEC: International Electrotechnical Commission
- ISO 2372 Group G: Mechanical vibration of machines with operating speeds from 10 to 200 rev/s—Basis for specifying evaluation standards
- ISO 2954: Mechanical vibration of rotating and reciprocating machinery – Requirements for instruments for measuring vibration severity
- ISO 3740: Acoustics—Determination of sound power levels of noise sources —Guidelines for the use of basic standards
- ISO 8573-1: Compressed air—Part 1: Contaminants and purity classes
- PED: Pressure Equipment Directive PED 97/23/EC with Annexes I to VII
- TEMA: The standards of Tubular Exchanger Manufacturers' Association

5.4 Procurement

5.4.1 Procurement strategy

Prior to making a call for tender to procure the TMCP, ESS engineers made a budgetary cost estimate for the TMCP cost using historical cost data from similar size plants and scaling factors. This was done to inform the reasonableness of proposals received. When the expected overall configuration was known, ESS contacted several suppliers to get budgetary proposals. These budgetary numbers were used to refine the TMCP technical requirements.

One item to note is that the TMCP design heat load increased from an initial 25 kW at 15 K to 32 kW at 15 K between the time that budgetary proposals were received and the final technical specification was drafted. This increase was due to changes in the cryogenic moderator design. This of course resulted in an increase in the expected cryoplant cost.

The procurement of the TMCP was initiated after the ACCP procurement, and as such was able to benefit from the policies and procedures developed from that project for posting call for tenders, evaluating proposals, and awarding contracts. The ESS cryogenics team worked with members of the ESS procurement office (Luis Ortega, Meredith Shirey, Mirko Menninga, and Malcolm De Silva) and the ESS Legal Office (Ohad Graber-Soudry) to ensure that the procurement went smoothly. Thanks to the ACCP experience and the ESS procurement team, the TMCP procurement was executed essentially on schedule and without any significant challenges or issues.

As with the ACCP, ESS chose to procure the TMCP via competitive bids in response to a detailed technical specification and statement of work (SOW) based on specific TMCP performance requirements. The TMCP technical specification was written largely as a functional specification, i.e., the bidder was responsible in their proposal for defining the thermodynamic cycle design and the associated equipment and controls (both hardware and software) required to implement a design that met the specification performance requirements. Besides the TMCP cryoplant, there were two additional items specified in the call for tender—the helium/hydrogen heat exchanger for the CMS, and the CTL test jumper spool box (JSB).

While there was no restriction on the bidder's country of origin, the ESS call for tenders required that bidders meet certain requirements including business size, financial stability, and previous experience with designing, fabricating, and installing helium cryoplants of a similar scale and scope.

The ESS's use of in-kind contributions as a funding approach was described in chapter 1. Since all the cryoplants (ACCP, TMCP, and TICP) were planned as commercial purchases, the procurement of these plants was never seen as suitable for in-kind contributions and the cryogenic plants were funded directly by the ESS laboratory.

5.4.2 TMCP technical specification

The technical specification along with an accompanying SOW defined the top-level requirement for procuring the TMCP. As with the ACCP and TICP, the TMCP specification required sufficient detail to assure that the product delivered met all the technical requirements but also left sufficient room for suppliers to play to their strengths and propose innovative and efficient designs. The TMCP specification was written after the ACCP specification, and drew largely from that document to assure consistency in how ESS defined the requirements. The TMCP specification was a collaborative effort, written principally by John Jurns, with contributions from Philipp Arnold and John Weisend. The document was reviewed by ESS cryogenic team members, CMS engineers, in-kind partners, and ESS procurement division personnel before drafting in its final form.

Key topics in the TMCP technical specification included:

- Background and aims.
- Codes and norms.
- ESS site conditions and utility interface.
- Performance of the TMCP.
- Key components:
    - Warm compressor station (WCS).
    - Cold box.
    - Helium/Hydrogen heat exchanger.
    - Vacuum jacketed test JSB.
- Spare parts.
- TMCP control system.
- Instrumentation, interlocks, and electrical design.
- Mechanical design, manufacture, and mounting requirements.
- Tests at the manufacturer's facilities.
- Tests at the ESS site.
- Quality management.
- Documentation.

The performance requirements section of the technical specification detailed the various operating modes that the cryoplant had to run under. These modes (described in the Technical Requirements section) are as follows:

- Nominal design mode.
- Nominal low power mode.
- Nominal turndown mode.
- Minimum turndown mode.

Two items included in the scope of the TMCP purchase in addition to the cryoplant were the helium/hydrogen heat exchanger for the CMS and a vacuum jacketed test JSB.

The He/H_2 heat exchanger was procured with the TMCP because it was anticipated that the TMCP supplier would source all the cold-box heat exchangers from the same supplier, and including the CMS He/H_2 heat exchanger would provide cost and schedule savings. The He/H_2 heat exchanger would then be provided to the CMS project to incorporate into their system. Figure 5.7 shows the interface between the helium and hydrogen sides of the heat exchanger.

The vacuum jacketed test JSB was required for TMCP acceptance and performance tests because it was anticipated that the TMCP and CTL would be complete before the CMS was available. The JSB would also provide an interface between the CTL and the CMS once the CMS was installed.

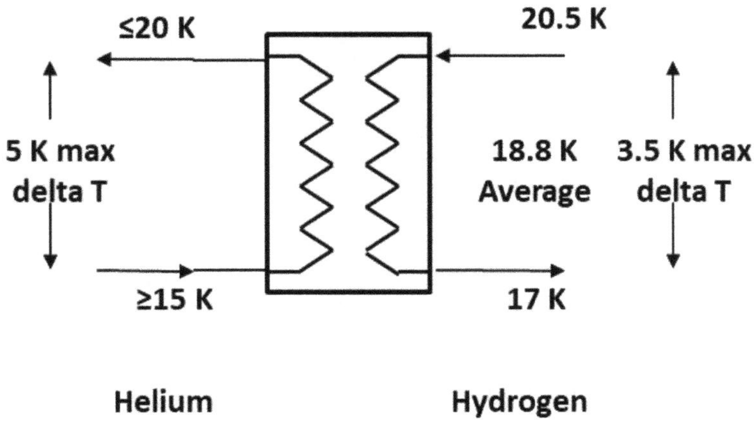

Figure 5.7. He/H_2 heat exchanger.

5.4.3 TMCP SOW

Whereas the TMCP technical specification detailed all the technical requirements for the TMCP and associated equipment, the SOW defined the contract deliverables, what would be provided by ESS, what reviews would be required, documentation requirements, approval criteria, and delivery terms.

5.4.4 Call for tender and bid evaluation

A call for tender was issued for the TMCP in October 2015. Qualified bidders were provided with the following set of tender documents:

- Tender Management.
- Company Qualification.
- Instruction to Bidders.
- TMCP Statement of Work.
- TMCP Technical Specification.
- Proposal Evaluation Criteria.
- Draft Supply Agreement and Performance Guarantee.
- General Conditions of Tender.

The TMCP project was awarded based on assessing bids received against the evaluation criteria provided to bidders. Bidders had to meet qualification criteria to be considered. Factors including past performance on similar size projects, key employees, confirmation with applicable codes and standards, compliance with ESS documentation requirements, ability to provide aftersales service, capacity to perform the work, and meeting quality assurance and sustainability standards were used to determine if ESS would consider the bid.

Once qualified, bidders were evaluated on the following criteria:

- CAPEX—Capital cost of the bid.
- OPEX—Operating cost of the cryoplant.
- Quality factors.

Table 5.4. TMCP proposal evaluation weights.

Award criterion	Maximum score
Technical award criterion	**25**
Quality score of technical proposal	25
Commercial award criteria	**75**
CAPEX	45
OPEX	30
Total score	**100**

Evaluation factors were weighted according to table 5.4.

As with the ACCP, bids were not only evaluated on capital cost but also on operating cost and quality factors. Significant weight was given to the operating cost criteria to emphasize to bidders the value ESS placed on minimizing environmental impact.

It is worth noting that the criteria give almost equal weight to the initial capital cost of the plant and the plant's operating cost. Thus, a vendor could propose a more expensive plant and still win the bid if that plant would be more energy efficient, and thus have a lower operating cost. This willingness of ESS to spend money up front to gain long term energy savings is consistent with ESS's commitment to sustainability.

Bids responding to the formal TMCP call for tender were received by December 2015. After reviewing and evaluating qualified bids, ESS awarded the TMCP contract to LKT with signing of the contract June 2016. It is worth noting that bids for the TMCP were extremely close. The deciding factor ended up being the advantage LKT offered in the OPEX evaluation.

One particular aspect of the bid evaluation was noteworthy to report here. The ESS TMCP specified high reliability for the equipment to assure greater than 99% operational reliability. LKT proposed using helium compressors manufactured by GEA Grasso. These compressors did not have a significant operating history with helium, but were a good match to TMCP operating requirements. Linde recognized this potential risk, and included a spare compressor in their proposal to mitigate this risk. We note that to date, the GEA Grasso compressors have operated with no issues.

5.4.5 Vacuum insulated piping

5.4.5.1 Procurement strategy

Vacuum insulated piping was required for the TMCP system. The CTL is the interconnecting piping between the TMCP cryoplant and the test JSB located in the target building. The Hydrogen Cryogenic Transfer Line (HyTL) is the interconnecting piping between a cryostat housing circulating pumps and heat exchangers and the Target neutron moderators. Since part of the CTL was planned to be installed in a below-grade duct between the compressor building and cold-box building, the

CTL procurement was split into two parts. The first procurement was for the piping installed from the TMCP cryoplant to a point in the below-grade duct where it terminated. A second procurement was made for the HyTL piping and the balance of the CTL piping from the termination point in the duct to its final tie-in point at the JSB.

The CTL procurements followed the same process as other systems. Namely, ESS developed a functional specification that defined the technical requirements for the CTL and HyTL, drafted SOW that defined the scope of the work, issued calls for tender soliciting bids from qualified suppliers, evaluated bids based on pre-defined evaluation factors, and awarded contracts.

The first portion of the CTL procurement was for the piping spools only. Final installation and testing of the vacuum insulated piping were planned as a separate activity as ESS scope to be performed by an on-site contractor. The second procurement of the remaining CTL and all of the HyTL piping was procured as a turnkey project, with the supplier manufacturing, installing, and testing the piping systems.

As with the TMCP, there was no restriction on the bidder's country of origin. The ESS calls for tender required that bidders meet certain requirements including business size, financial stability, and previous experience with designing and fabricating vacuum insulated systems of a similar scale and scope.

5.4.5.2 CTL technical specifications and SOW

The technical specifications defined the top-level requirement for the vacuum insulated piping procurements, outlining the temperature, flow, heat leak, and pressure drop requirements, as well as the interface points. It also defined what codes and norms were required for manufacture and testing, required piping routing, and ESS site conditions. It also defined contract deliverables, ESS provided scope, review and documentation requirements, approval criteria, and delivery terms. The vacuum insulated piping technical specifications and SOWs for each procurement were combined in a single document, as opposed to the TMCP, which had separate technical and SOW documents.

5.4.5.3 Calls for tender and bid evaluation

A call for tender was issued for the first CTL procurement TMCP July 2016. A second call for tender was issued for the balance of the CTL and the HyTL procurement October 2018. Suppliers were required to meet the following selection criteria to have their offers evaluated:

- Candidate's size and personnel—List key employees demonstrating sufficient and experienced manpower in the relevant fields.
- Codes and standards—Conformation with all applicable European Directives regulating safety aspects.
- References—List of current or recent projects in the relevant field.
- Additional Technical Criteria—Compliance with ESS documentation requirements, ability to provide aftersales service, and sufficient workshop capacity to perform the work.

Table 5.5. CTL proposal evaluation weights.

Award criterion	Maximum score
Technical award criterion	**25**
Quality score of technical proposal	25
Commercial award criteria	**75**
Capital cost of the bid (CAPEX)	75
Total score	**100**

Table 5.6. Quality score criteria. Courtesy of W. Hees and P. Arnold ESS

Rating criteria	Weight
Completeness of proposal with regard to requested documentation	5
Understanding and compliance with SOW and technical specification	5
Conformity to commercial terms and conditions as set forth in ESS draft contract	5
Project management plan	5
Additional features implemented in the proposed design that are technically and/or commercially advantageous for ESS	5
Total maximum possible score	**25**

- Quality Assurance and Sustainability—Work to be performed according to a quality management system (i.e., ISO 9001 and ISO 14001).

Qualified bidders were evaluated according to table 5.5.
The bidders' CAPEX score was calculated as follows:

$$CS = CS_{MAX} \times \left(\frac{CAPEX_{MIN}}{CAPEX} \right)$$

Where:
CS_{MAX} = Maximal Score for CAPEX
$CAPEX$ = Price of bidder in EUR
$CAPEX_{MIN}$ = Lowest price of all received bids in EUR

The bidders' quality score was determined according to table 5.6.

5.5 Project execution

5.5.1 Design reviews

Executing the TMCP project required a steady stream of meetings between stakeholders to effectively manage progress, schedule, and budget. The contract with LKT included specific milestone reviews, as listed in table 5.7.

Table 5.7. TMCP design reviews.

Review	Participants	Review dates
Kick-off meeting	ESS, FZJ, LKT, TUD, MAX IV[1]	June 2016
Preliminary design review	Cold box: ESS, LKT, TUD	September 2016
	CTL: ESS, Kriosystem	April 2020
HAZOP review	ESS, LKT, TUD, MAX IV	November 2016
Critical design review 1	ESS, LKT, TUD, MAX IV	May 2017
Logic design review	ESS, LKT, TUD	July 2017
Critical design review 2	Cold Box: ESS, LKT, TUD	November 2017
	CTL: ESS, Kriosystem	August 2020, April 2021
Installation readiness review	ESS, LKT, JOBSAB[2]	April 2018
Site acceptance test	ESS, LKT	December 2022

[1] MAX IV Synchrotron Light Source.
[2] JOBSAB Installation contractor.

Table 5.7 lists the contractually obligated reviews with LKT for the project. Each meeting was set with an agenda containing specific project elements to be addressed, and followed up with written meeting notes and action items assigned to various parties. Participants in each review were present based on the agenda elements. For example, participants from Technische Universität Dresden (TUD; responsible for CMS design) were present when discussing the interface between the TMCP and CMS. ESS also called in outside experts to support these meetings. Dr Hans Quack (TUD) provided expert advice on cryogenic aspects of the design and Andreas Thiel (MAX IV) provided expert advice on helium compressor technology.

In addition to these review meetings, ESS, LKT, and other parties were in constant communication by teleconferences and emails to follow-up on action items.

The following paragraphs briefly summarize the scope of each of these review meetings:

- The TMCP kick-off meeting (KOM) agenda included a project status summary, review of the overall target station strategy, discussion on technical topics and communication interfaces, and setting a time and agenda for the preliminary design review.
- The preliminary design review (PDR) included a discussion on the following topics: tentative dates for HAZOP, piping layout for the TMCP cold box, process flow diagram overview, process and instrument diagram (P&ID) overview, T–S diagram review, plate-fin heat exchanger (PFHX) specification sheets review, process control description review, 20 K cold adsorber review, warm compressor review, control valve review, CMS test JSB, cold-box room layout, KOM follow-up items, and a preliminary discussion on-site acceptance testing.
- The HAZOP review included a follow-up on PDR action items, an overview of the HAZOP procedure/approach, a review of specific system nodes (CMS

interface, warm compressor, turbine, cold box, and warm compressor system), and review of the failure modes and effects analysis (FMEA) for cold box and warm compressor system. Following the HAZOP meeting, the review board delivered an evaluation of the review, determination of whether the HAZOP and FMEA studies met requirements for safety, personnel and machine protection, and gave recommendations for action items.

- Critical design review number 1 (CDR-1) included a review of the warm compressor P&ID, a review of the overall system P&ID, the utility list review, the acceptance test procedure review, a discussion on controls (CMS interface and signal exchange), a review of warm compressor system 3D models, a review of the compressor hall layout and cold-box hall layout, a review of the cold box internal piping design, and a review of the electrical overview diagrams.
- The logic design review included a review of software architecture, an introduction to compressor logic, a review of gas management controls, a review of permissives and shutdowns for TMCP and TJS cold modes, an introduction to basic control loops and step chains for the TMCP and TJS, and an introduction to control loops and step chains of dynamic operations for the TMCP and TJS.
- CDR-2 (held at ESS) was a follow-up to CDR-1 topics and included a review of PID changes, further discussions on the TMCP/CMS interfaces, a review test JSB design, a review on updates to the compressor hall and cold-box hall layouts, a review on more details of the cold-box external and internal piping design, a review of the purge panel design, a review of the 20 K adsorber regeneration skid design, a review of electrical overview diagrams and interfaces, discussion on QA/QC issues, and a site visit for LKT personnel.
- After the main components had been delivered on-site and before the installation started, an Installation Readiness Review (IRR) was conducted at ESS. This review evaluated the readiness of LKT to start installation at ESS, with an emphasis on the interfaces between components and subsystems, controls integration, and a detailed look at the plans, staff, and tooling required for the installation work itself. The IRR covered review of the compressor hall layout, review of the installation plan, a hazards analysis review, a review of the compressor and cold box placement, a review of the work and safety coordination plan, a review of the electrical and control system design, a review of compressor factory acceptance testing (FAT), and review of commissioning tasks and installation schedule.
- The site acceptance test review took place on December 8, 2022. The TMCP has passed all acceptance tests as specified and ESS received all final documentation and a punch list on minor items that still needed to be fixed. A provisional acceptance certificate has been issued and signed by both the LKT and ESS.

In addition to meetings with external stakeholders, the TMCP engineering team also held regular review meetings with other internal ESS organizations and stakeholders. Regular meetings were held with:

- The ESS Conventional Facilities Division to coordinate issues related to the TMCP installation and utility requirements.
- ESS Target Division, Forschungszentrum Jülich, and TUD to coordinate design and interface requirements between the TMCP and CMS.
- ESS Accelerator Division Instrument and Controls group to coordinate interface between TMCP controls and the overall accelerator control architecture.

Summaries of all meetings and activities were provided to management in a monthly work package report.

5.5.2 Inspections and factory acceptance tests

The TMCP supplied by LKT consisted of a number of components manufactured by suppliers which were then integrated into the complete cryoplant. During manufacture, ESS personnel were invited to witness fabrication and FAT of these components. A summary of site visits follows:

- *Heat exchangers*—The aluminum plate-fin heat exchangers for the cold box and CMS cryostat were fabricated by Linde Schalchen (Tacherting, Germany). ESS personnel visited the site August 2017 and witnessed the final pressure tests. Linde Schalchen provided quality assurance documentation following the site visit and testing.
- *Cold box*—The TMCP cold box was fabricated by Simic S.p.a. (Camerana, IT). ESS personnel were on-site for leak testing of the cold box internal piping. Photographs of the cold box under construction and during testing are shown in figures 5.8–5.10.

Figure 5.8. TMCP cold box shell under construction.

Figure 5.9. TMCP cold box internal piping.

- *Helium compressors*—The TMCP warm helium compressors were manufactured by GEA Refrigeration Germany GmbH (Berlin, Germany). ESS visited the site July 2017 to witness a FAT. The testing included mechanical and vibration testing. One item to note was that although ESS had stipulated acceptance testing at full load (21.4 bara discharge pressure), the GEA test stand only had the capability of testing at a lower pressure and flow rate due to a limitation on the size of the compressor drive motor at the test stand. This requirement was unfortunately not passed from LKT to the compressor manufacturer. After some discussion, ESS agreed to have the FAT performed at the 'minimum turndown' operational case—1.13 bara suction pressure, 5.57 bara discharge pressure, and 453 g s^{-1} nominal flow rate. The compressors passed FAT at these conditions. Figure 5.11 shows one of the compressors on the GEA test stand.

- *Helium compressor skid*—Once the helium compressors passed FAT, they were shipped to Enerproject S.A. (Mezzovico-Vira, Switzerland) to be integrated into a skid package that included the compressor, motor, controls, piping, and oil separator. ESS visited Enerproject two times—once in September 2017 to witness helium leak tests of the skid, and once in October 2017 to witness a FAT for the assembled compressor skid. Helium

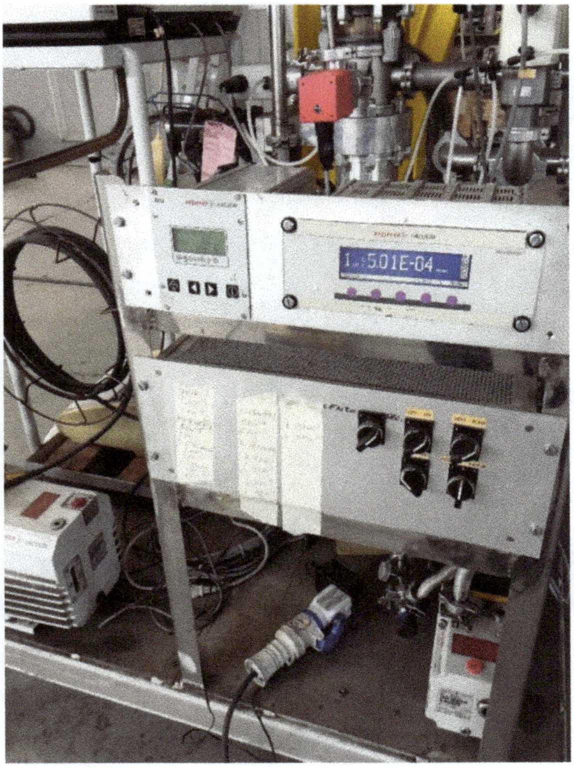

Figure 5.10. Cold-box piping leak testing in progress.

leak testing was performed by a third party in accordance with standard EN 1779. The compressor skids successfully passed leak test. FAT included field signal functionality and malfunctions, pneumatic valve activation, lubricating oil circuit functionality with oil pump in service, and compressor slide movement. The compressor motor was not operated due to power limitations at the Enerproject facility. FAT goals were largely achieved, with a small list of punch items identified to be addressed. Figure 5.12 shows the compressor skids during FAT at Enerproject.

- *Oil recovery system*—The oil recovery system (ORS) consisting of three oil coalescing filters, a final oil removal vessel, and interconnecting piping was manufactured by KASAG Swiss AG (Langnau, Switzerland). ESS visited KASAG March 2018 to inspect equipment and review the quality assurance and test documentation. The vessels were fabricated and tested in accordance with the PED directive. Figure 5.13 shows the ORS during assembly at KASAG.

- *Gas management panel*—The gas management panel (GMP) was fabricated by BERO Technik AG (Sirnach, Switzerland). ESS visited BERO in March

Figure 5.11. Helium compressor on the GEA Grasso test stand.

Figure 5.12. Two helium compressor skids during test at Enerproject S.A.

Figure 5.13. ORS being assembled in the KASAG facility.

2018 to inspect equipment and review the quality assurance and test documentation. Figure 5.14 shows the GMP under construction at BERO Technik.

- *Test Jumper Spool cryostat*—The test JSB cryostat was manufactured by Cryo World B.V. (the Netherlands).

5.6 Description of the delivered TMCP

A simplified schematic of the TMCP circuit is shown in figure 5.15. There are several features to point out, demonstrating the capacity flexibility of the cryoplant. First to be noted is the presence of two warm helium compressors and two cold end expansion turbines (T2a and T2b). With only one of the compressors and expansion turbines operating, the TMCP can operate at half capacity. Operating at the load points shown in table 5.3 is accomplished by varying the compressor HP and LP pressures and compressor slide valve.

Additional power savings are realized by utilizing an ambient heat exchanger instead of electric heaters for heat load balancing.

The cooling capacity required for the TMCP is around five times higher than those of existing facilities (SNS or J-PARC). Figure 5.16 shows a more detailed

Figure 5.14. GMP under construction.

overview of the TMCP. The WCS consists of two oil injected screw compressors in parallel (562 g s^{-1} and the input power of 1062 kW for one compressor), an oil recovery system (an oil separator, three oil filters, and a charcoal adsorber), and the GMP. There are four helium buffer tanks with a volume of 70 m^3 each. The TMCP cooling capacity will be controlled by the so-called floating pressure process [6], where matching higher heat loads are achieved by loading mass into the system and by increasing the pressure levels while the pressure ratio over the rotating machinery remains roughly constant. The two buffer tanks (HPB: high pressure buffer volume and LPB: low pressure buffer volume) are set to different pressures and one of them will be selected by the GMP, which controls the loading or unloading of helium to maintain the HP at the set point, dependent on the set pressure of the HP. On the other hand, the LP pressure is controlled by the bypass valve on the GMP. The HP helium stream enters the cold box at ambient temperature.

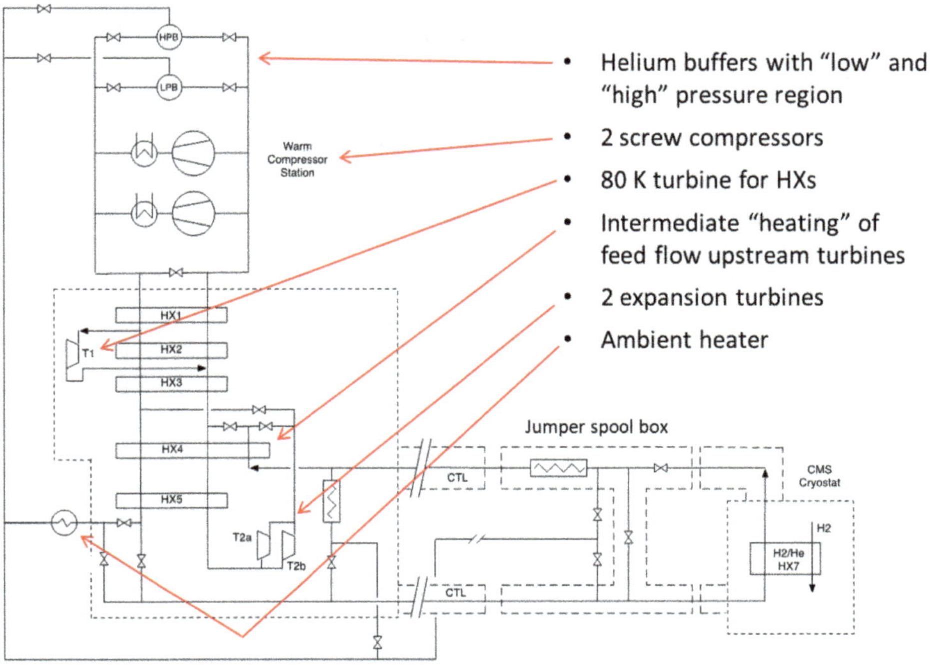

Figure 5.15. TMCP simplified schematic.

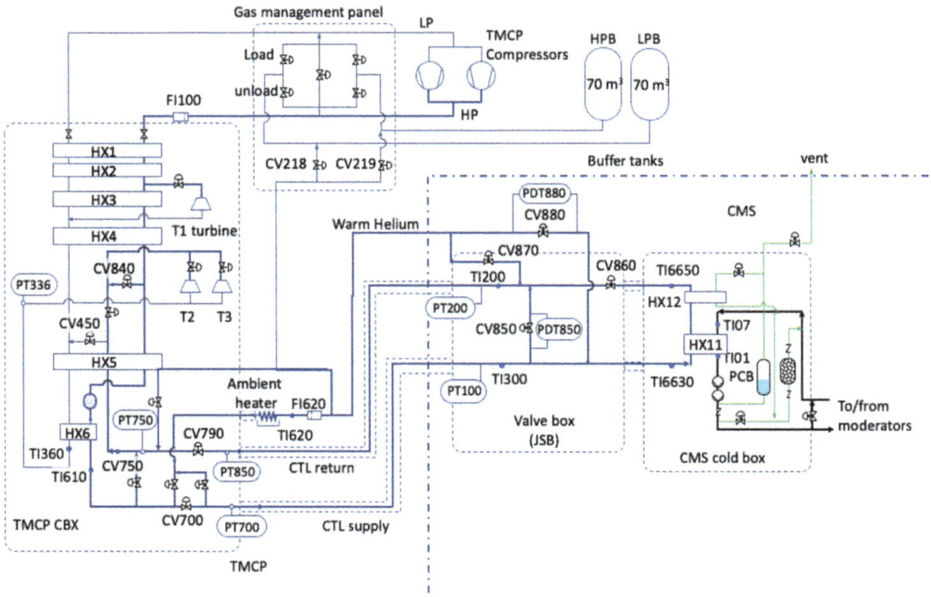

Figure 5.16. Overview of the TMCP. Reproduced from [8].

There are three expansion turbines in the cold box. One is located between the second and third heat exchangers. Part of the HP stream is led to the first turbine (T1) and is expanded into the LP stream at between the third and fourth heat exchangers, where the temperatures are around 60 K. The HP stream is cooled by a counter flow passing through the third heat exchanger and is further cooled down to around 20 K by the LP return stream going through the fourth and fifth heat exchangers. A 20 K adsorber is installed downstream to collect impurities of neon and hydrogen. The HP stream is cooled down to around 15–16 K through the sixth heat exchanger, where the LP stream cooled by the other cold turbines (T2 and T3) enters. The CTL supply temperature (TI610) is regulated by a temperature controller that controls two cold turbine bypass valves (CV450 and CV840) by means of a split range. The 16 K-HP stream flows through the approximately 300 m long vacuum insulated CTL into the JSB. Part of the HP helium flow (< 25 g s^{-1}) feeds through an ambient heater (aluminum star fin tube vaporizer) at the cold end of the cold box and is transferred to two mixing control valves (CV870 and CV880) at the JSB. The CMS supply temperature is adjusted by injecting warm helium via CV880. The cooling capacity required for the CMS is adjusted changing the feed flow rate by CV860. The return temperature (TI200) is always maintained at 21.2 K by injecting the warm helium via the other valve (CV870). The return helium stream passes through the three-stream heat exchanger (HX5) and is expanded by two cold parallel expansion turbines. The LP stream warms up through all the heat exchangers and returns to the suction side of the compressor.

5.6.1 Compressor system

The compressor system mainly consists of two helium screw compressors in parallel; the ORS, including a coarse oil separator, an integrated coalescer, and a charcoal adsorber; and the GMP, as shown in figure 5.17. During operation, the slide valve is closed and is only used during startup of the compressor. The high pressure is regulated by the operation conditions, whereas the low pressure is floating. The oil injected in the compressor block has to be separated in the ORS located downstream of the warm compressors. The oil is removed from the helium in the first coalescer located in the bulk oil separator of each compressor skid. The bulk oil separator accumulates oil droplets entrained within the helium stream. The collected oil flows back into the compressor. The ORS is designed for the final separation of the compressor oil from the helium gas stream. The three coalescer vessels are all mounted on one skid. The residual oil content in the gas will finally be less than 10 ppb(w) range. The separator cartridge can be replaced.

The final removal of oil and oil vapor takes place in the charcoal oil adsorber vessel. The remaining impurities are removed from the helium below 10 ppb by weight. The gas analyzer is connected downstream of the charcoal vessel to analyze impurities in helium gas, such as moisture, nitrogen, and hydrocarbons. The LP pressure is set ranging from 1.7 to 4.9 bar. The sampled gas does not return to the suction of the compressor but releases to the gas bag.

The GMP is required for the pressure control (HP and LP pressure control). The GMP is also required for loading and unloading of helium to/from the four warm

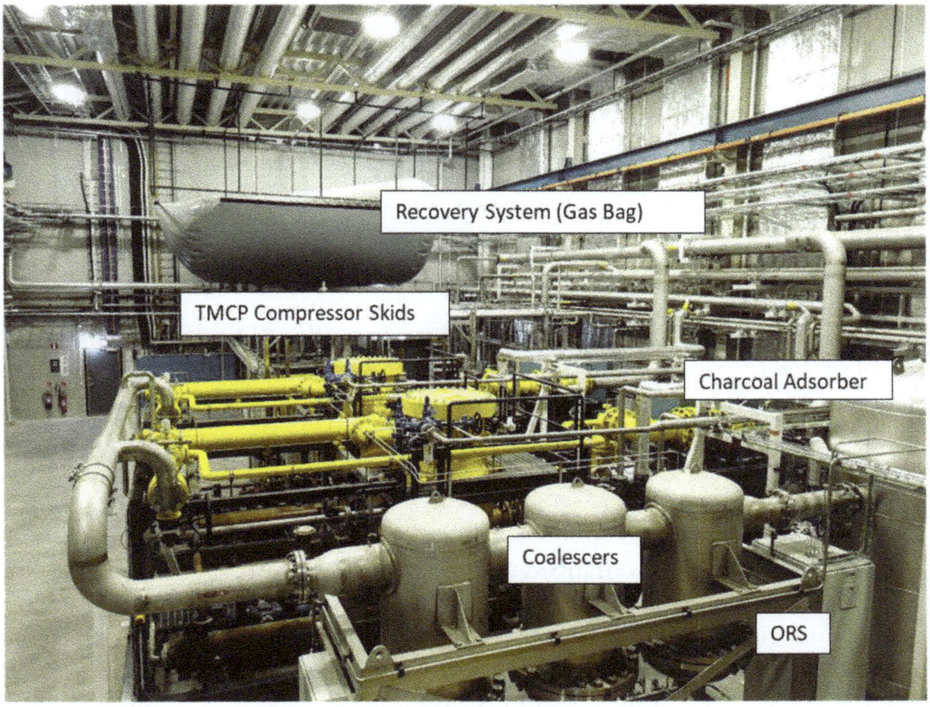

Figure 5.17. TMCP compressor system.

Figure 5.18. TMCP cold box.

helium buffer tanks. The loading and unloading valves allow the control of the discharge pressure of the compressor. The compressor bypass valves allow the control of the suction pressure of the compressor during dedicated operation steps.

5.6.2 Cold box

Figure 5.18 shows the TMCP cold box. The vacuum vessel is always evacuated by a vacuum system composed of a combination of a turbomolecular pump and a rotary roughing vacuum pump. The vacuum is maintained by a continuous pumping down and the vacuum pressure is maintained below 2×10^{-6} mbar during the cryogenic operation. There are six heat exchangers, three expansion turbines, and a 20 K adsorber.

5.6.2.1 Plate fin heat exchangers

The plate-fin heat exchangers are made of aluminum alloy. The process pipe is made of stainless steel, with the inlet and outlet pipe of the heat exchanger jointed by friction welding. The temperature difference between the HP and the LP stream should not exceed an allowable temperature of 40 K at both ends to avoid cracking. The protection is included in the TMCP control logic.

5.6.2.2 Expansion turbines

The expansion turbines are equipped with dynamic gas bearings and gas inlet filters. The speed of the expansion turbine is controlled by the brake valve. Regulation is done by adjusting the flow rate in the compressor brake circuit. Each turbine has its own dedicated controller that calculates the optimal speed set point of the turbine depending on the operating conditions. The turbine speed is controlled by the turbine brake valve, which is located in the brake circuit. For the warm stage turbine, the outlet temperature is controlled by adjusting the inlet valve to avoid decreasing lower than 60 K.

The CTL supply temperature is controlled by a temperature controller that controls two cold turbine bypass valves (CV450 and CV840) by means of a split range. The cooling capacity was regulated by adjusting the flow rate entering into the cold parallel turbines. If the CTL supply temperature is higher than the set point, CV450 decreases to zero and CV840 starts to open. Part of the HP stream bypasses the CTL and is directly diverted to the parallel cold turbines. On the other hand, if it is lower than the set point, CV840 closes first and CV450 opens. Part of the HP stream returning from the CTL bypasses the cold turbines and directly flows into the LP side.

5.6.2.3 20 K adsorber

A 20 K adsorber (figure 5.19) is installed vertically downstream of the HX5, where it is around 20 K at the nominal condition. It is filled with a charcoal to remove impurities of hydrogen and neon. When the 20 K adsorber's performance is impaired during the operation, the adsorber can be isolated and be regenerated by a heater in it and vacuum pump unit located downstream of an ambient heater. Furthermore, there is a precooling line that is connected to the LP between the HX3 and HX4.

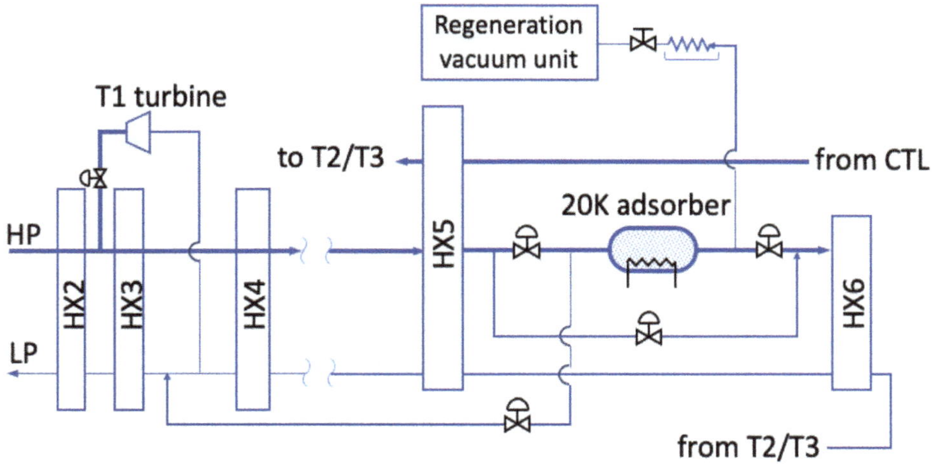

Figure 5.19. 20 K adsorber.

5.7 CTL

The 16 K-HP stream is delivered from the cold box in the cold-box hall to the JSB in the target building through the CTL, as shown in figure 5.20. The HP helium stream can reduce the piping size due to helium's higher density than on the LP side. This results in a smaller size pipe and reduced heat load. The CTL has been designed under the condition displayed in table 5.8. The diameter of the process pipe is 88. 9 mm. The lengths of the supply and the return CTLs (S-CTL and R-CTL) are 305 m and 307 m, respectively. Each CTL has 15 elbows and 19 metal flexible hoses and is divided into 11 vacuum segments. The pressure drops of the S-CTL and R-CTL are estimated to be 242 mbar and 350 mbar. The designed heat load is 1.27 kW. At the cold end of the cold box, part of HP cold helium (< 20 g s^{-1}) is warmed by an ambient heat exchanger (figure 5.21) and is delivered to the two mixing valves on the JSB (CV870 and CV880, see figure 5.8) through a warm helium pipe with the diameter of 60.3 mm.

A more complete description of the HyTL cryogenic piping for the CMS is given in chapter 6.

5.7.1 JSB

There is a valve box called the JSB at the end of the CTL in the hydrogen room on the fourth floor of the target building. The JSB has four control valves (CV850, CV860, CV870, and CV880), as shown in figures 5.16 and 5.22.

The transient heat load of 17.2 kW is suddenly applied to the CMS when the proton beam is turned on. The temperature controller for the CTL supply temperature (TI610) is not suitable for controlling a transient temperature variation and the effect persists for a long time. This results in the CTL supply temperature fluctuation. One of the JSB functions is to compensate the transient heat load. The CTL return temperature is always regulated at 21.2 K by injecting the warm

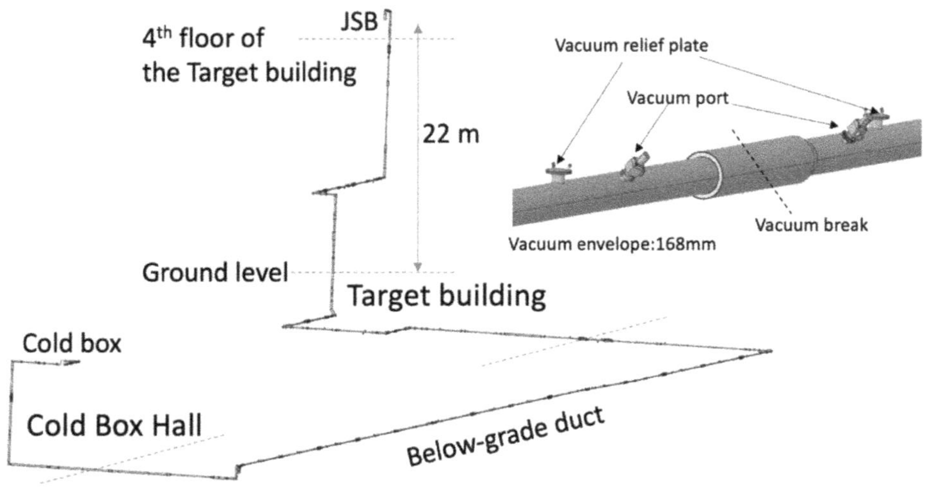

Figure 5.20. Overview of the CTL.

Table 5.8. CTL design requirement.

Process line	Allowable pressure drop (mbar)	Flow rate (g s^{-1})	Operational pressure (bar)	Operational temperature (K)
S-CTL	250	1125	20.5	15
R-CTL	375	1125	19.9	20
Warm He supply	35	20	21	300

helium via CV870. The heat load applying to the cold box would be able to be kept constant. The other mixing valve (CV880) regulates the CMS supply temperature, which is set to 0.5 K higher than the CTL temperature (TI300). The cooling capacity of 17.2 kW will be immediately adjusted by changing the feed helium flow rate using CV860 when the proton beam is turned on or off. The JSB bypass valve (CV850) will be controlled at the same pressure drop in conjunction with CV860 in order to avoid changing the flow rate delivered to the JSB.

5.8 TMCP operation

5.8.1 Maximum normal design mode

In the maximum nominal design mode, the TMCP will perform at peak capacity of 30.2 kW. This mode will occur with the beam on. In this mode, the discharge pressure after the WCS is 22.4 bar and both compressors will work simultaneously delivering the required mass flow of 1125 g s^{-1} [7]. The HP helium at 21.5 bar enters the cold box at 1020 g s^{-1} and is cooled against the returning low pressure helium stream. The suction pressure is controlled at 5.2 bar with the compression ratio of

Figure 5.21. The ambient heat exchanger.

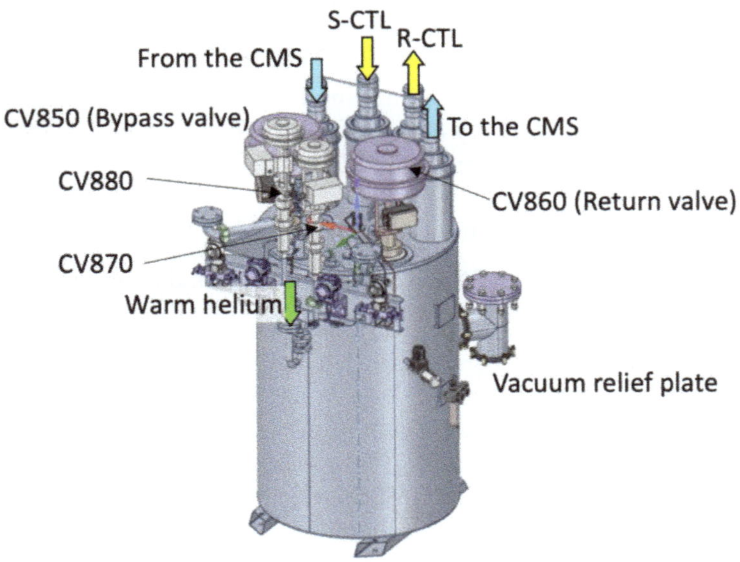

Figure 5.22. Overview of the JSB.

4.1. Part of the HP helium stream is expanded in the turbine (T1) and is mixed with the returned LP stream in order to enhance the heat exchange in the first three heat exchangers. The rest of the HP helium stream is further cooled down in the following heat exchangers to 20 K by means of the returning LP helium. Finally, the HP stream is cooled to 15 K in the last exchanger HX-6 and it is supplied to the S-CTL.

At the JSB, the feed flow rate to the HX-11 in the CMS cold box is adjusted by the JSB bypass valve (CV850) and the return valve (CV860). The HP helium stream is returned to the cold box at about 20 K and 20.6 bar.

The returned helium stream is passed through the three-stream heat exchanger HX-5 and is expanded by means of two identical, parallel turbines (T2 and T3). At the outlet of the turbines, the two LP streams are mixed again into just one LP stream, which is returned at around 14 K and 5.4 bar.

5.8.2 Normal low power mode

When the capacity of the TMCP is reduced to the one corresponding to the low power mode, only one compressor and one out of the two parallel turbines are operated. This mode will occur with the beam on but operating at less than full power.

5.8.3 Normal turndown mode

The mode will occur with the beam off. For short duration interruptions of the beam, appropriate means will be used to allow the heat load to the cryogenic TMCP to remain constant. For longer duration interruptions, the cryogenic TMCP capacity will be 'turned down' via the control system to allow for this reduced heat load requirement in the most energy efficient manner.

5.8.4 Long term switching from nominal to turndown

The TMCP will transition from nominal mode to turndown mode by unloading helium inventory from the CTL by means of specific bypasses in the TMCP through a heat exchanger to the ESS warm helium HP and LP buffer tanks.

5.8.5 Long term switching from turndown to nominal

The TMCP will transition from turndown mode to nominal mode by loading helium inventory to the CTL from the ESS warm helium HP and LP buffer tanks.

5.8.6 Beam trip

Beam trip—In the event that proton beam to the target trips off when the TMCP is operating in 'nominal design mode', the cooling required for the CMS drops significantly. The result is a significant change in the load seen by the TMCP. To deal with this situation, the cold helium supply to the CMS will be partially bypassed and warm helium injected into the return stream back to the TMCP sufficient to produce the same load as that lost due to the beam trip. Thus, the TMCP will see a constant load through the whole event. As the beam to target is re-established, the warm helium will simultaneously throttle back, thereby maintaining a constant load. In the event that the beam cannot be re-established in a reasonable period of time, the TMCP settings can be adjusted to 'nominal turndown mode' by throttling warm helium flow and permitting the TMCP control system to adjust the TMCP settings so that it can benignly follow the ramp down.

5.9 Installation

The commissioning and acceptance testing of the cryoplant is split into two phases, with over two years in-between them. Phase 1 comprised the commissioning and testing of the compressor skids and the cold box alone, and was finished by Q3 2019, as illustrated in figures 5.9 and 5.10. Phase 2 integration of the helium transfer lines and the JSB and has been completed in the summer of 2022, as shown in figure 5.23.

After the installation of the CTL, a helium leak test was implemented under the condition of pressurized to the maximum operational pressure of 20 bar using a mixture of 70% of nitrogen and 30% of helium. The entire CTL was pressurized by the mixing gas from the cold end of the cold box and each vacuum segment was tested in turn from the cold-box hall to the hydrogen room. It took a few days until all the leak tests were completed.

Before the commissioning, the helium leak test was conducted again using 100% helium, which was pressurized by the compressor. A small leak of around 10^{-6} mbar l s^{-1} was detected in the L-shaped segment 9, which had a horizontal part (32 m) and a vertical one (11.5 m) located in the target building as shown in figure 5.24. Figure 5.25 shows the leak test result of the process line at the segment 9 where a leak detector was connected to the vacuum envelope. For the process pressure lower than 6 bar, the leak rate was almost on the background level. The leak rate was rapidly increased from 10^{-10} to 10^{-7} mbar l s^{-1} with increase in the pressure to 13 bar. Above 13 bar, the leak rate was rapidly increased to 10^{-6} mbar l s^{-1} and was increased up to 3.7×10^{-6} mbar l s^{-1} at the maximum operational pressure of 20 bar. With a decrease in the pressure to atmospheric pressure, the leak rate got back to the initial background level of 1.2×10^{-12} mbar l s^{-1}. We have confirmed that the maximum leak rate was not increased, even though the pressure cycle was repeated five times. Eventually, we have found out the origin of the leak which was located at

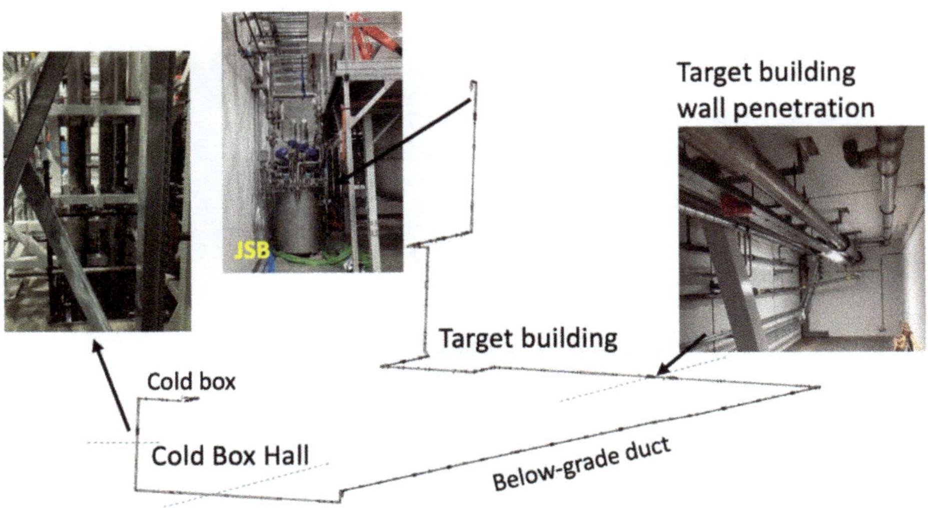

Figure 5.23. Installation of the CTL, the warm helium line and the JSB.

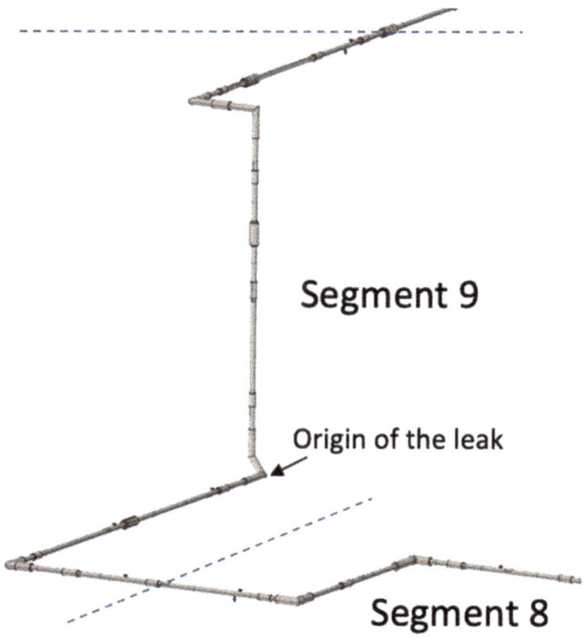

Figure 5.24. Small Leak found at the CTL segment 9.

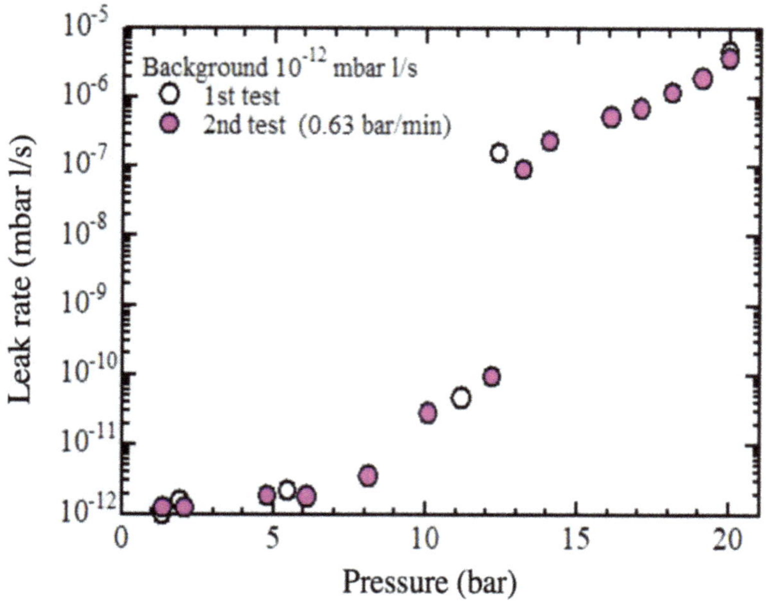

Figure 5.25. Leak test result of the segment 9. Reproduced from [8]. © IOP Publishing Ltd. CC BY 4.0.

a bend part from the horizontal to the vertical pipe. According to the result, it would be difficult to find such a small leak at the low-pressure difference between the process line and vacuum envelope so the helium leak test was implemented at the maximum operational pressure. Furthermore, the mixing test gas would not be uniform in the vertical long pipe after lapse of time from the injection of the mixing gas because helium has moved upward.

Finally, the CTL segment 9 was repaired by replacing it in February 2023. The installation of the final segment between the JSB and the CMS cold box was also completed at the same time.

5.10 Commissioning

5.10.1 Preparation

There is no dedicated external purification for the TMCP. The TMCP is purified by being diluted with the loading helium from the buffer tank via a GMP and being released to the recovery system. The impurities are always being monitored by a gas analyzer downstream of the charcoal vessel. The oil concentration is approximately 10 ppb and nitrogen is below 5 ppm. The sampling gas is delivered to the recovery system.

Figure 5.26 shows the trend of the impurities in the TMCP during the purification process, where the HP was 8 bar and the circulation flow rate was 28 g s^{-1}. After the CTL installation, pumping and purge were repeated three times and the purification was started. The nitrogen analyzer is available below 50 ppm. It took 4 days until it decreased down to 7 ppm. The total amount of the purged helium was 275 kg, which was estimated using the total pressure reduction of the buffer tank the volume of 70 m^3.

5.10.2 Compressor station

A one-stage oil injected screw compressor station with two identical screw compressor skids is installed in parallel. The WCS was tested with the ORS and GMP to check the function and pressure stability at reduced mass flow and pressure.

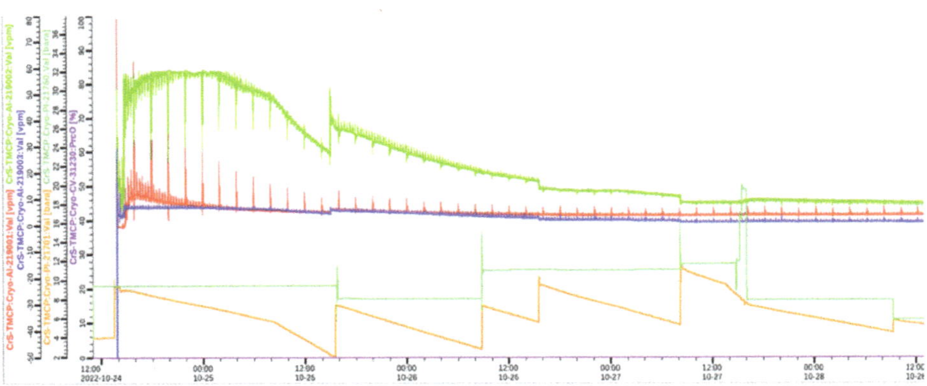

Figure 5.26. Purification operation.

Table 5.9. Commissioning result of the compressors.

Parameters	Unit	Design value	Comp-1	Comp-2
Mass flow	$g\ s^{-1}$	562.5	604	602.5
Intake temperature	°C	36	32.8	31.2
Intake pressure	bara	5.1	5.1	5.1
Discharge pressure compressor	bara	22.4	23.1	23
Discharge pressure WCS	bara	21.48	22.5	22.4
Discharge temperature compressor	°C	105	105.3	105.5
Discharge temperature WCS	°C	40	40	40
Consumed power	kW	1200	1105	1104
Main rotor speed	RPM	2980		
Oil injection temperature	°C	51.4	42.5	35.6
Isothermal skid efficiency, η_{iso}	%	45.0	53.6	53.1

A 100 h-reliability run test of the two compressors was successfully completed without connecting to the TMCP cold box after the installation. The test results are displayed in table 5.9.

5.10.3 Expansion turbine performance

It is generally known that a turbomachine characteristic can be expressed using dimensionless expressions of a head coefficient, ψ, a discharge coefficient, ϕ, and the wheel speed, u_s. Adiabatic efficiencies, η, of the turbines were estimated using ϕ

$$\psi = \frac{\Delta P}{\rho {u_S}^2} \tag{5.1}$$

$$\phi = \frac{\dot{m}}{\rho u_S D^2} \tag{5.2}$$

where ρ is the density, $\dot{m}$ is the mass flow, u_s is the wheel speed, and D is the impeller diameter.

The adiabatic adiabatic efficiencies are also estimated using the discharge coefficient. Figures 5.27–5.29 show the nondimensional mechanical properties and adiabatic efficiency for each turbine.

The impeller sizes for all the turbines are the same. It is verified that the mechanical properties among all the turbine are on the same curve.

On the other hand, the adiabatic efficiencies change depending on the flow coefficient independent of the pressure and temperature. At the beginning of the cooldown, the adiabatic efficiency for T1 was around 60% and those for T2 and T3 were around 42%. With decrease in the temperature, the efficiency was increased. At the nominal condition, the adiabatic efficiency increased to around 70% for T1 and more than 80% for T2 and T3. It was verified that the turbines were operated at the maximum adiabatic efficiency point.

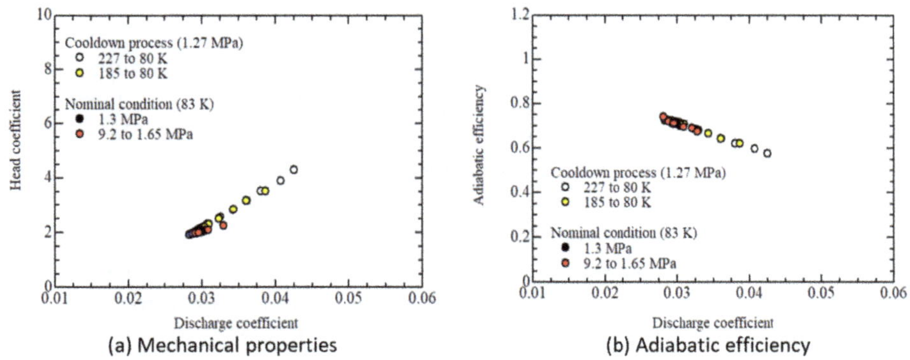

Figure 5.27. Nondimensional performance and adiabatic efficiency for T1 turbine. Reproduced from [8].

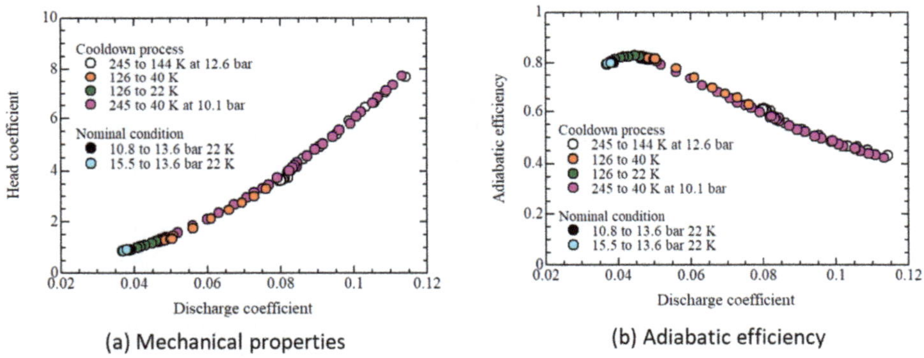

Figure 5.28. Nondimensional performance and adiabatic efficiency of T3 turbine.

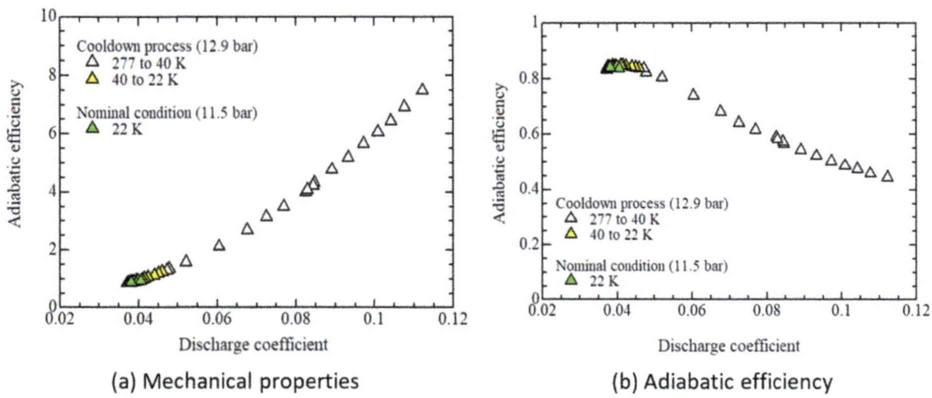

Figure 5.29. Nondimensional performance and adiabatic efficiency of T3 turbine.

5.10.4 CTL heat load measurement

The heat load of the overall CTLs including the JSB was estimated using a temperature difference between the supply (TI610) and the return temperatures (TI750) at the cold end of the cold box under the condition that only the T2 turbine was operated, CV450 was opened, and CV870 was closed. In order to reduce the cooling capacity and the flow rate, the operational pressure was decreased to 0.79 MPa. The flow rate passing through the CTL is the same as that measured by FI110. In advance, the temperature sensor of TI750 has already been calibrated compared with TI610 when the TMCP cold box was operated without connecting to the CTL. Figure 5.30 shows the measured CTL heat load at the flow rate of 0.235 kg s^{-1} for more than 200 min. The overall heat load was estimated to be 1.52 kW, which was 21% higher than the design value of 1.27 kW [8]. The maximum cooling capacity was predicted to be 30.9 kW at 2.0 MPa, as mentioned in section 5.10.6, and was 9 kW higher than the CMS heat load of 21.9 kW for the four-moderators arrangement at the nominal condition.

5.10.5 CTL pressure drop measurement

The pressure drops for the S-CTL and R-CTL were calculated to be 25.1 and 34.7 kPa at the flow rate of 1.125 kg s^{-1} and the maximum operational pressure of 2.0 MPa. The supply and return temperatures were 16 K and 21 K, respectively. The JSB is located at 22 meters above the ground. The effect of the head pressure should be considered when the pressure drop is calculated using the two pressure transmitters in the cold box and JSB. In this study, the pressure drops of the overall CTL, including the JSB, were estimated using two pressure transmitters (PT700 and PT850) located at the cold end of the cold box and a differential pressure transmitter (PDT850), which measured a pressure drop via CV850. The effect of the head pressure was able to be negligible. The pressure drop was measured under the condition where CV450 is opened. The return temperature (TI200) was controlled at 21.2 K by injecting warm gas through CV870, whose flow rates, $\dot{m}_w$, ranged from 4 to 14.5 g s^{-1}. The flow rates through the S-CTL and R-CTL can be estimated by

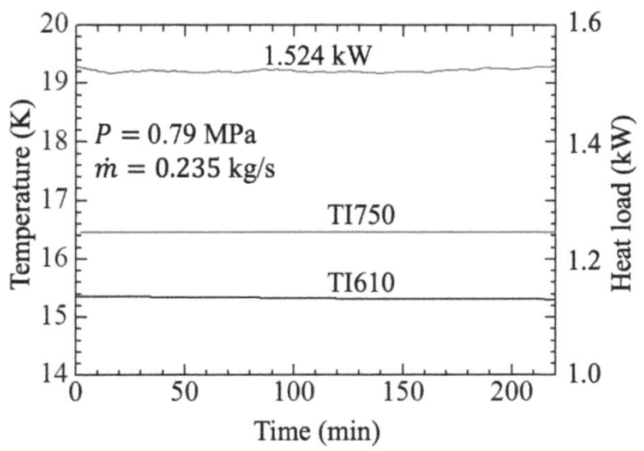

Figure 5.30. CTL heat load test result. Reproduced from [8]. © IOP Publishing Ltd. CC BY 4.0.

$\dot{m} - \dot{m}_{T1} - \dot{m}_w$ and $\dot{m} - \dot{m}_{T1}$. Figure 5.31 shows comparison of the measured pressure drops with the calculated ones for the supply temperatures (TI610) of 16 and 17 K [8]. The measured pressure drops over the overall CTL agree with the calculated ones within 6%. It is verified that the fabrication and installation of the CTL has been carried out as designed.

5.10.6 TMCP cooling capacity

The CTL supply temperature (TI31610) is adjusted by the split range controller by CV31450 and CV840. Figure 5.32 shows dependence of the positions of CV31450

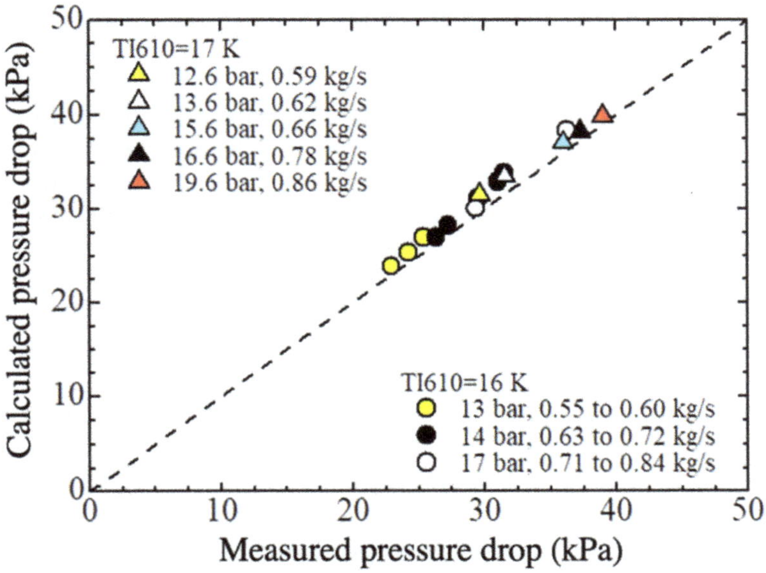

Figure 5.31. CTL pressure drop test result. Reproduced from [8]. © IOP Publishing Ltd. CC BY 4.0.

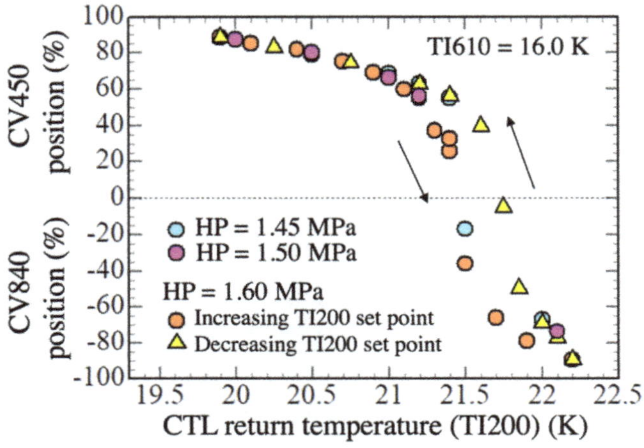

Figure 5.32. Dependence of CV450 and CV840 on TI200 set point. Reproduced from [8]. © IOP Publishing Ltd. CC BY 4.0.

and CV31840 on the CTL return temperature (TI51200) set point for TI31600 = 16.0 K [8]. The valve position seems to be independent of the HP. For the set point higher than 21.2 K, the valve positions were widely changed. It took a long time until the TMCP reached steady-state. We found that the CTL return temperature had to be set to maintain the CV31450 position more than 55% from the viewpoint of the TMCP stable operation. When CV31450 opens, the flow rates through the S-CTL can be calculated by $\dot{m} - \dot{m}_{T1} - \dot{m}_w$. The CTL flow rate can be expressed as follows when CV840 is opened.

$$\dot{m}_s = \frac{h_w - h_R}{h_R - h_S}\dot{m}_w \tag{5.3}$$

where h_w is the enthalpy of the mixing warm helium, h_R is the enthalpy at TI200, and h_S is the enthalpy at TI300.

The TMCP cooling capacities were measured for operating (i) the T1 and one of the cold parallel turbines, and (ii) the T1 and two parallel turbines (T2 and T3) at the CTL supply temperature of 16.0 K and the CTL return temperature of 21.2 K, changing the HP. The heat load given by injecting the warm helium via CV51870, $Q = \dot{m}_w(h_w - h_S)$, was considered as the cooling capacity. Figures 5.33 and 5.34 show the dependence of the TMCP cooling capacity and the S-CTL flow rate on the HP. Although the maximum operational HP was 2.0 MPa, the HP could only be increased to 1.8 MPa because of the limitation by the leak rate, as mentioned in 5.9 in the commissioning [8]. The CV700 position needed to be reduced to 36% because the pressure drop driven by the flow via the S-CTL was too small to inject the required warm flow through CV51880 and CV51870, especially during the cool-down process. The cooling capacities and the feed flow rates seemed to be proportional to the HP. For CV31700 = 36%, the maximum cooling capacities, which appear at 2.0 MPa, can be predicted to be 14.6 kW and 26.7 kW for case (i) and (ii)

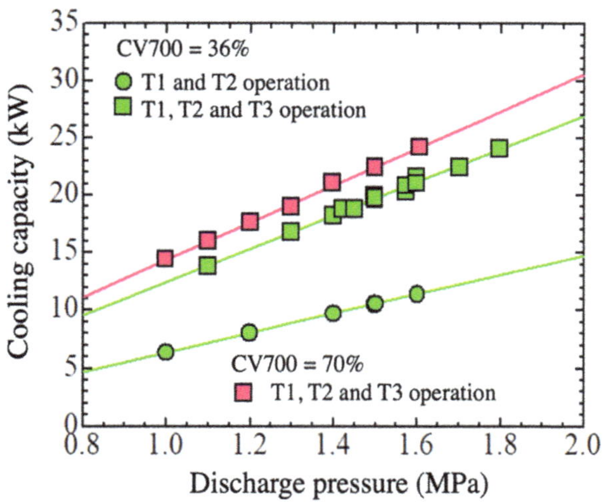

Figure 5.33. Effect of the TMCP cooling capacity on HP. Reproduced from [8]. © IOP Publishing Ltd. CC BY 4.0.

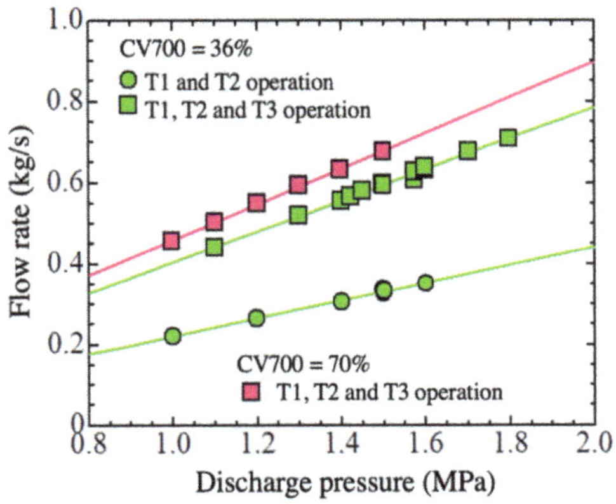

Figure 5.34. Effect of the flow rate on HP. Reproduced from [8]. © IOP Publishing Ltd. CC BY 4.0.

by being extrapolated from the measured ones. It turns out that the operation of the T1 and one of the two cold turbines at the HP of 1.25 MPa meets the requirement for the two-moderator arrangement at the proton beam power of 5 MW. For the four-moderator arrangement in the future, which is required for the cooling capacity of 21.9 kW at the 5-MW proton beam power, all the turbines of T1, T2, and T3 need to be operated at the HP more than 1.65 MPa. For CV31700 = 70%, the measured cooling capacities were also proportional to the HP and were higher than those for CV700 = 36%. The maximum cooling capacity would be predicted to be 30.6 kW based on an extrapolation of the measured data. It was verified that the cooling capacity met the design requirement of 30.2 kW.

5.10.7 Two cold turbine and one warm turbine operation.

Figure 5.35 shows the test result of the cooldown process from 284 K in Phase I to 29.5 K in Phase II at the pressure of 1.5 MPa for operating the two compressors, the two cold parallel turbines (T2 and T3), and one warm turbine (T1). The HP was temporarily changed to 1.1 MPa and the compression ratio was reduced to 3.6 to indirectly reduce the turbine expansion ratio, as well as the case of the T1 and T2 turbines operation described in section 5.6. The temperature difference at the cold end of the HX6, ΔT_{HX}, did not exceed the allowable value of 40 K. The cooling speed for TI610 was set to 0.62 K min^{-1}, which was almost the same as that used in the simulation. The cooling speed of the CTL return temperature (TI51200) was set to be the same as that for TI31610. Given the simulation result, the CTL supply temperature was temporarily maintained at 120 K for an hour, which is called holding state of 120 K. The CV450 was always acting, except for the early stage of the holding state. The CTL flow rate was around 230 g s^{-1}, which is 1.8 higher than that for the one-cold turbine operation at 1.7 MPa. With a decrease in the temperature, the cooling capacity, Q_c, was increased. The measured value of Q_c was 8.1 kW below 150 K and was 1 kW higher than the required Q_c given by the simulation.

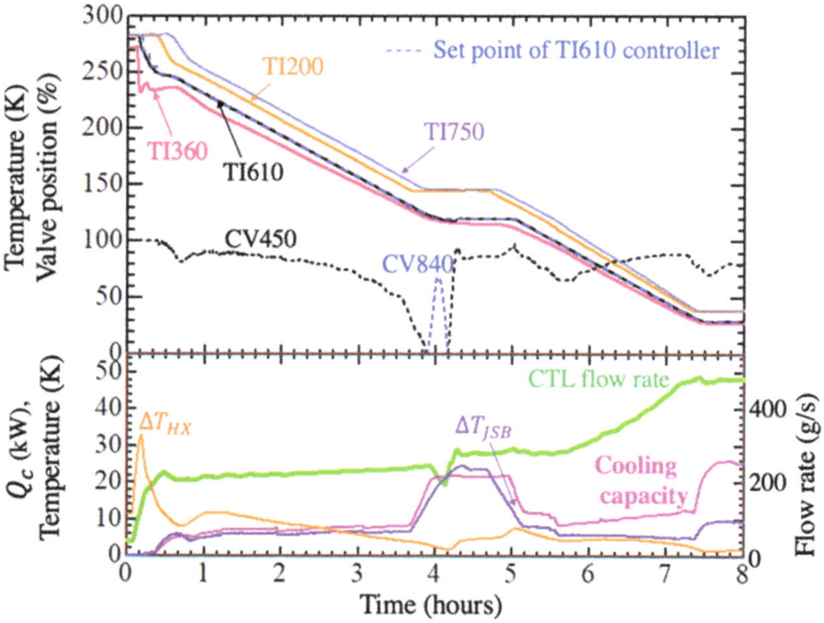

Figure 5.35. TMCP cooldown test results in Phase I and II using the two cold turbines. Reproduced from [7].

When the set point of the temperature controller reached 120 K, the CTL temperature undershot for 12 min and was decreased to 118 K. It was clarified from the TMCP commissioning result that the temperature controller was not suitable for suppressing a transient phenomenon. During the CMS cooldown process, hydrogen is supplied to maintain the pressure at the set point of 11 bar in Phase I due to its condensation. The temperature rise of the feed helium flow would bring about a large pressure fluctuation in the hydrogen loop, especially in Phase II and III. Therefore, the CMS supply temperature (TI6630) should be set to a certain set point higher than TI300 and be fine-tuned by injecting the warm helium via CV880 in order to achieve a more stable CMS cooldown operation

The CTL supply temperature was decreased to 29.5 K, which is lower than a critical temperature, T_c, of hydrogen at the same cooling speed. On the other hand, the CTL return temperature (TI200) was decreased to 39 K, which was higher than T_c. The CV450 was acting in the range of 66 to 88% and the available cooling capacity was more than 9.0 kW in the range of 120 to 29 K, and was much larger than the required one given by the simulation. During the holding state of 29 K at the final stage of Phase I, the available cooling capacity was 26.7 kW at $\Delta T_{\mathrm{JSB}} = 9.5$ K. We found that the two cold turbine operation at the pressure of 1.5 MPa met the Phase 1 CMS cooldown process requirement.

At the final stage of Phase I, the CMS supply temperature (TI6630) will be maintained at 34.5 K, which is higher than T_c. In Phase II, only the CMS supply temperature will be decreased at the cooling speed of 0.012 K min^{-1} given by the

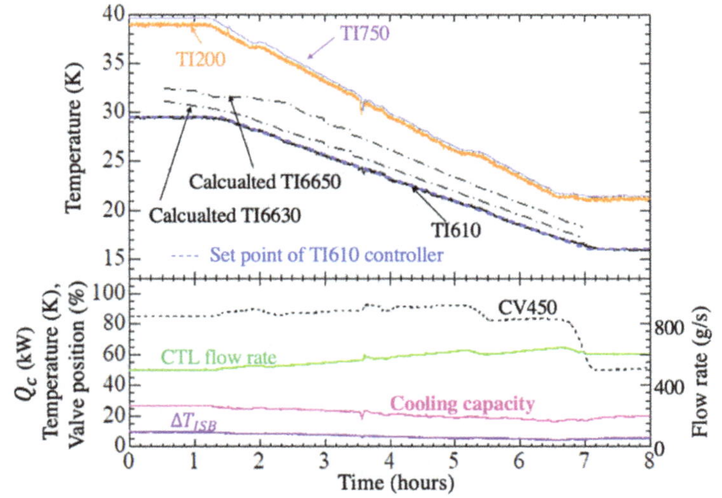

Figure 5.36. TMCP cooldown test results in Phase III with the two cold turbines. Reproduced from [7]. © IOP Publishing Ltd. CC BY 4.0.

simulation result, controlling the warm helium via CV880 under the condition that TI610 and TI200 are always maintained at 29.5 K and 39 K, respectively. This is because a slight supply helium temperature would bring about a huge pressure oscillation temperature during the condensation process around T_c where hydrogen density drastically changes. In this study, it was verified that the CTL supply (TI610) and return temperatures (TI200) were able to be maintained stably at the set points of 29.5 K and 39 K for 12 h and the TMCP operation met the requirement of Phase II operation.

Figure 5.36 displays the test result of Phase III. The supply temperature was decreased from 29 K to 16 K at 0.04 K min^{-1}, which was slightly slower than that of the simulation. The CTL return temperature was decreased to 21.2 K at 0.06 K min^{-1} and the CV450 position tried to be maintained above 90% in this study. The CTL flow rate was increased to 600 g s^{-1} at the nominal condition of 16 K. The temperature difference at the JSB, ΔT_{JSB}, was decreased from 9.6 K to 5.0 K during the cooldown process. The cooling capacity, Q_c, was always more than 17 kW in Phase III, which was much higher than the required one of 5.5 kW, although it was 26 kW at the early stage of Phase III. It was verified that the TMCP has a sufficient available cooling capacity to meet the Phase III of the CMS cooldown condition. The optimized operational parameters have been determined through the commissioning.

5.10.8 Transient heat load test.

In this study, the transient heat loads, Q_c, of 17.5 kW, which correspond to those for the two- and four-moderator arrangements at the proton beam power of 5 MW, was rapidly applied and removed. Propagation of the temperature fluctuation caused by the transient heat load over the TMCP was clarified. The transient heat load was simulated by injecting the warm helium using the other mixing valve, CV51880 with the Kv of 7.8. The CV51870 was always controlled to keep the CTL return temperature at 21 K. Therefore, the flow rate measured by FI31620 described that injected via CV51880 and CV51870. The flow rate via CV51880 was needed in order to estimate the heat load given by CV51880, Q_c. There is a differential pressure transmitter, PDT51880, to measure a pressure drop through CV51880, ΔP_{in}. In advance, a correlation between ΔP_{in}, the flow rate, $\dot{m}_{in}$, and the position of CV51880 had been calibrated. The heat load of Q_c was estimated by $\dot{m}_{in}(h_{620} - h_{300})$, where h_{620} and h_{300} were the enthalpies at TI31620 and TI51300.

When the transient heat load was applied or removed, we studied how much the temperature fluctuation was be able to be mitigated using a combination with a feed-forward and the PID control. Figure 5.37 shows the transient heat load test result for $Q_c = 17.5$ kW where the CTL return temperature was controlled by a combination of the feed-forward and the PID control. As the initial condition, the CTL supply and return temperatures were kept at 16.8 K and 22.0 K, respectively. The HP helium stream was delivered through the CTL supply at the flow rate of 761 g s^{-1}. The heat load of 17.5 kW was changed by injecting or stopping the warm helium through CV880 at the ramping speed of 7.9 kW s^{-1}. The valve position of CV870 (from 41% to 60%), which corresponded to the heat load of 17.5 kW, was moved within 1 s when the temperature rise of more than 1 K detected at TI200. Although a

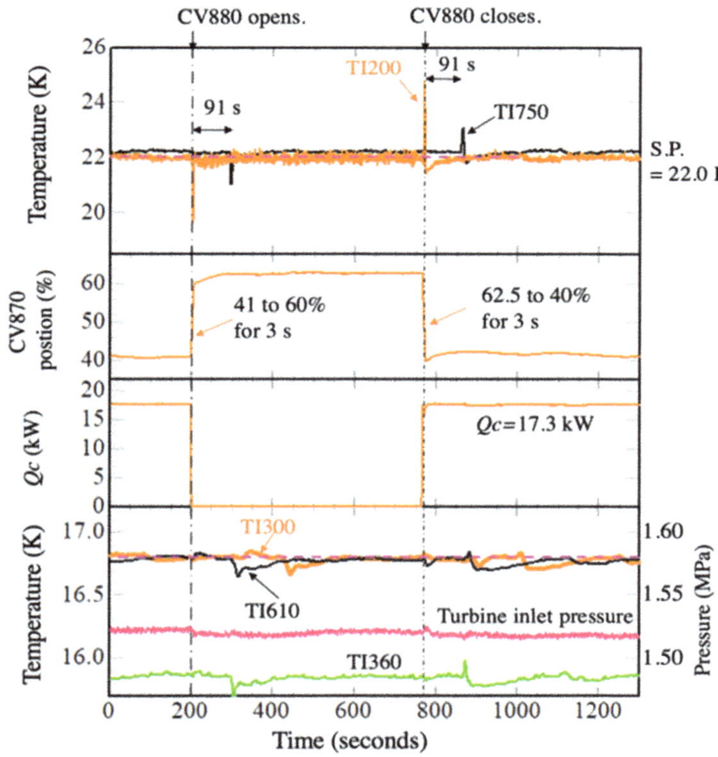

Figure 5.37. Thermal compensation using a PID and feed-forward control when the heat load of 17.3 kW is applied and removed. Reproduced from [9]. © IOP Publishing Ltd. CC BY 4.0.

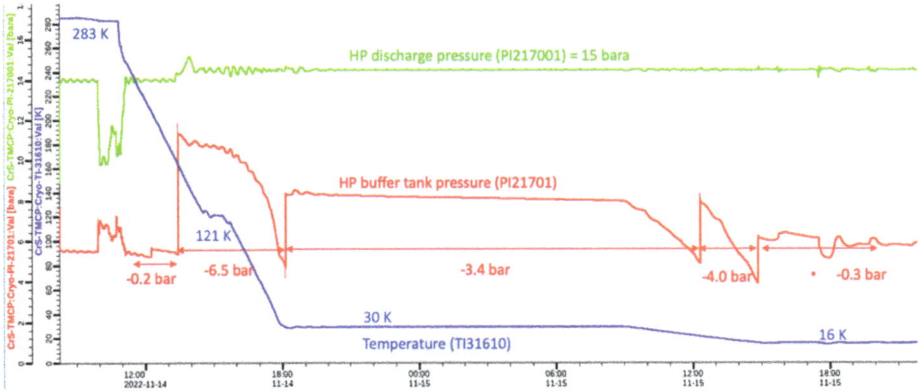

Figure 5.38. TMCP inventory estimation by reducing the buffer tank pressure.

temperature change of around 2 K appeared during a short duration of less than 2 s before the feed-forward control executed, the fluctuations of the temperature and the pressure became much smaller than those for only the PID control. In addition, although the temperature fluctuation propagated to the CTL supply side, it was below 0.1 K, which was much lower than the margin of the TI6630 controller. According to the test results, we can expect that the temperature fluctuation can be further reduced by optimizing the timing for starting the feed-forward control.

5.10.9 Estimation of TMCP inventory

Figure 5.38 shows the pressure change of the buffer tank during the cooldown operation. The total pressure reduction at HP=15 bar is 14.4 bar. When the HP discharge pressure is increased/decreased by 1.0 bar at the nominal condition (TI31610 = 16 K), the change pressure of the buffer tank is measured to be 1.2 bara. Therefore, for the maximum operational pressure of 20 bar, the required helium volume of the HP buffer tank is estimated to be $14.4 + 5 \times 1.2 = 20.4$ bar.

5.11 Summary

The TMCP has been designed to cool the CMS and remove the transient heat load caused by the beams on or off. The first phase of the installation (the compressor station and the cold box) had been completed in 2019. In the second phase, the CTL and JSB were installed and it was verified from the TMCP commissioning without connecting the CMS that the TMCP met the requirement. The final installation of the U-type transfer lines between the JSB and the CMS was completed in February 2023. The commissioning of the TMCP–CMS together started in February 2024.

5.12 Lessons learned

- Address challenges due to multiple suppliers involved, while making sure that information is transmitted to and understood by all parties.

- Recognize and address challenges in coordinating work between manufacturers and on-site installation contractors, especially in an international environment with risks of language and cultural barriers to overcome.
- Be aware of issues of integrating TMCP into overall facility—plan for flexibility to accommodate schedule and scope changes that one may not have control over. Closely coordinate with project management to address items such as timing and availability of interfaces, and planning for intermediate activities such as phased testing and commissioning.
- Plan for visits to sister facilities to benefit from their lessons learned and experiences.
- Have personnel with the right qualifications to evaluate early design decisions (hazardous environments and equipment location).
- Early engagement with suppliers is important to incorporate their expertise and insights into the technical specifications—make sure that what one specifies is something that can actually be built!

References

[1] Peggs S *et al* 2013 *ESS Technical Design Report* https://europeanspallationsource.se/sites/default/files/downloads/2017/09/TDR_online_ver_all.pdf

[2] Garoby R *et al* 2018 The European spallation source design *Phys. Scr.* **93** https://iopscience.iop.org/article/10.1088/1402–4896/aa9bff

[3] Mason T E, Gabriel T A, Crawford R K, Herwig K W, Klose F and Ankner J F *The Spallation Neutron Source: A Powerful Tool for Materials Research* (Oak Ridge, TN: Spallation Neutron Source)

[4] Oyama Y 2003 Present status of J-PARC high intensity proton accelerator project in Japan *Proc. ICANSXVI* ed G Mank and H Conrad (Germany: Forschungszentrum Jülich GmbH, Jülich) pp 7–11

[5] Bessler Y, Henkes C, Hanusch F, Schumacher P, Natour G, Butzek M, Klaus M, Lyngh D and Kickulies M 2017 Final design, fluid dynamic and structural mechanical analysis of a liquid hydrogen Moderator for the European Spallation Source *IOP Conf. Ser: Mater. Sci. Eng.* **171** 012131

[6] Ganni V and Knudsen P 2010 Optimal design and operation of helium refrigeration systems using the ganni cycle advances in cryogenic engineering *Adv. Cryo. Engr.* **55** 1057–71

[7] Tatsumoto H, Arnold P, Horvath A and Boros M Study of operational modes of cool-down and warm-up process for the ESS large-scale 20 K helium refrigeration system *IOP Conf. Ser: Mater. Sci. Eng.* **1301** 13011126

[8] Tatsumoto H, Arnold P, Boros M, Horvath A, Segerup M, Tereszkowski P and Huber R Commissioning of the ESS large scale 20 K helium refrigeration system for the cryogenic moderator system *IOP Conf. Ser: Mater. Sci. Eng.* **1301** 13011131

[9] Tatsumoto H, Arnold P, Horvath A and Boros M Effect of a thermal fluctuation caused by a proton beam injection on the ESS large-scale 20 K helium refrigeration system *IOP Conf. Ser: Mater. Sci. Eng.* **1301** 13011123

Chapter 6

The cryogenic moderator system

H Tatsumoto

The cryogenic moderator system (CMS) is a forced-flow liquid hydrogen cooling system that cools high-energy neutrons down to cold neutrons in two cryogenic hydrogen moderators (four in the future), where an estimated nuclear heating of 6.7 kW (17.3 kW in the future) is generated in the moderators at 5 MW proton beam power. The European Spallation Source (ESS) CMS has been designed by the ESS in-kind partner, Forschungszentrum Jülich GmbH (FZJ), in cooperation with Technische Universität Dresden. The installation into the ESS Target building started in November 2021 and the commissioning started in February 2024.

6.1 Introduction

At the ESS target, high-energy spallation neutrons are produced by impinging 2 GeV protons (at nominal current of 62.5 mA) with an average beam power of 5 MW on the high-Z material, tungsten. The proton beam is pulsed with a repetition of 14 Hz and a pulse length of 2.86 ms. The spallation neutrons are moderated to cold and thermal energies by the moderators. The moderator system consists of a water pre-moderator and two liquid hydrogen cold moderators. The design of a flat butterfly-shaped hydrogen moderator vessel, which is made of a high-strength aluminum alloy (AL6061-T6), is optimized to achieve a high cold neutron brightness [1]. The typical operating temperature of the liquid hydrogen moderators is 18 K, where the parahydrogen fraction is 99.9% in equilibrium. At this cryogenic temperature, the neutronic performance of the liquid hydrogen moderator depends on the quantum states of the bi-atomic molecule, the spin triplet ortho state, and the spin singlet para state because the total neutron scattering cross-section of the parahydrogen is almost two orders of magnitude lower than that of the normal hydrogen in the cold neutron energy range. The neutronic performance of the cold moderators degrades rapidly with the decreasing parahydrogen fraction below 99.5% [2]. The neutron collisions in the cold moderators increases the orthohydrogen fraction [3] if it is not compensated by the orthohydrogen to parahydrogen conversion driven by

doi:10.1088/978-0-7503-3223-1ch6

the catalyst system. Therefore, it is important to keep the parahydrogen fraction close to 99.5% in the liquid hydrogen moderators, which is the theoretical value derived by the Boltzmann distribution function at 20 K.

At the beginning, the ESS will install two hydrogen moderators above the target wheel. The current plan is to replace them with four (two above and two below, respectively) in the future. The calculated nuclear heating generated in the two hydrogen moderators will be 6.7 kW for the proton beam power of 5 MW, while for the four moderators will be 17.2 kW [2].

6.2 Hydrogen

Under normal conditions, hydrogen is a colorless and odorless gas. Hydrogen is highly flammable and explosive because the flammable range is approximately 4% to 75% by volume of hydrogen in air at atmospheric pressure and the ignition energy is very low (0.017 mJ, which is 10 times lower than that for hydrocarbons) [4].

Figure 6.1 shows a phase diagram of hydrogen [5]. At standard atmospheric pressure (101.3 kPa), liquid hydrogen boils at 20.28 K and freezes at 13.83 K. Hydrogen has the second lowest boiling point of all known substances. Saturated liquid hydrogen at boiling point has a density of 70.8 kg m^{-3}, which is 0.57 times lower than that of 125.0 kg m^{-3} for liquid helium at 4.22 K and is 0.088 times lower than that of 808.5 kg m^{-3} of liquid nitrogen at 77.4 K. Liquid hydrogen is the

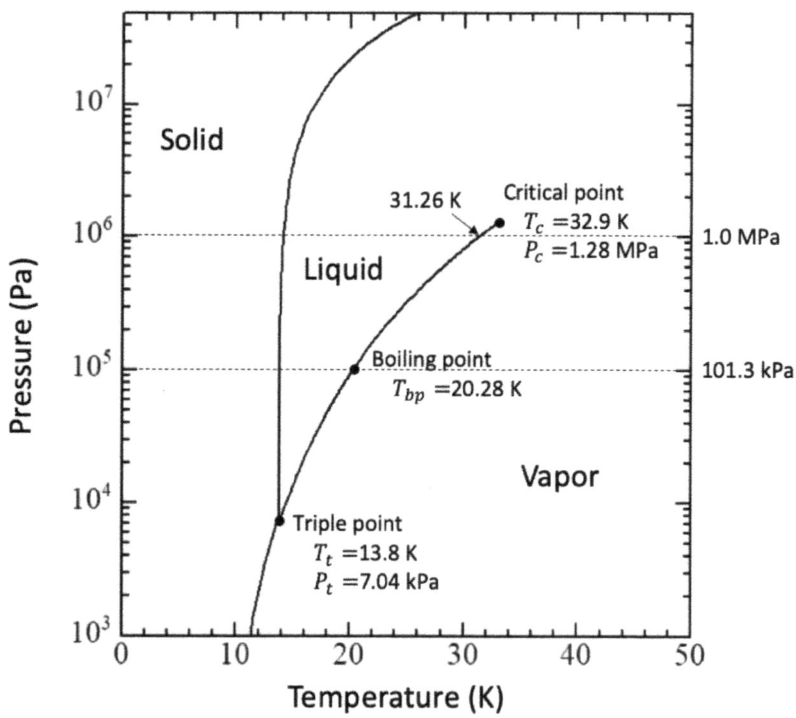

Figure 6.1. Phase diagram of hydrogen.

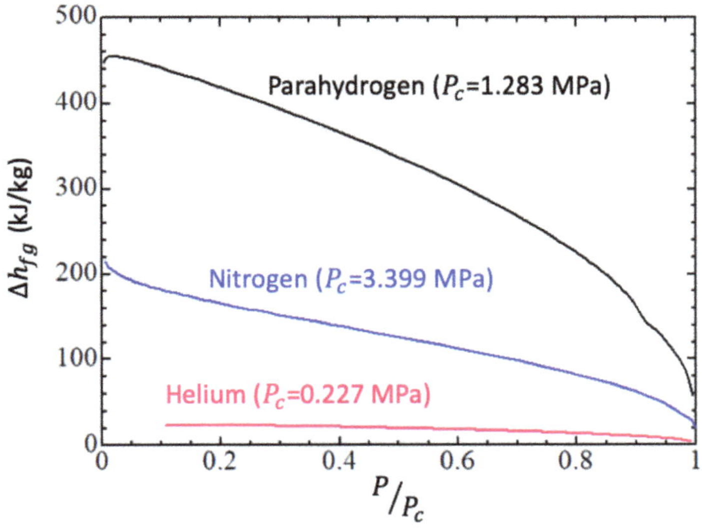

Figure 6.2. Latent heats of cryogenic fluids [5].

lightest liquid. Temperature and pressure at the triple point are 13.8 K and 7.04 kPa. The saturated temperature increases to the critical temperature of 32.9 K (T_c) when increasing the pressure to the critical pressure of 1.38 MPa (P_c). As shown in figure 6.2, latent heat decreases to zero as the pressure approaches the critical pressure, P_c. The latent heat of liquid hydrogen is 445.4 kJ kg^{-1}, which is 2.24 times larger than that for liquid nitrogen (P_c = 3.40 MPa and T_c = 126.26 K) and 21.5 times larger than that for liquid helium (P_c = 0.228 MPa and T_c = 5.20 K).

Molecular hydrogen (H_2) has two different forms: orthohydrogen and parahydrogen. The hydrogen molecule consists of two electrons and two protons that possess spin. When the nuclear spins are in the same direction, the angular momentum vectors are in the same direction. This is called orthohydrogen. Hydrogen with the nuclear spin in opposite directions is called parahydrogen. Hydrogen exists as a mixture of 75% orthohydrogen and 25% parahydrogen by volume at room temperature and is called normal hydrogen. The equilibrium ratio of orthohydrogen to parahydrogen, which is called equilibrium hydrogen, is given as a function of temperature, as shown in figure 6.3 [6]. As the temperature is decreased, orthohydrogen converts to parahydrogen and the equilibrium orthohydrogen fraction decreases 75% to 0.2% at the normal boiling point. This conversion is conducted over a definite period of time because the change is made through energy exchanges by molecular magnetic interactions. Orthohydrogen drops to a lower molecular energy level during the conversion and release a quantity of energy, called the heat of conversion. The heat of conversion is 703.3 kJ kg^{-1} at normal boiling point and is higher than the latent heat of vaporization of 445 kJ kg^{-1}. Figures 6.4 and 6.5 shows the specific heat and thermal conductivity of parahydrogen and normal hydrogen. Parahydrogen has a higher thermal conductivity and specific heat than normal hydrogen, unlike density and viscosity (table 6.1).

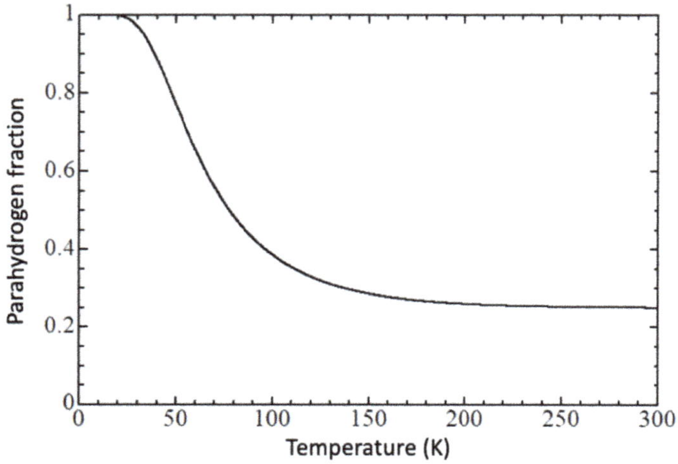

Figure 6.3. Temperature dependence of equilibrium parahydrogen fraction [6].

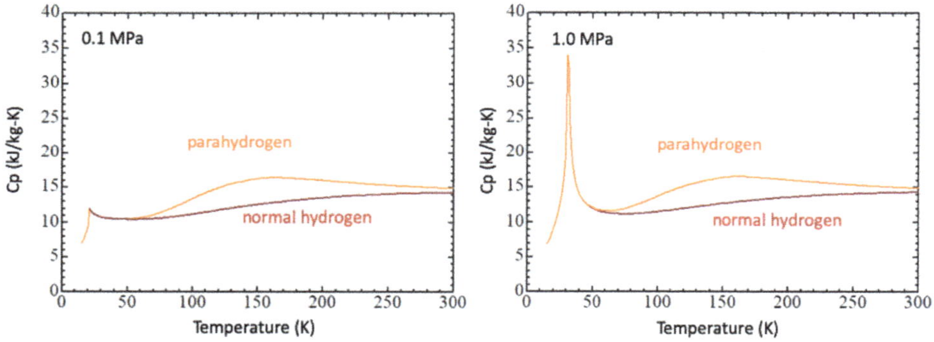

Figure 6.4. Specific heats of parahydrogen and normal hydrogen [5].

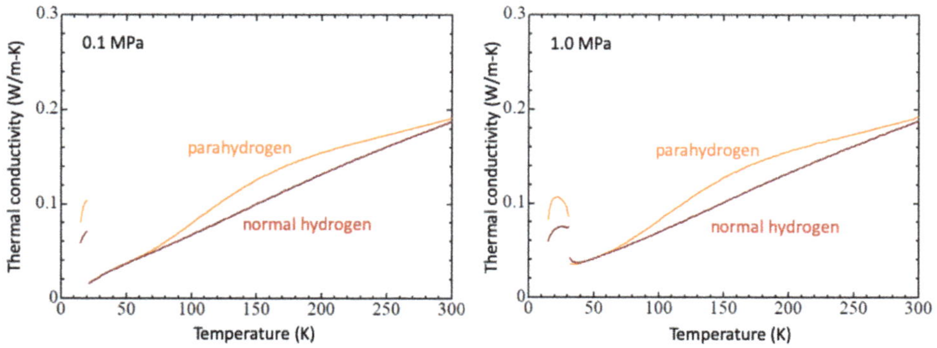

Figure 6.5. Thermal conductivities of parahydrogen and normal hydrogen [5].

Table 6.1. Equilibrium concentration of parahydrogen [6].

Temperature (K)	Parahydrogen fraction
17.0	0.9996
17.3	0.9995
17.5	0.9994
18.0	0.9993
19.0	0.9989
20.0	0.9982
22.7	0.9950
25.0	0.9900
77.8	0.5001
300.0	0.2507

Table 6.2. CMS features of the SNS and J-PARC [9, 10].

Conditions	SNS	J-PARC
Cold moderator material	Supercritical H_2	Supercritical H_2
Number of cold moderators	3	3
Number of CMS loops	3	1
Design pressure (bar)	19	21
Operational pressure (bar)	14–15	15
Supply temperature (K)	17	18
Hydrogen inventory (kg)	6.8	18.7
Ortho-parahydrogen catalyst	—	$Fe(OH)_3$
LH2 pump	Ball bearing	Dynamic gas bearing
Pressure control	Accumulator + Feed He flow control	Accumulator + heater
Helium refrigerator	7.5 kW at 17 K	6 kW at 17 K [14]

6.3 CMS design

At MW-class pulsed neutron sources such as the Spallation Neutron Source (SNS) at Oak Ridge National Laboratory [7] and the Japan Proton Accelerator Research Complex (J-PARC) [8], cryogenic hydrogen at a supercritical pressure (15 bar and 18 K) has been adopted as a cold moderator material. The pressure is higher than the critical pressure of 12.9 bar and the temperature is much lower than critical temperature of 32.9 K, as shown in figure 6.1. Therefore, the supercritical hydrogen exists as a single phase, which means there is no clear liquid-vapor interface, in the hydrogen loop and has a high density, as well as subcooled liquid hydrogen at the same temperature. The features of the CMSs at SNS [9] and J-PARC [10] are summarized in table 6.2. Figure 6.6 shows the overview of the SNS CMS. Each moderator has its own hydrogen loop, which has a hydrogen pump (ball bearing), an

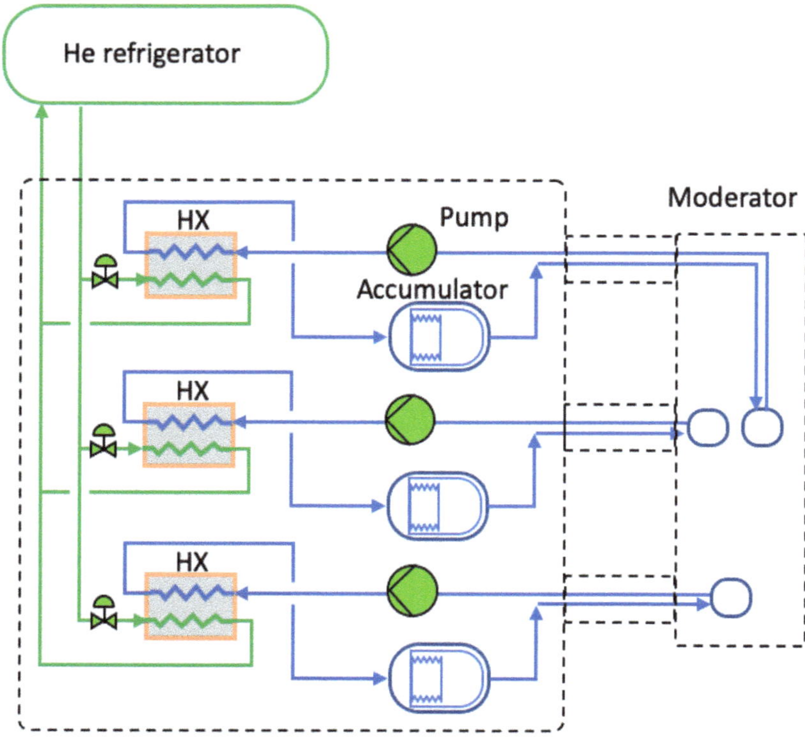

Figure 6.6. Overview of the SNS CMS [9].

accumulator, and a heat exchanger. The accumulator's function is to mitigate a pressure fluctuation. It has a bellows filled with helium gas because the bellows passively compresses and expands as the pressure changes in the hydrogen loop. There is no control valve in the hydrogen loop and the hydrogen feed flow rate is adjusted by the pump's rotation speed. A thermal disturbance applied to the hydrogen loop is adjusted by a helium flow control valve.

Figure 6.7 gives an overview of the J-PARC CMS [10]. A hydrogen loop circulates supercritical cryogenic hydrogen at a flow rate of 0.185 kg s^{-1} using two hydrogen pumps (dynamic gas bearing) [11] placed in parallel, and three moderators are arranged in parallel. An ortho-parahydrogen (OP) converter (Fe $(OH)_3$) is placed to provide parahydrogen concentration of more than 99% to the moderators. A He-H_2 heat exchanger is placed downstream of the pump and OP converter to maintain a constant hydrogen temperature to the moderators. At J-PARC, an accumulator for volume control (allowable volume change of 6 l) [12] and a heater for thermal compensation (maximum capacity of 7 kW) [13] are used to mitigate the pressure fluctuation caused by a temperature rise by the nucleate heating of 6.7 kW at the moderators, which is estimated to be 2.3 K for the 1 MW proton beam operation. The heater makes it possible to always provide the supercritical hydrogen within 0.1 K of thermal disturbance, even when a transient heat load is suddenly applied [10].

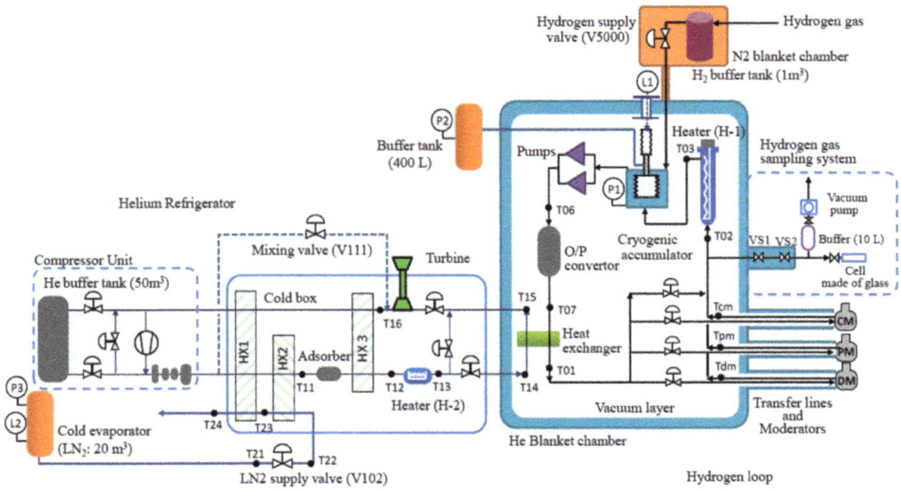

Figure 6.7. Overview of the J-PARC CMS. Reprinted from [10], with the permission of AIP Publishing.

Unlike SNS and J-PARC, the ESS CMS was selected to use subcooled liquid hydrogen instead of supercritical hydrogen at a temperature of around 17 K as the cold moderator material for the cold neutrons [15]. Therefore, a complicated accumulator is not necessary for the ESS CMS. Subcooling, which is the difference between saturated temperature and liquid temperature, is decreased by 1.5 K compared with the case of SNS and J-PARC. A higher feed flow rate is required in order to avoid the occurrence of nucleate boiling with a low hydrogen density in the moderators.

The following design conditions are required for the ESS CMS to satisfy the ESS's goals of providing high brightness cold neutrons for science [15]. In order to meet the first requirement, the moderators have to be arranged in parallel to ensure the feed liquid hydrogen temperature of around 18 K and an OP catalyst is prepared to keep a desirably high parahydrogen fraction in the cold moderators. The required feed flow rate is more than 250 g s^{-1} for each moderator. Therefore, a total circulation flow rate of 1 kg s^{-1} is required for the four-moderator configuration.

(1) Liquid hydrogen with a temperature of around 18 K and a parahydrogen fraction of more than 99.5% shall be provided to the cold moderators.
(2) The average temperature rise over the cold moderators shall be below 3 K.

Figure 6.8 shows an overview of the ESS CMS, which is cooled by a helium refrigerator called the target moderator cryoplant (TMCP) [16] via heat exchangers. Figure 6.9 shows the layout of the CMS in the Target building. The CMS design pressure is 16 bar.g. The most fragile component is the cold moderator in the twister. The TMCP design pressure is 25 bar.g. The ESS CMS consists of two turbo pumps in series, a pressure control buffer (PCB) in a bypass and an ortho-parahydrogen (OP) converter in a bypass, and two kinds of heat exchanger (HX)—a plate-fine type (HX-61100) and a finned tube type (HX-61200)—, vacuum insulated cryogenic

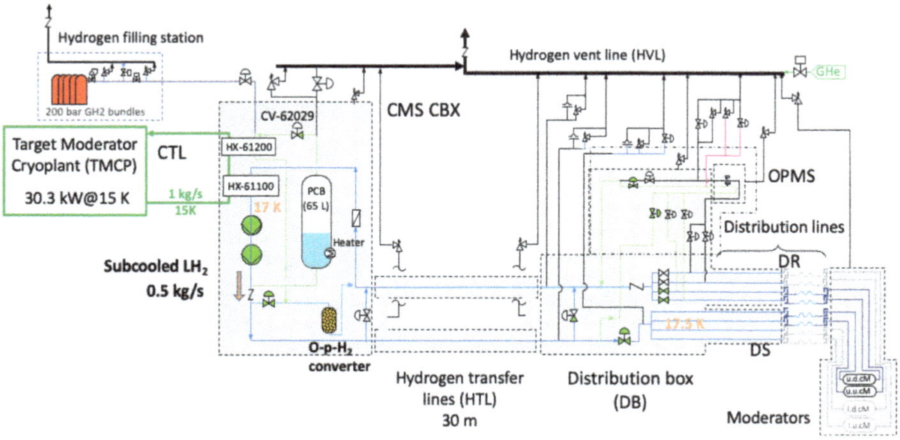

Figure 6.8. Overview of the ESS CMS [15].

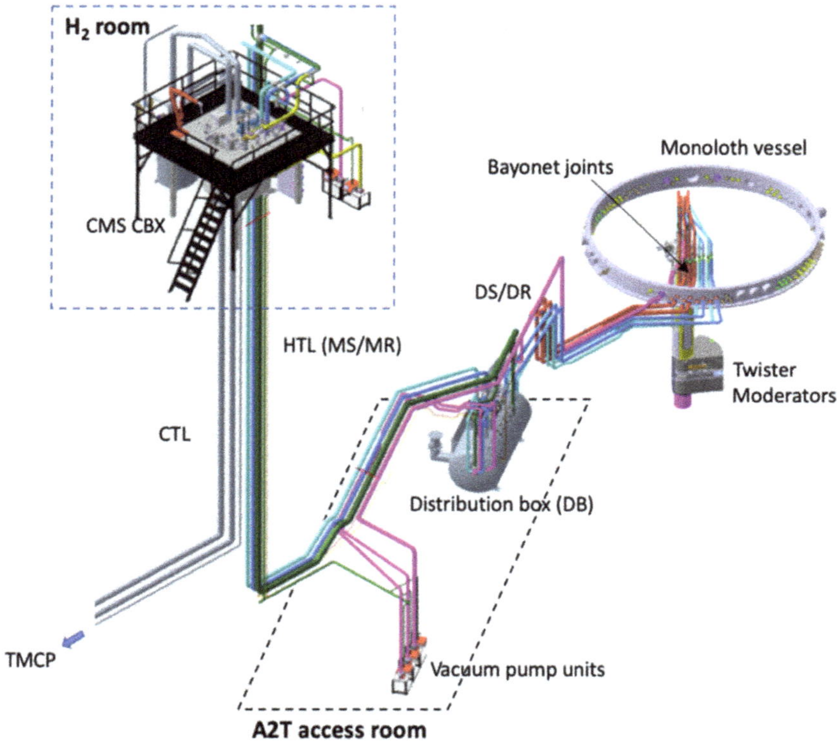

Figure 6.9. Layout of the CMS in the ESS target building.

hydrogen transfer lines (HTLs), a distribution box (DB), an in situ OP measurement system using a Raman spectroscopy (OPMS) in the DB, distribution lines (DLs) to the cold moderator vessels, and a hydrogen vent line (HVL). The HTL is split into four DLs in the DB. Removable u-shaped bayonets connect to each DL with the pipe inside the twister. The parallel flow allows each moderator to operate at the same temperature and results in less pressure loss. There are two cold moderators above and two below the target wheel (in the future) inside the twister. There are five separate vacuum spaces: the CMS coldbox (CBx), the HTL, the DB including the DL, the OPMS, and the moderators.

The TMCP has two warm oil-lubricated screw compressors, the high pressure varies from 0.6 to 2.1 MPa and the feed flow rate is 0.12 to 0.97 kg s^{-1} at 15 K. The TMCP can change the cooling power of 5 to 30.2 kW at 15 K. TMCP CBx located in another building that is 350 m away from the CMS CBx. The helium temperature leaving the HX-61100 is 20 K, while the hydrogen temperatures at both of its ends are 17 K and 20.5 K.

A hydrogen filling station (design pressures: 200 barg and 25 barg) is located outdoors to reduce the hydrogen inventory in the Target building as much as possible. All the hydrogen will be released to air after the scheduled proton beam operation. The filling station stores seven 200 bar high-pressure hydrogen bundles. High-purity hydrogen with a pressure of 20 bar is supplied at a maximum flow rate of 120 Nm^3 h^{-1} to the CMS through the GH2 feed line with a length of 123 m during the cooldown operation.

6.4 Hydrogen safety measures

The CMS's safety design has been considered based on the following hydrogen safety principles.

- The CMS has been designed and fabricated according to Directive 2014/68/EU (PED) (Design pressure: 16 barg).
  - Pressure relief device (PRD) is installed to avoid exceeding the design pressure. The required size is calculated according to EN-ISO 21013-3 design code guidelines [17], considering the case of the most serious case of 'vacuum failure' due to air ingress, which will be described in 6.7.1.
  - Additionally, control release valves are installed to protect from over pressure when the pressure rise is relatively slow.
  - The control valve is a fail-open function, and hydrogen is released and depressurized in case of the instrument air failure.
  - Safety devices for the vacuum envelope are also installed to be maintained below the design pressure of 0.5 barg. The sizes are determined for the case where liquid hydrogen is leaked to the vacuum space [18, 19], which will be described in section 6.7.2.
- All the cryogenic process lines are fully welded and have a vacuum envelope, which is considered as a second barrier. There is no possibility of a hydrogen leak to the outside of the cryogenic piping.
  - The vacuum is maintained and is always monitored by a vacuum pressure transmitter.
  - There are bayonet joints with a CF flange and a metal gasket on the DS and DR on the top of monolith vessel, which enable the moderator-reflector-plug, including moderators, to be replaced every year due to

radiation damage. The bayonet part includes a monolith vessel (vacuum environment), which works as a safety barrier.
  - All the GH_2 feed lines between the filling station and H2 room are fully welded (design pressure of 25 barg).
- The hydrogen inventory in the Target building is designed to be as small as possible.
  - The GH_2 filling station where 10 200 bar-bundles are stored is located outdoors.
  - All the hydrogen will be released after the CMS operation is finished and is replaced by an inert gas such as nitrogen or helium. During the maintenance period, a hydrogen-free environment is formed.
  - After the CMS cool-down operation, the feed line from the GH_2 filling station to the CMS CBx is depressurized and is kept at a positive pressure, e.g., 2 bar.
  - A pneumatic shut off valve is installed into the feed line at the GH2 filling station. If something happens in the Target building, the hydrogen supply will be immediately shut off and hydrogen in the GH2 filling station and the feed line will be released by pneumatic release valve via the HVL of the filling station.
- Hydrogen is released to atmosphere through a HVL, where inert gas is filled at a positive pressure and a check valve is installed at the end to prevent air leak.
  - The HVL is designed according to *CGA G-5.5, Hydrogen Vent Systems* [20].
  - The pressure in the HVL is always monitored. If the inert gas pressure falls below a set pressure, then it will be automatically refilled.
- High ventilation for the H2 room and A2T access room is applied in order to dilute potentially leaked hydrogen below 1% of LEL.
  - Stationary hydrogen leak sensors are installed on the ceiling in the H2 room and A2T access room because there are mechanical joints where warm hydrogen exists, e.g., a pipe fitting, thread fitting for a pressure transmitter, or a top flange of the hydrogen pump. The hydrogen leak sensors are placed above the mechanical fittings. If a hydrogen leak is detected, then hydrogen is immediately released by a control valve and electrical power can be partly shut off.
- All the instruments and equipment are connected to ground.
- An area classification and risk assessment are taken into account according to Directive 2014/34/EU (also known as ATEX 114 or the ATEX Equipment Directive).
- Failure monitoring function and failure action mode are developed in the CMS control system (vacuum failure, high-pressure warming, and hydrogen leak). When the failure action mode is activated, motors are stopped, hydrogen is released via the control valve, and electrical power is shut off.

6.5 Main components

6.5.1 Liquid hydrogen pump

A ball-bearing type hydrogen pump was designed to circulate 1 kg s^{-1} at 17 K for the four-moderator arrangement at the design phase. The hydrogen pump impeller is made of aluminum alloy and has a diameter of 95.4 mm. For now, the required flow rate is only 0.5 kg s^{-1} for the two moderators. Two hydrogen pumps are placed in series because of redundancy and are operated at the same rotation speeds. If one pump fails, the other is ramped up to get the required flow rate. It is known that the performance curve can be arranged using dimensionless expressions of a head coefficient, ψ, discharge coefficient, ϕ, and the wheel speed, u_s [21].

$$\psi = \Delta P(\rho u_S{}^2)^{-1} \tag{6.1}$$

$$\phi = \dot{m}(\rho u_s D^2)^{-1} \tag{6.2}$$

$$u_s = \pi ND/60 \tag{6.3}$$

where ΔP is the pump head, $\dot{m}$ is the mass flow rate, ρ is the density, D is the impeller diameter, N is the rotation speed (rpm), and $g = 9.8$ m s^{-2}.

The non-dimensional performance curve given by a measurement using LN_2 is shown in figure 6.10.

Figure 6.11 shows the hydrogen pump performance at the nominal condition of 17 K and 10 bar for two-series and single operations. For comparison, the pressure drops for two- and four-moderator configurations are also displayed. For the two pumps, the required rotation speeds are 7500 rpm for the two-moderator arrangement and 10,000 rpm for the four-moderator arrangement because the required pump heads are 93 kPa for 0.5 kg s^{-1} and 167 kPa for 1 kg s^{-1} to overcome the

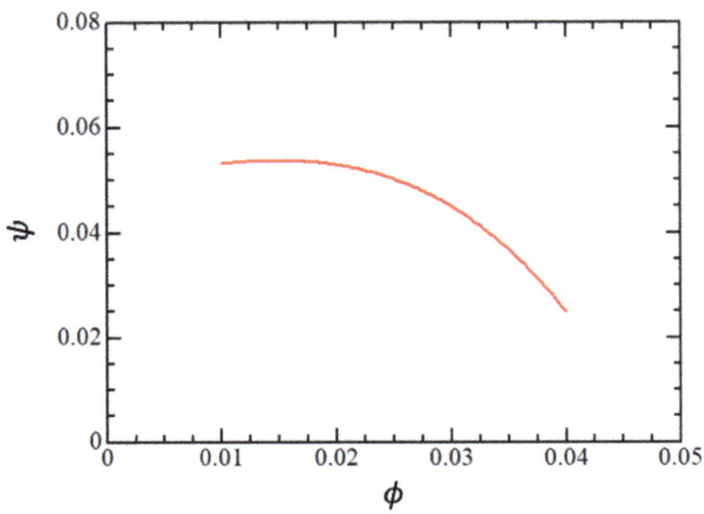

Figure 6.10. Hydrogen pump design performance curve.

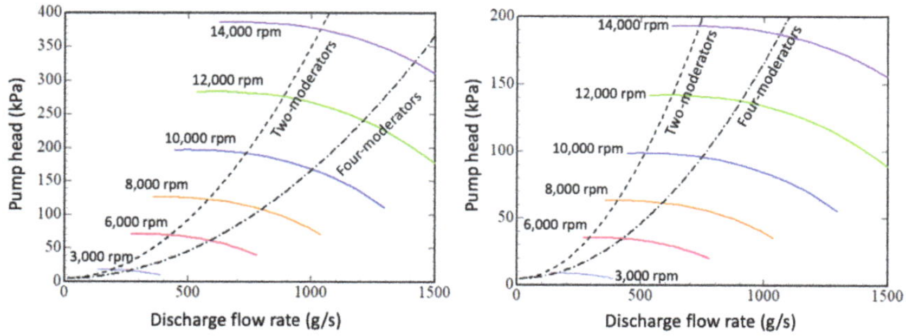

(a) Two-pump in series operation. (b) Two-pump in series operation.

Figure 6.11. Predicted hydrogen pump performance at nominal conditions (17 K and 1.0 MPa).

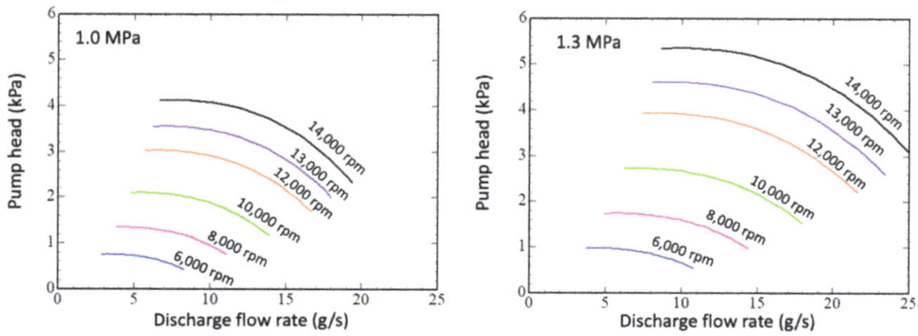

(a) P=1.0 MPa (b) P=1.3 MPa

Figure 6.12. Predicted hydrogen pump performance at ambient temperature for two-series operations.

pressure drop, as mentioned in 6.6.1.1. It turns out from figure 6.11 that even if one pump fails the revolution speed required for the remaining pump can be lower than the maximum speed of 13,200 rpm.

Figure 6.12 shows the predicted hydrogen pump performances for two-series operations at ambient temperature at the pressures of 1.0 and 1.3 MPa. The pump head is predicted to be 4 kPa at most at the maximum rotation speed of 14,000 rpm and ambient temperature because of the lightest hydrogen density. At a pressure of 1.3 MPa, the pump head is 1.3 times higher than that at 1.0 MPa because the density is proportional to the pressure. This would be just one option to increase the pressure if you need a slightly higher pump head at the beginning of the cool-down operation (around ambient temperature).

6.5.2 OP converter

The natural conversion mechanism during and after the cool down below 20 K is rather slow, taking months to enrich the parahydrogen to a level higher than 99.5%. Orthohydrogen (oH_2) could decrease the moderation performance significantly because scattering cross-sections for orthohydrogen are more than 40 times higher

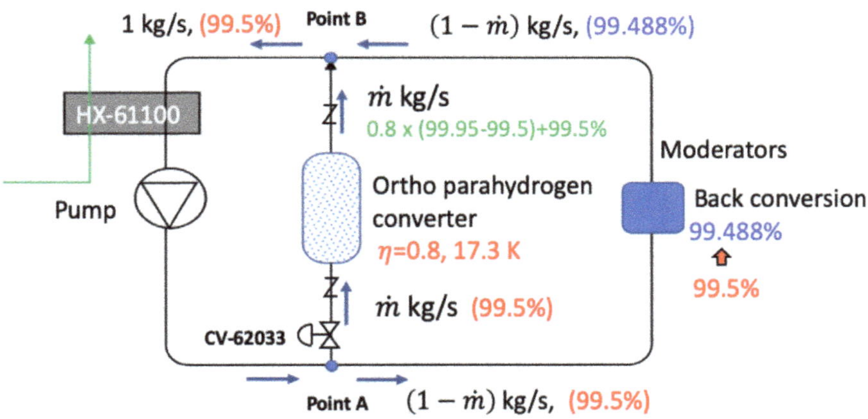

Figure 6.13. Flow and parahydrogen fraction balances in the CMS loop.

than those for parahydrogen below 10^{-2} eV. Iverson and Carpenter [22] predict that a neutron irradiation induces conversion of 0.003% of the parahydrogen volume to orthohydrogen per pulse. Based on this, a parahydrogen of 0.0114% via the moderators would convert to orthohydrogen for the proton beam at 14 Hz. For 99.5% of the parahydrogen fraction at the inlet of the moderators, it decreases to 99.488% downstream of the moderator. The ESS CMS is equipped with an OP catalyst vessel in a bypass of the main loop to convert hydrogen from the ortho state to the para state, to get equilibrium parahydrogen during the cool-down process, and to keep a desirably high parahydrogen fraction in the cold moderators. This is because the pressure drop over the o-parahydrogen converter would be too large, e.g., 79 kPa at 1 kg s^{-1}, if it was arranged in the main loop.

At nominal operation, liquid hydrogen with a temperature of 17.3 K, where the equilibrium parahydrogen fraction is 99.95%, is provided to the OP converter because of the heat load of the hydrogen pump. Figure 6.13 shows the flow and parahydrogen fraction balances in the CMS loop. Assuming a conversion efficiency, η_p, of 80%, the OP converter would give a parahydrogen fraction of 99.868%, if the parahydrogen fraction at the inlet is 99.5%.

$$\eta_p = \frac{x_p - x_{pf}}{x_{pe} - x_{pf}} \tag{6.4}$$

where x_{pe} is the parahydrogen fraction in equilibrium, and x_p and x_{pf} are the parahydrogen fraction at the outlet and inlet of the catalyst, respectively.

Considering the mass balance at the confluence point (Point B), the required bypass flow rate is calculated to be 38.5 g s^{-1}.

In the presence of a catalyst, the reaction approaches a first order reaction. The rate of change of the orthohydrogen is given by

$$-\frac{dx_o}{dt} = kx_o - k'(1 - x_o) \tag{6.5}$$

where $x_o(=1 - x_p)$ is the parahydrogen fraction in equilibrium, k is the reaction rate constant converted from orthohydrogen to parahydrogen, and k' is the reversed reaction rate constant.

At the equilibrium condition $(\partial x_o/\partial t = 0)$, the orthohydrogen fraction, x_o, is equal to x_{oe}. The reversed reaction rate constant is given by

$$k' = k\frac{x_{oe}}{1 - x_{oe}} \tag{6.6}$$

Equation (6.5) can be rearranged as follows

$$-\frac{dx_o}{dt} = \frac{k}{1 - x_{oe}}(x_o - x_{oe}) \tag{6.7}$$

The space velocity (min^{-1}), SV, is calculated under the condition of $x_o = x_{of}$ at $t = 0$.

$$SV = \frac{1 - x_{oe}}{t} = -\frac{k}{\ln\left(\frac{x_O - x_{Oe}}{x_{Of} - x_{Oe}}\right)} \tag{6.8}$$

Equation (6.7) can be rearranged using equation (6.4).

$$SV = -\frac{k}{\ln(1 - \eta_p)} \tag{6.9}$$

The space velocity is calculated to be 2174 for $\eta_p = 0.8$ and $k = 3500$ [23, 24]. The required catalyst bed size is calculated to be 9.6 l. The catalyst bed size is conservatively chosen to be 35 l. The diameter of the OP converter was determined to be 273.05 mm and a commercial catalyst IONEX® Type OP with an average grain size of 0.5 mm was selected. A 60 μm mesh filter is installed into the bottom of the converter vessel. The design parameters of the OP catalyst vessel are summarized in table 6.3 [15]. The pressure drop over the bypass line can be calculated to be 20.8 kPa at the flow rate of 38.5 g s^{-1}, which includes the two check valves and piping.

Table 6.3. Design parameters of the OP catalyst vessel [15].

Catalyst	IONEX® Type OP (Ferric oxide)
Conversion efficiency (%)	80
Catalyst particle size (mm)	0.5
Catalyst bed size (L)	35 (9.6 l: calculated value)
Porosity (-)	0.4
Space Velocity (min^{-1})	3500
Flow rate (g s^{-1})	31.5
Pressure drop (kPa)	0.5
Temperature (K)	17.3
Vessel diameter size (mm)	273

This is much smaller than the pump head of 93 kPa for 0.5 kg s^{-1} for the two-moderator arrangement, as shown in figure 6.11.

6.5.3 PCB tank

The function of the PCB is to keep the pressure in the CMS loop stable during steady-state operation and to mitigate transient pressure fluctuations within $\pm$ 0.5 bar caused by a proton beam injection or stop [25]. The CMS forms a closed loop, which is filled with subcooled liquid hydrogen with a pressure of 10 bar and a temperature of around 17 K that behaves like an incompressible fluid. The PCB is arranged in a bypass, which is routed from the outlet of the hydrogen pumps to the inlet of the OP converter via the release valve (CV-62029) and the small heat exchanger (HX-61200), as shown in figure 6.8. There is a vapor–liquid surface in the PCB and expansion or compression of the vapor phase can mitigate pressure fluctuation. This is an advantage of selecting subcooled liquid hydrogen, unlike supercritical hydrogen in SNS [9] and J-PARC [10]. The PCB, with a volume of 65 l and a diameter of 323.9 mm, is mounted vertically in the CMS CBx, as shown in figure 6.14. Electrical heaters are wrapped on it to produce pressure caused by vaporized hydrogen. If the CMS pressure increases too much, the vapor is released by the control release valve (CV-62029, $Kv = 1.1$) placed on the top. The released vapor is liquefied by HX-61200, as mentioned in section 6.5.4. To achieve faster responsiveness, it is useful and effective to keep the liquid hydrogen within the PCB at a saturated temperature of 31.26 K at 1.0 MPa. Figure 6.15 shows measured pool boiling curves of saturated liquid hydrogen from a wire, as reported by Tatsumoto *et al* [26]. It is found from the figure that the heaters should be designed in

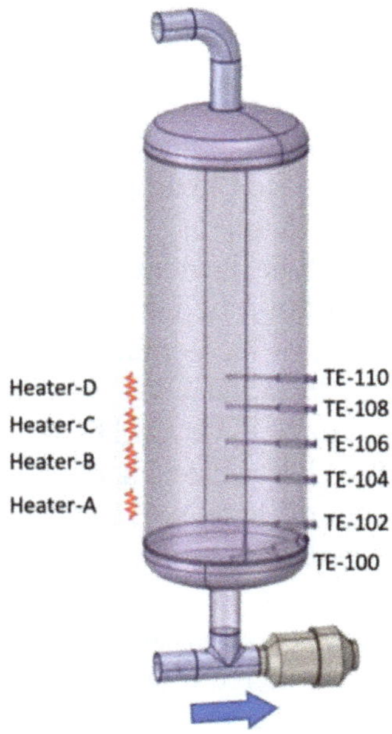

Figure 6.14. Schematic of the PCB tank.

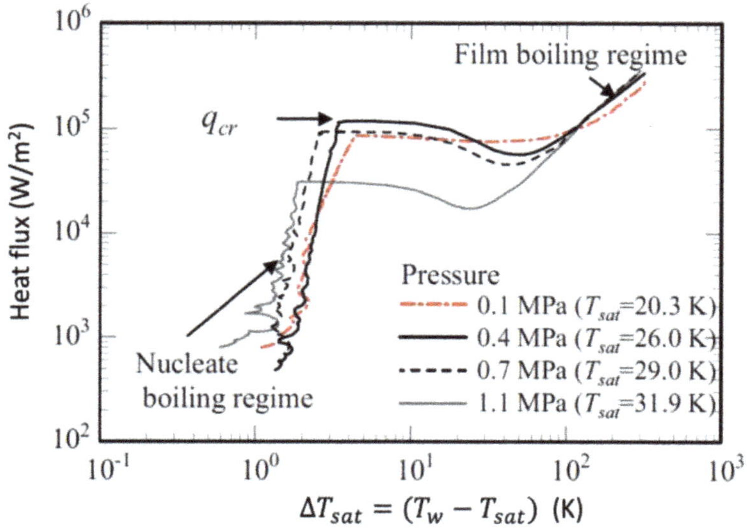

Figure 6.15. Saturated liquid hydrogen boiling curves. Reprinted by permission from Springer Nature Customer Service Centre GmbH: [26], Copyright (2015)

such a way that the heat flux to the liquid hydrogen is from 2×10^3 to 3×10^4 W m^{-2}. The liquid hydrogen level in the PCB is detected in two ways: a hydrostatic head of liquid hydrogen and temperature sensors (Si-diode) placed vertically at six different altitudes corresponding to the liquid volume of 3%, 11%, 21%, 29%, 36%, and 43%.

6.5.4 Heat exchangers

There are two heat exchangers. The first is a plate fin heat exchanger (L500 × W250 × H228 mm), HX-61100, placed in the main CMS loop in order to remove the heat load in the CMS and maintain the feed hydrogen temperature at 17 K. The design parameters are described in (table 6.4) [15].

The second is a finned tube type heat exchanger, HX-61200, which has the function of condensing the vapor released from the PCB to maintain the CMS pressure at around 1.0 MPa [15]. For the nominal operation, the hydrogen flow rate of 0.31 g s^{-1} released from the PCB via CV-62029, whose temperature is 110 K, should be condensed and cooled to 22 K via the HX-61200, as described in 6.6.3. The removal heat load via the HX-61200 is calculated to be 459 W. This is because the HX-61200 is located downstream of the HX-61100, where the helium outlet temperature is 20 K. The fin tube is coiled helically around a central mandrel and is housed in a surrounding stainless-steel vessel, as shown in figure 6.16. The length and the diameter of the copper finned tube where hydrogen is flowing have been determined to be 12 m and 16.5 mm to meet the requirements, while the length of the annular space through which helium is passing as cross flow is set 0.33 m. A CFD analysis shows that the pressure drop of the helium flow is 0.8 kPa under the conditions of 1 kg s^{-1}, 20 K, and 2.1 MPa and satisfies the design criteria of maximally 3 kPa.

Table 6.4. Design parameters of the heat exchanger (HX-61100) [15].

HX-61100	Plate fin type	
Size (mm)	L 500 × W 250 × H 228	
Fluid	H2	He
Flow rate (kg s^{-1})	1.0	0.99
Pressure (MPa)	1.0	1.9
Temperature (K)	20.5/17.0	15.12/19.88
Pressure drop (kPa)	10	13
Layers	15	14
Heat exchange (kW)	28.8	

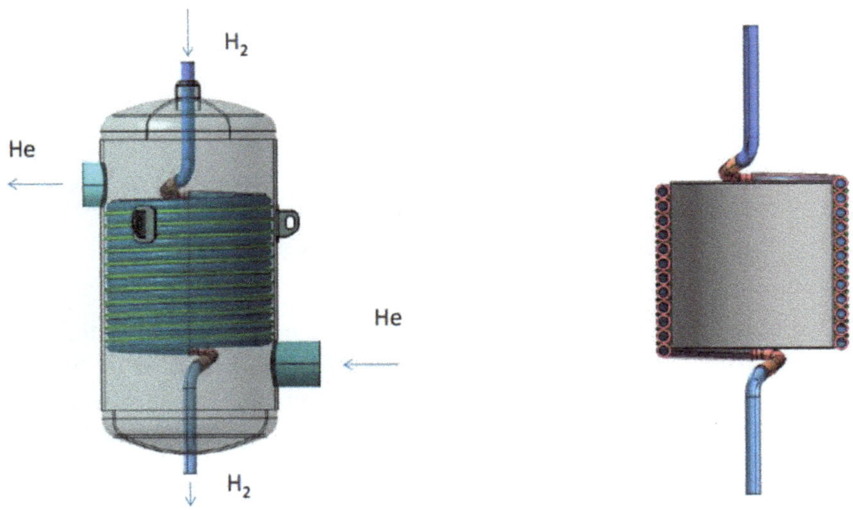

Figure 6.16. Design of HX-61200 and the finned tube helix around inner flow blocker.

6.5.5 In situ real-time measurement system for ortho- and para-fractions of LH_2 (OPMS)

An in situ measurement system for ortho- and para-fractions of liquid hydrogen (OPMS) is being developed in order to ensure a parahydrogen fraction of 99.5% or more at the moderators and to study the effect of the neutron scattering driven para-to-ortho back conversion. The details have been reported in [19]. A Raman spectroscopy will be used to identify the orthohydrogen and parahydrogen fractions. The required measurement precision is 0.1% to detect an undesirable shift towards a high orthohydrogen fraction. A highly reliable sapphire window that endures in low temperature and high-pressure environments was developed so that Raman laser beams could be directly injected into the CMS loop. As a safety measure, the

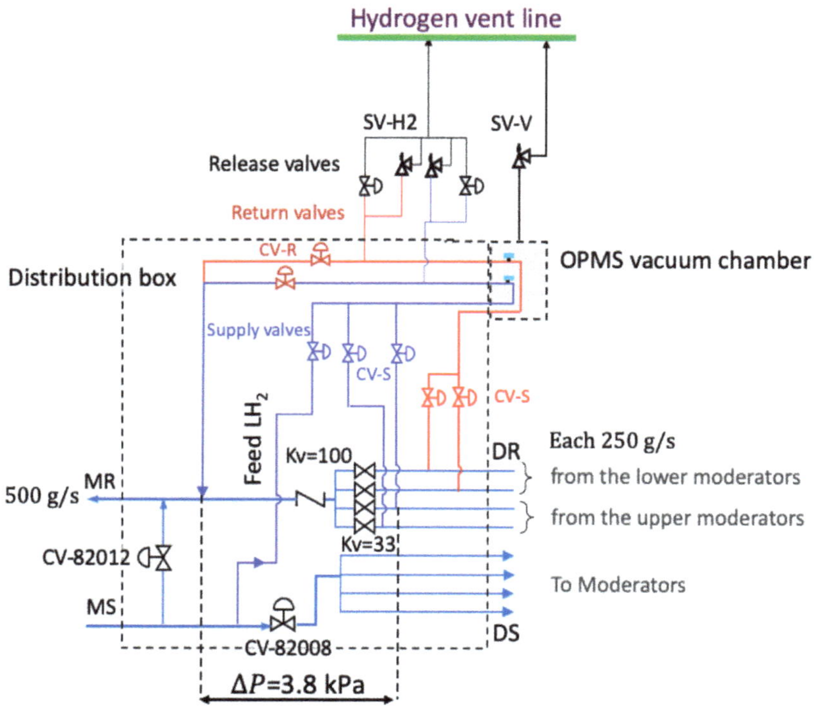

Figure 6.17. OPMS dedicated sampling line with the sapphire window [19].

sapphire window part is not placed in the process line but in a dedicated sampling line, which has been designed as a bypass in the DB, as shown in figure 6.17. This is because the hydrogen inventory is reduced as much as possible and the sampling line is physically isolated from the main CMS process line if a failure in the sapphire window would ever happen.

The feed flow rate to each moderator is adjusted to be 240 g s^{-1} by a manual hand valve downstream of its own return distribution line (DR). Two dedicated sampling lines with a sapphire window with a diameter of 15.0 mm for the upper moderators and the lower ones are placed in the DB. The sampling piping for the upper moderators is routed from each DR and the main supply (MS) HTL back to the main return (MR). The total inventory of each sampling line between the control valves of CV-S and CV-R is 0.62 l. A vacuum space around the sapphire window is also physically separated from the DB, taking into account the case of the hydrogen leak due to a sapphire window failure. The CMS would be able to be continuously operated without the OPMS until the end of a scheduled user program.

6.5.5.1 ESS Raman spectroscopy system and the sapphire window section design

Figure 6.18 shows an overview of the ESS Raman spectroscopy system. A Raman laser optics system consists of a Raman laser (532 nm), Raman spectroscopy and an optical probe box which includes focusing and de-focusing lenses, a mirror, a high

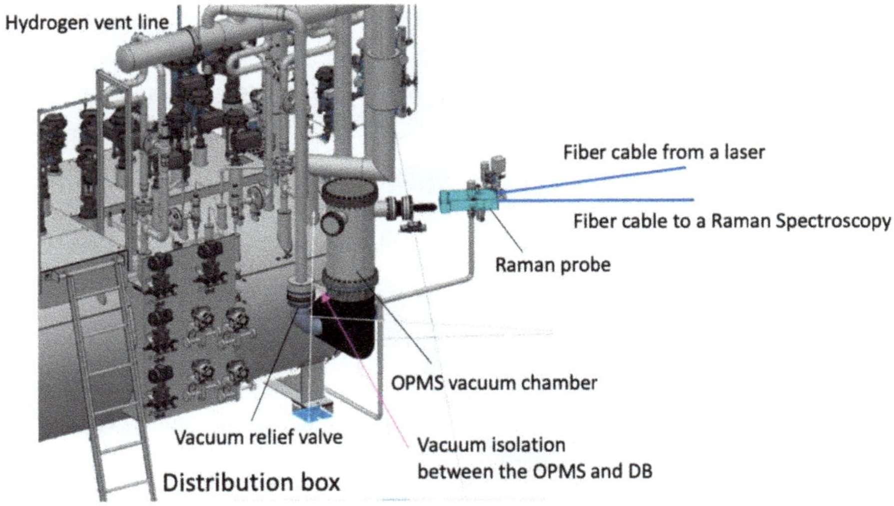

Figure 6.18. Overview of the ESS Raman spectroscopy system.

pass filter, and a band pass filter. The laser and Raman spectroscopy are connected to the Raman optical probe box via a fiber cable and are installed into a cabinet that is 4 m away from it.

6.5.5.2 Sapphire window section design

Figure 6.19 shows an overview of the sapphire window section of the OPMS sampling line. The details have been reported in [19]. Two sampling pipes are mounted in a vacuum chamber with an inner diameter of 300 mm, which is physically isolated from the DB. A CF flange is selected as a safety measure against a liquid hydrogen leak. The sapphire window part is exchangeable by a VCR fitting. As shown in figure 6.20, a Raman laser probe will be mounted on a side port on a welded bellows with a z-translation stage to be able to adjust a focal point. There is another sapphire window with a diameter of 30 mm on the flange that creates a boundary between the atmosphere and vacuum. The laser guide tube is maintained as the same OPMS vacuum space and has a focal lens that makes the focal point at between 23 and -5 mm from the inside of the sapphire window. A support jig made of G10 is designed to easily and correctly setup the sapphire window and the laser probe on the same optical axis. The steady-state heat load to the sapphire window through the G10 jig was calculated to be 0.81 W, which is lower than the criteria, using the finite element analysis software ANSYS 2020 R1.

The sapphire window with a diameter of 15 mm and a thickness of 2.0 mm is brazed to a pipe with a thickness of 0.8 mm made of Kovar (Fe-Ni-Co alloy) in a vacuum environment. The sapphire is 99.99% purity single crystal Al_2O_3 and maintains a transmittance of more than 80% in a wavelength range higher than 290 nm. Fluid analyzes were performed using a CFD code, FLUENT, at the inlet temperature of 19 K and inlet flow range of 2 g s^{-1} in order to optimize the distance

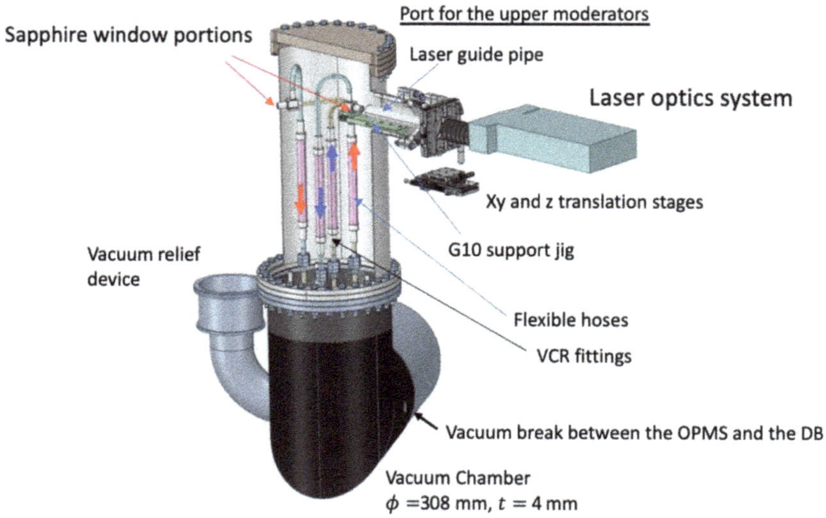

Figure 6.19. Overview of the sapphire window section of the OPMS sampling line. Reproduced from [19].

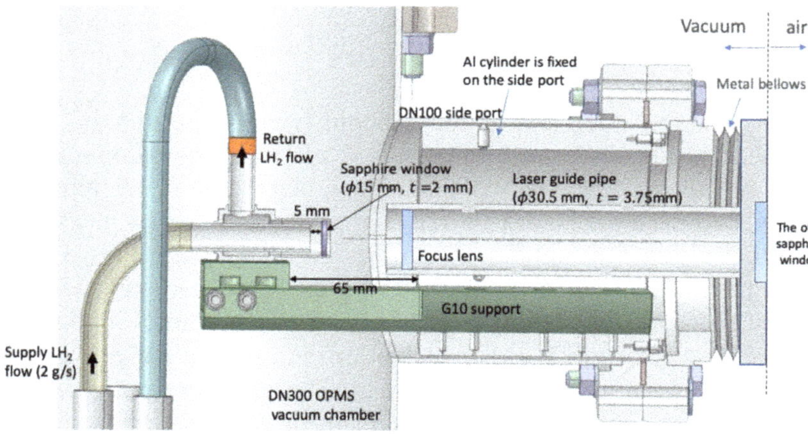

Figure 6.20. Sapphire window support in the OPMS and a Raman laser focal point. Reproduced from [19].

between the sapphire window and the inlet pipe, L_g, with an outer diameter of 12 mm and a thickness of 2 mm to prevent the liquid hydrogen from accumulating around the focal point. Figure 6.21 shows the liquid hydrogen flow analysis results for $L_g = 5$ and 15 mm. For $L_g = 5$ mm, the flow uniformly turns around the sapphire window and smoothly flows out along the outside of the inlet pipe, unlike that for $L_g = 15$ mm. There is no stagnation point around the focal points. The calculated pressure drop is 15.4 Pa and is vanishingly small. The gap size, L_g, is determined to be 5 mm based on the CFD simulation results.

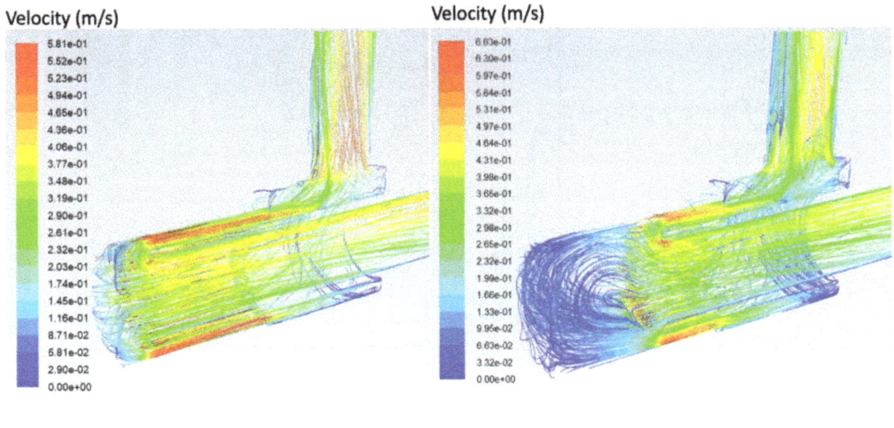

(a) L_g = 5.0 mm (b) L_g = 15.0 mm

Figure 6.21. LH_2 flow patterns through the sapphire section at 2 g s^{-1}. Reproduced from [19]. © IOP Publishing Ltd. CC BY 3.0.

6.5.5.3 Sapphire window development

The ESS CMS plans to be cooled down twice a year. The parahydrogen fraction should be confirmed using the OPMS at the final cool down operation stage. The time required for the OP fraction measurement by Raman spectroscopy would be 25 min, which includes the precooling time of 20 min after switching the sampling line and 5 min for the Raman measurement. It is thought to be sufficient to measure the orthohydrogen and parahydrogen fraction measurement once a week during the stable operation. The OPMS will be isolated from the CMS main loop after the measurement. Forty cycles of the cool-down and warm-up will be carried out per year. The control valves of CV-S and CV-R enable the cool-down speed to be adjusted in order to avoid an extreme thermal shock. It was important for the ESS OPMS to develop a highly reliable sapphire window that endures in low temperature and high-pressure environments. The ESS is proceeding the R&D in collaboration with the Japan Aerospace Exploration Agency (JAXA), which possesses a liquid hydrogen tank internal visualization method using a 60 mm-diameter observation sapphire window at 0.6 MPa. In a preliminary experimental study, a pressure-cycle test in LH_2 was conducted using a standard sapphire window with a diameter of 20 mm and a thickness of 2 mm manufactured by Kyocera (GMM-90167-1). The details have been reported in [19]. An x-ray inspection of the sapphire brazing in the Kovar sleeve that was made in a dry GN_2 environment showed the existence of some small cavities in the brazing. A prototype test stand for the pressure cycling was also fabricated, as shown in figure 6.22. One side of the sapphire window part was to be connected to a 1/2-inch pipe and was evacuated. The opposite side was to be connected to a 1/8-inch pipe to be pressurized or depressurized between 0.1 and 1.7 MPa by a small amount of GH_2 within 1 s. The sapphire window section was immersed in saturated LH_2 at 0.1 MPa. Figure 6.23 shows a pressure-cycle test result for four cycles of 0.5 MPa, 100 cycles of 1 MPa, and 300 cycles of 1.7 MPa. After the

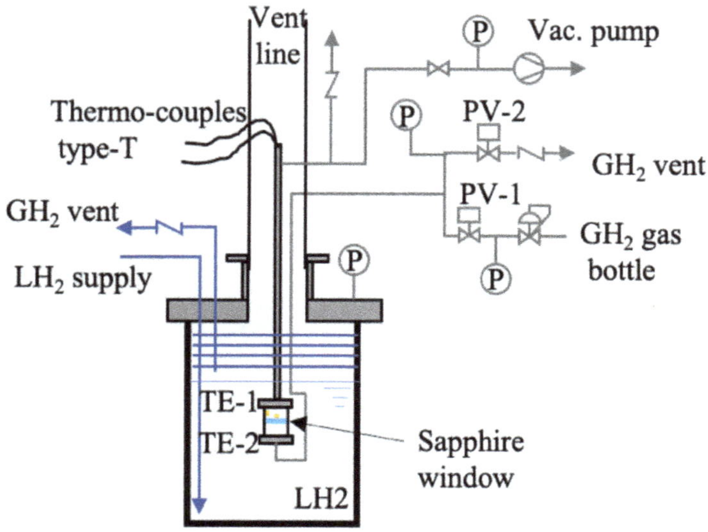

Figure 6.22. Pressure-cycle test stand for a sapphire window immersed in LH_2. Reproduced from [19]. © IOP Publishing Ltd. CC BY 3.0.

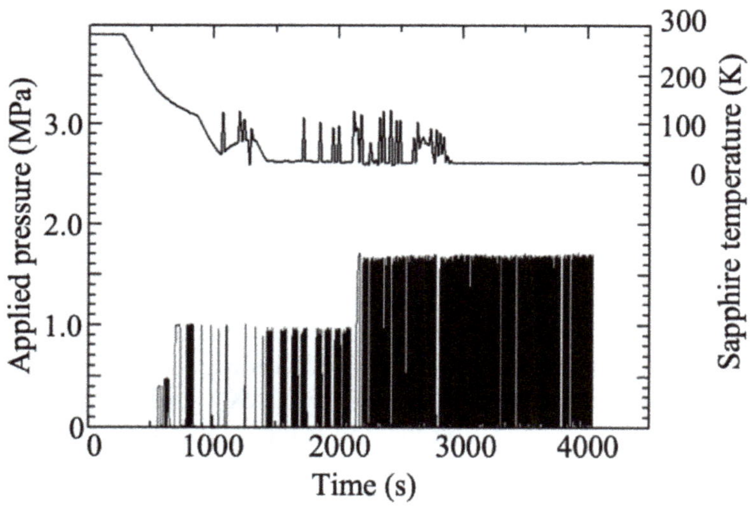

Figure 6.23. Pressure-cycle test result. Reproduced from [19]. © IOP Publishing Ltd. CC BY 3.0.

pressure-cycle test, a helium-leak test was implemented under the background level of 2.7×10^{-10} Pa m^3 s^{-1}. It was verified that there are no harmful leaks through the brazing part or the sapphire itself.

6.5.6 HVL

The HVL, whose design pressure is 0.15 MPa, is routed from the A2T access room to the roof top through the hydrogen room, where the CMS CBx is placed, as

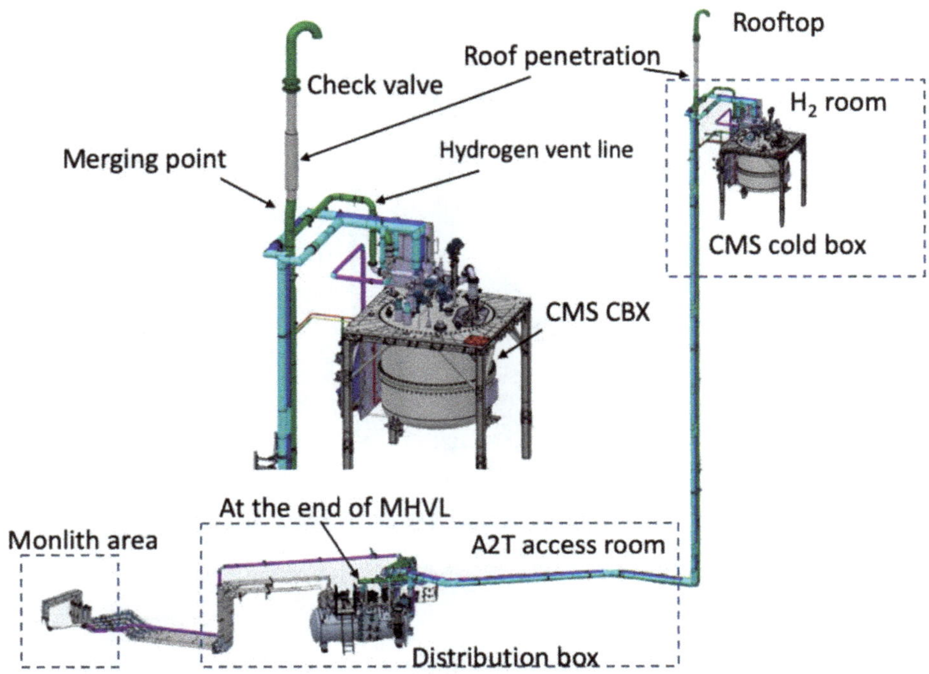

Figure 6.24. HVL layout in the target building [27].

Table 6.5. Released hydrogen flow rate for each case. Reproduced from [27]. © IOP Publishing Ltd. CC BY 3.0.

	Operation mode	Flow rate (g s^{-1})	Release time
Nominal case	Normal warm-up	6.4 (average)	60 min
	Quick warm-up	38 (average)	10 min
Failure case	Pneumatic failure (CV-62001)	300 (maximum)	3 s
	Air leak into the moderator vacuum space	650 (maximum)	60 s

shown in figure 6.24. The total length is 36 m and the outer diameter is 168.3 m. The details have been reported in [27]. As shown in figure 6.8, there is a merging point from the sub-HVL for the CMS CBx in the hydrogen room. The HVL outlet is located 3 m from the roof top and is guarded by a lightning rod. There is a check valve with $Kv = 760$ and a cracking pressure of 3.7 kPa on the roof top. The check valve maintains the HVL in a helium environment at a positive pressure in order to avoid an explosive environment forming in it. The HVL has to be sized large enough to limit the backpressure below 0.15 MPa during the release of hydrogen. The released hydrogen flows for each postulated case are summarized in table 6.5. During the normal warm-up operation, the warm-up speed is to be controlled by the helium refrigerator and hydrogen is to be always released via CV-62001,

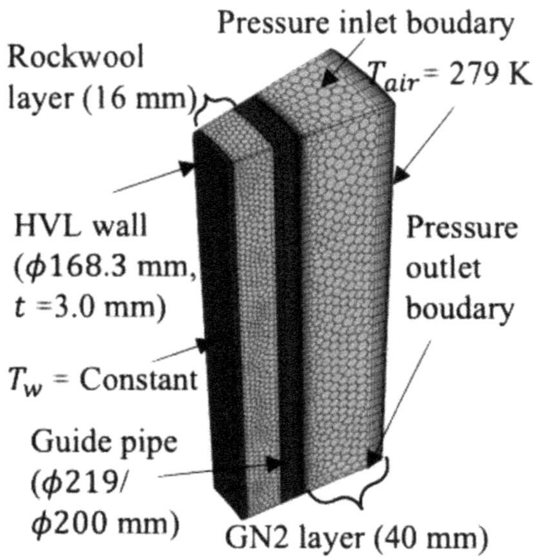

Figure 6.25. Transient thermal analysis model at the penetration. Reproduced from [27]. © IOP Publishing Ltd. CC BY 3.0.

maintaining the pressure at 1 MPa. It would take an hour until the release of the liquid hydrogen is completed. The quick warm-up mode will be used when the liquid hydrogen needs to be released within 10 min by warm GH_2 purge without being circulated by the two hydrogen pumps. The average release flow rate is estimated to be 38 g s^{-1}. A vacuum loss around the moderators due to air ingress is considered as one of the severe failure events mentioned in section 6.7. The maximum release flow rate is estimated to be 650 g s^{-1} when temperature increases to around critical temperature.

The released hydrogen expands in isenthalpic process and the HVL would be partially cooled down to cryogenic temperatures. The roof penetration temperature should be maintained above 253 K. A guide pipe with an inner diameter of 200 mm and a thickness of 9.5 mm has already been embedded. Rockwool will be stuffed into the gap between the guide pipe and the HVL. The allowable lowest HVL wall temperature is estimated by transient thermal analyzes using a CFD code, ANSYS FLUENT, using a simplified analysis model, as shown in figure 6.25. A GN_2 layer is added instead of air in order to consider the effect of the natural convection and the temperature on the outside of the air layer, T_{air}, is kept to 279 K, which is an average temperature between indoor and outdoor temperatures in winter in Lund. The details have been reported in [27]. It turns out from figure 6.26 that the HVL wall temperature, T_w, should be maintained higher than 175 K. However, even if the HVL wall temperature is decreased to 50 K, the guide pipe temperature is maintained above 253 K within 41 minutes. Most of liquid hydrogen for any failure cases has been released within 60 s, as described in table 6.5. The wall temperature of the HVL has to be maintained above 100 K while the liquid hydrogen is being released.

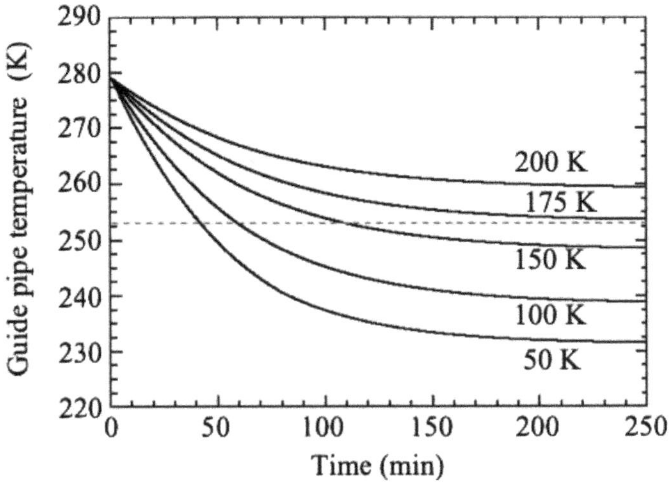

Figure 6.26. Roof penetration guide pipe temperature reduction while liquid hydrogen is being released. Reproduced from [27].

6.5.7 Hydrogen filling station

For hydrogen safety measures, the hydrogen inventory is designed to be as small as possible. All the hydrogen will be released after the scheduled beam operation and is replaced by inert gas, such as nitrogen or helium. During the maintenance period, the ESS Target building forms a hydrogen-free environment. Therefore, CMS cool-downs are expected to happen approximately two times per year. The CMS operational pressure is 10 bara in nominal steady-state operation. The total inventory of liquid hydrogen is 0.344 m^3 and that of gaseous hydrogen is 0.05 m^3, as mentioned earlier. The required hydrogen mass is 26.2 kg for the filling and cool-down operation. Additionally, 2 kg of gaseous hydrogen (GH_2) will be used for the purge process before the cool-down operation. Therefore, we need about 29 kg of GH_2 for one cool-down operation cycle.

The hydrogen filling station is located on the outside of the Target building where 200 bar-GH_2 bundles are stored and its own HVL is prepared. The filling station area has a floor space of 6 m × 12 m (concrete floor) and an asphalt floor with a width of 1.5 m around a concrete area, and is enclosed by a fence with a height of 2 m. The GH2 filling station is placed on the left-hand side of the concrete floor (6 m × 7.5 m). All the equipment and pipes are grounded with a resistance to earth of no more than 25 ohms [20]. An electrical cabinet is placed outside the fence. As shown in figure 6.8, GH_2 will be supplied from the GH_2 bundle, which has 12 standard hydrogen bottles with a volume of 50 l each at 200 barg (8.7 kg of GH_2). The rack size of one bundle is 1020 × 780 × 1930 mm. The pressure is reduced to 20 bar by a pressure regulator ($Cv = 0.13$) and the GH_2 is provided to the CMS CBx in the Target building through a 125-long feed line, whose design pressure is 25 barg. During a condensed process of the cool-down operation, the maximum flow rate is estimated to be 120 $Nm^3\ h^{-1}$. All the hydrogen is released through a HVL for the filling station to atmosphere. The HVL, whose design pressure is 0.5 barg, has been designed according to CGA G-5.5 Hydrogen Vent Systems [20]. The exit of the vent line is located 10 m away from the ground to prevent the released GH_2 from coming back to the operational area. There is a check valve close to the exit as well as the HVL in the Target building and it is filled with dry nitrogen while being maintained at a positive pressure. The postulated maximum release flow rate is 28 g s^{-1}, considering the case of the pressure regulator failure. For the size of DN80, the pressure drop is calculated to be 4 mbar. The check valve with a *Kv* of more than 130 and a cracking pressure lower than 0.1 bar is selected not to exceed the design pressure.

6.6 CMS process design

6.6.1 Pressure drop

6.6.1.1 Pressure drop over the CMS process line

The pressure drop over the main liquid hydrogen circuit is estimated at the nominal operation condition for both the two- and four-moderator arrangements. The fluid properties of parahydrogen at 17 K and 1.0 MPa given by GASPAK [5] have been used. The friction factor of a pipe is calculated using the Colebrook equation [28], where a surface roughness of 0.05 mm is used,

$$\Delta P = f\frac{L}{D}\frac{\rho}{2}u^2 \tag{6.10}$$

$$\frac{1}{\sqrt{f}} = -2\log\left(\frac{\varepsilon}{3.7D} + \frac{2.51}{\mathrm{Re}\sqrt{f}}\right) \tag{6.11}$$

where f is the friction factor, ε is the surface roughness, L is the length, and D is the diameter.

The pressure drops of the moderators are calculated by a CFD simulation [2]. The feed control valve ($Kv = 65$) and the return manual control valve ($Kv = 30$) are set opened by 100% and 100%, respectively. There are two check valves, which have a flow resistance given by $Kv = 100$. The cracking pressure is also considered. HX-61100 has been specified for a pressure drop of 10 kPa at 1 kg s^{-1}. A summary of the pressure drop calculation is given in tables 6.6 and 6.7, where the pressure drops for

Table 6.6. Pressure drop over the main liquid hydrogen circuit for the two-moderators arrangement at a nominal operation condition [15].

Pressure (MPa)	Temperature (K)	$\dot{m}$ (g s^{-1})	Total pressure drop, ΔP (kPa)	Pressure drop from point A to B in figure 6.13, ΔP_{op} (kPa)
1	17	250	27.1	22.5
		500	92.6	81.8

Table 6.7. Pressure drop over the main liquid hydrogen circuit for the four-moderators arrangement at a nominal operation condition [15].

Pressure (MPa)	Temperature (K)	$\dot{m}$ (g s^{-1})	Total pressure drop, ΔP (kPa)	Pressure drop from point A to B in figure 6.13, ΔP_{op} (kPa)
1	17	500	45.9	35.0
		1000	166.9	131.2

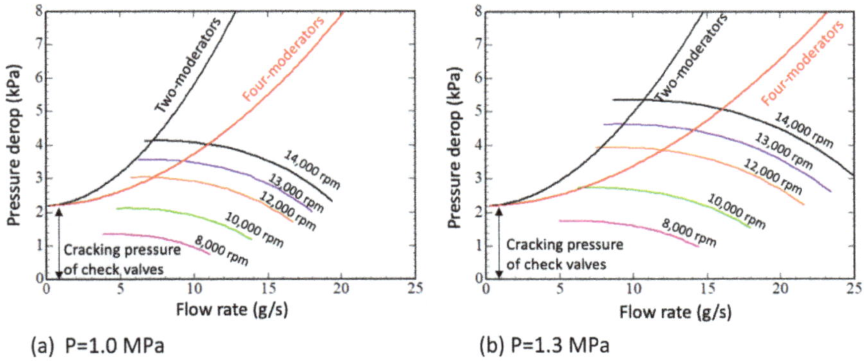

Figure 6.27. Pressure drop over the main liquid hydrogen circuit at 300 K.

not only the piping but also equipment, e.g., a valve, an orifice flow meter, and a filter are included. The circulation flow rates required for the two- and four-moderator arrangements are 0.5 kg s^{-1} and 1.0 kg s^{-1}, respectively, and the calculated pressure drops are 92.6 kPa and 166.9 kPa under the nominal operation condition. The dependence of the pressure drop at 17 K and 1.0 MPa are also shown in figure 6.11. As mentioned in section 6.5.1, the pressure drops at the required flow rates are lower than the hydrogen pump performance.

Figure 6.27 shows pressure drops over the main CMS loop at temperatures of 300 K and the pump performance curves operated by two pumps in series. At ambient temperature and pressure of 1.0 MPa, the maximum pump head is 4.1 kPa at the maximum speed of 14,000 rpm. The cracking pressure of the check valves occupies 52% of the total pressure drop. For the two-moderator arrangement, the pressure drop curve at the pressure of 1.0 MPa exists in the surging region. If the pressure is increased to 1.3 MPa, the pump performance is improved and the pump with the speed of more than 13,000 rpm can circulate ambient hydrogen gas over the CMS loop. For the four-moderator arrangement, the pump speed of more than 10,000 rpm can circulate the hydrogen gas, even if the pressure is even 1.0 MPa. Figure 6.28 shows pressure drops over the main CMS loop at temperatures of 100 K. For lower temperatures, the pump performance is improved. Even for the two-moderator arrangement, the pump can be operated stably for speeds of more than 8000 rpm.

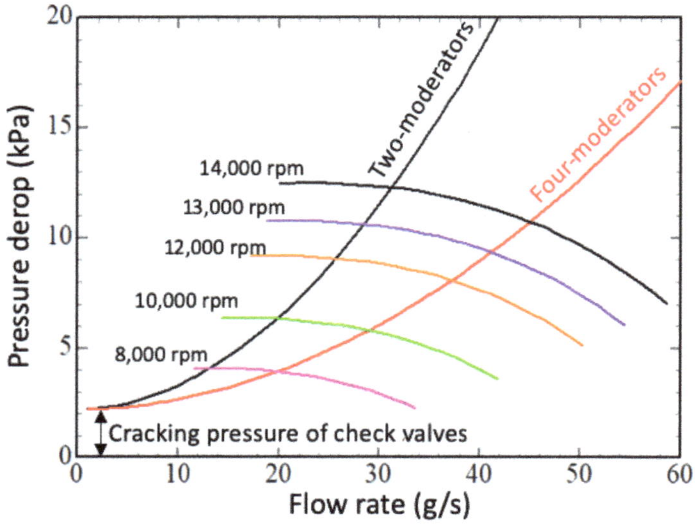

Figure 6.28. Pressure drop over the main liquid hydrogen circuit at 100 K and 1.0 MPa.

6.6.1.2 Pressure drop over the OP converter bypass line

The required flow rate through the OP converter is 38.5 g s^{-1}, as mentioned in section 6.6.2. The pressure drop through the catalyst bed is calculated by a well-known Ergun equation [29].

$$-\frac{\Delta P}{L} = 150\frac{(1-\varepsilon_p)^2}{{\varepsilon_p}^3}\frac{\mu}{{d_p}^2}U + 1.75\frac{(1-\varepsilon_p)}{{\varepsilon_p}^3}\frac{\rho}{d_p}U^2 \tag{6.12}$$

where L is the catalyst bet length, ε_p is the porosity, d_p is the particle diameter, μ is the fluid viscosity, and U is the superficial fluid velocity.

Figure 6.29 shows the pressure drop through the OP converter at 17.3, 100, and 300 K and at the pressure of 1.0 MPa. The pressure drop is 0.38 kPa at the required flow rate of 38.5 g s^{-1} at the nominal condition and is much less than the pressure drop over the main loop.

The *Cv* values of CV-62033 and two check valves at the inlet and outlet of the catalyst vessel are 1.2 and 16, respectively. Figure 6.30 shows the pressure drops over the bypass line at the nominal condition (17.3 K and 10 bar). In comparison, the pressure drops of the main loop from point A to B shown in figure 6.13, which correspond to that of the OP converter bypass line, are also displayed in this figure. The pressure drop via the bypass line at 38.5 g s^{-1} for CV-62033 = 100% is much lower than that of the main loop. For the two-and four-moderator arrangements, CV-62033 positions are 81% and 76%, respectively, so that liquid hydrogen with a flow rate of 38.5 g s^{-1} passes through the catalyst.

6.6.1.3 Pressure drop over the OPMS sampling line

There is a manual control valve (*Kv* = =33) and a check valve (*Kv* = 100) in the process line, which are arranged in parallel with the OPMS sampling line, as shown

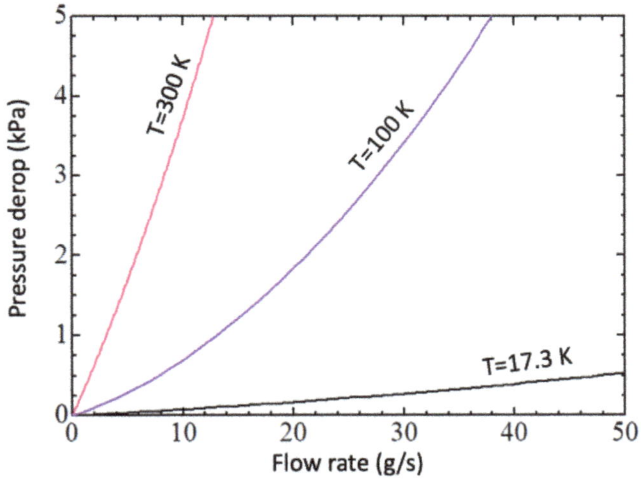

Figure 6.29. Pressure drop over the OP converter.

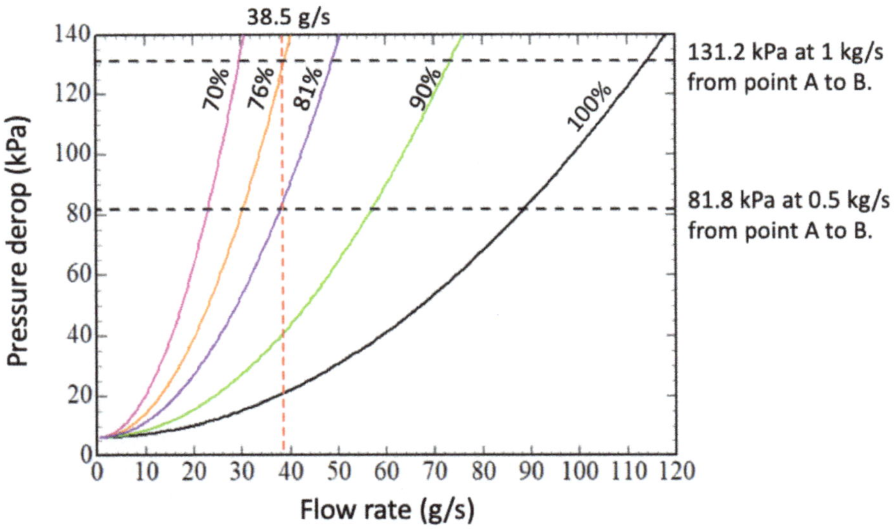

Figure 6.30. Pressure drops over the bypass line for various CV-62033 positions at the nominal condition (17.3 K and 10 bar).

in figure 6.17. The calculated pressure drops of the process line are summarized in table 6.8. The pressure drops for two and four-moderator configurations are estimated to be 3.84 kPa and 6.45 kPa, respectively, for the 95% position of the manual control valves. The designed pressure drop of the sampling line is set to be 3.84 kPa. The dedicated sampling line has an outer diameter of 12 mm and a thickness of 1.0 mm. The nominal liquid hydrogen flow rates range from 1 to 2 g s^{-1}. Table 6.9 displays the calculated pressure drop over one sampling line at a flow rate of 2.0 g s^{-1}. The pressure drop through the sapphire window section is calculated to

Table 6.8. Calculated pressure drops of the process line arranged in parallel with the OPMS sampling line. Reproduced from [19].

Component in the process line	Two-moderator case		Four-moderator case	
	$\dot{m}$ (kg s^{-1})	ΔP (kPa)	$\dot{m}$ (kg s^{-1})	ΔP (kPa)
Manual control valve (Kv = 33)at 95% open	0.25	1.63	0.25	1.63
DN32 pipe (L = 4 m)		1.33		1.33
Check valve (Kv = 100)	0.50	0.45	1.00	1.79
DN50 pipe (L = 4 m)		0.43		1.70
Total		3.84		6.45

Table 6.9. Calculated pressure drop over the OPMS sampling line. Reproduced from [19].

Component	ΔP (kPa) CV-S = 100% CV-R = 100%	ΔP (kPa) CV-S = 50% CV-R = 80%
1. Supply line from the DR to the sapphire window (L = 5.6 m)	0.09	0.09
2. Supply valve (CV-S), Kv = 1.1	0.06	5.91
3. Sapphire window section	0.02	0.02
4. Return line from the sapphire window to the MR (L = 5.0 m)	0.08	0.08
5. Return valve (CV-R), Kv = 1.1	0.06	0.37
Total	0.31	6.51

be 15.4 Pa by FLUENT, as mentioned in section 6.5.5.2. The pressure drop is calculated to be much lower than the design value in the case where both of the control valves are fully opened. At the nominal operation, the CV-S and CV-R positions need to be adjusted, e.g., the CV-R and CV-S should be set to 80% and 50% at the flow rate of 2 g s^{-1}, respectively, for the four-moderator configuration.

6.6.1.4 Pressure drop over the feed line from the hydrogen filling station

The maximum flow rate is estimated to be 120 Nm^3 h^{-1}, where the hydrogen is condensed during the cool-down process. The middle pressure header in the hydrogen filling station has a length of 3 m and the size of DN15. There is a pneumatic shut off valve (Cv = 1.0) and two isolation hand valves (Cv = 3.1) on the feed hydrogen line (DN25) with a length of 125 m. The pressure drop is calculated to be 0.6 bar.

6.6.1.5 Pressure drop over the HVL

When the safety relief valve works or the CMS warm-up operation starts, liquid hydrogen should be released to the HVL. A pressure drop of the HVL has been

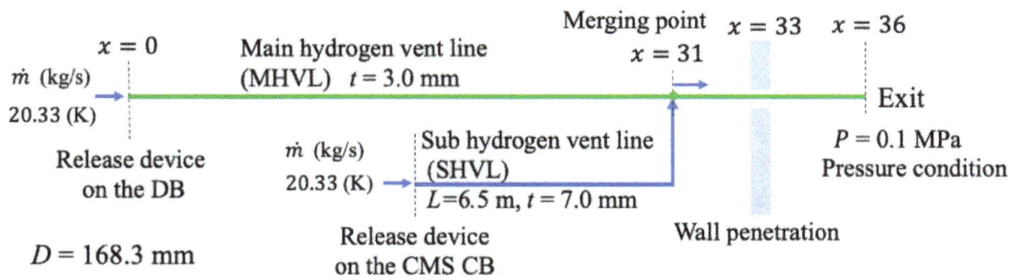

Figure 6.31. HVL pressure drop simulation model. Reproduced from [27]. © IOP Publishing Ltd. CC BY 3.0.

estimated, considering a transient process of releasing hydrogen. A one-dimensional process simulation code [27] has been developed based on the J-PARC CMS cool-down process simulation code [30]. It is proved that the average hydrogen and wall temperatures calculated by the developed code agree with those by FLUENT, i.e., within 11%. The details have been reported in [27]. The simulation code procedures will be briefly explained below.

A one-dimensional horizontal straight pipe with an outer diameter of 168.3 mm is considered as the analytical model, as shown in figure 6.31. The main hydrogen vent line (MHVL) has a thickness of 3.0 mm and a total length of 36 m. The sub-HVL from the CMS CBx has a thickness of 7.0 mm and a length of 6.5 m, and is connected to the MHVL at 6.5 m away from the CMS CBx.

Transient transport is calculated using the following enthalpy equation. The viscosity dissipation term and pressure-volume work term are ignored in the analysis,

$$\frac{\partial(\rho h)}{\partial t} = -\frac{\partial(\rho u h)}{\partial x} + \frac{\partial}{\partial x}\left(\lambda\frac{\partial T}{\partial x}\right) + Q \tag{6.13}$$

where h is the enthalpy, u is the flow velocity, λ is the thermal conductivity, Q is the energy source, and T is the temperature.

Non-boiling heat transfer in forced-flow liquid hydrogen can be also expressed well by the Dittus–Boelter correlation [31],

$$Nu = 0.023\,\mathrm{Re}^{0.8}\,\mathrm{Pr}^{0.4} \tag{6.14}$$

where Nu is the Nusselt number, Re is the Reynolds number, and Pr is the Prandtl number.

The two-phase film boiling forced convection heat transfer is considered in this simulation code. A correlation developed by Giarratano and Smith [32] that utilizes the Martinelli parameter, χ_{tt}, is applied,

$$Nu_{FB} = \exp\left(0.22 + 0.16 - 0.008(\ln \chi_{tt})^2\right)Nu \tag{6.15}$$

$$Nu = 0.026\,\mathrm{Re}^{0.6}\,\mathrm{Pr}^{0.33}\left(\frac{\mu_v}{\mu_w}\right)^{0.14} \tag{6.16}$$

$$\chi_{tt} = \left(\frac{1 - X}{X}\right)^{0.9}\left(\frac{\rho_f}{\rho_l}\right)^{0.5}\left(\frac{\mu_l}{\mu_f}\right)^{0.1} \tag{6.17}$$

where Nu_{FB} is the Nusselt number for two-phase flow, μ is the viscosity, and X is the quality. The subscripts of w, l and f denote the heated wall, liquid, and gas, respectively.

The Nusselt number for natural convection from isothermal surface of a long horizontal cylinder is a function of the Grashof, Gr, and Prandtl numbers [33],

$$Nu = C(GrPr)^a \tag{6.18}$$

where $C = 0.53$ and $a = 0.25$ are used for the laminar regime, and $C = 0.14$ and $a = 0.33$ are used for the turbulent regime.

For the single-phase pressure drop calculation, the Colebrook–White equation, equations(6.10) and (6.11), is used. For the two-phase flow pressure drop, the following homogeneous flow model [34] is applied where the pressure is defined in terms of the average mixture velocity,

$$\frac{\Delta P}{\Delta x} = 2\frac{fG^2}{D\rho_{TP}} \tag{6.19}$$

$$f = 0.08\,\mathrm{Re}^{-0.25} \tag{6.20}$$

$$\rho_{TP} = (1 - \alpha)\rho_l + \alpha\rho_g \tag{6.21}$$

$$\alpha = \left(1 + \frac{1 - X}{X}\frac{\rho_g}{\rho_l}\right)^{-1} \tag{6.22}$$

where G is the mass flow flux and α is the void fraction.

Each grid size is 0.1 m. The enthalpy equation is solved by the finite volume method. The convection term is discretized by applying a central differencing scheme. Time integration is explicitly performed with a time step of 0.05 ms. The inlet temperature is set to 20.23 K because the released hydrogen temperature decreases down to saturated temperature at 0.1 MPa due to isenthalpic expansion. Initially, the HVL is maintained at a temperature of 300 K.

Figure 6.32 shows the pressure drops when liquid hydrogen is released due to a hydrogen leak into the vacuum space. Even if a severe vacuum loss failure happens, the pressure drops can be always kept below the design pressure.

6.6.2 Heat load

All the cryogenic equipment is insulated by high vacuum and is partially covered with 20 layers of MLI, where the thermal radiation heat load between 300 and 20 K is considered to be 1.3 W m^{-2} on the cold surface. However, the moderator vessels and their pipes in the moderator-reflector plug (MRP) is not covered with MLI

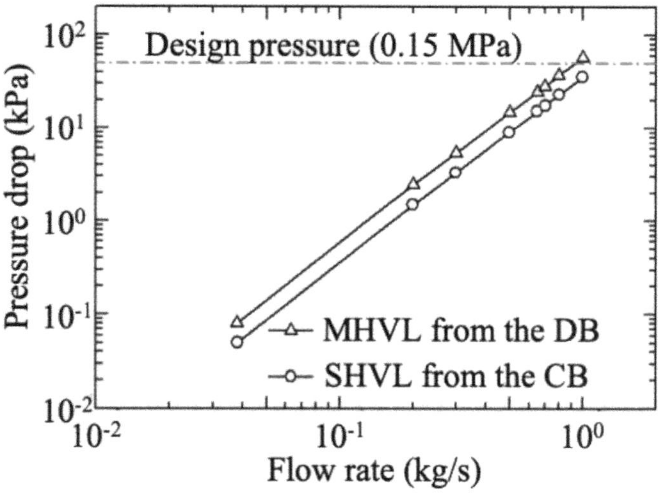

Figure 6.32. Pressure drop through the HVL. Reproduced from [27].

because of a high radiation field. The heat load is calculated using the well-known radiation heat transfer correlation with the Boltzmann constant [35]. The heat load of a pump, Q_p, is calculated using the pump head, ΔP, in table 6.6 as follows,

$$Q_p = \frac{\dot{V}\Delta P}{\eta_p} \tag{6.23}$$

where $\dot{V}$ is the volumetric flow rate and η_p is the isentropic efficiency, which is set to be the design value of 0.72.

Table 6.10 updates the heat loads for the two- and four-moderator arrangements. The heat loads of the pumps account for around 50% of all the static heat load in both cases. For the 5 MW proton beam, the total heat loads are 8.6 and 21.9 kW, respectively, which are 27% lower than the maximum TMCP cooling power of 30.3 kW.

The OPMS sampling pipe penetrates the vacuum break between the OPMS and DB. The heat conduction through the valves and the pipe for the release devices and pressure transmitter is estimated. Table 6.11 shows the calculation results of the heat load for the sampling line. The LH_2 temperature increases by 0.19 K at the sapphire window section at the flow rate of 2 g s^{-1}. It is numerically confirmed that the heat load line meets the design condition.

6.6.3 Pressure and temperature fluctuation caused by the proton beam being turned on or off

6.6.3.1 J-PARC CMS pressure control system

At the J-PARC, cryogenic hydrogen at a supercritical pressure of 1.5 MPa with a temperature of 18 K is circulated at a flow rate of 0.19 kg s^{-1} by the two pumps, as shown in figure 6.7. The nuclear heating generated in the three moderators is

Table 6.10. CMS heat load. Reproduced from [15]. © IOP Publishing Ltd. CC BY 3.0.

Component	Two moderator (kW)	Four moderator (kW)
Two liquid hydrogen pumps	0.935	2.909
PCB, HXs, and OP converter	0.037	0.037
Piping, valves, and auxiliary pipes in the CBx	0.015	0.015
Feed and return HTLs with valves	0.126	0.126
Feed and return HDTLs, valves, and auxiliary pipe	0.118	0.231
Moderator vessels and their pipes without MLI	0.227	0.454
Spacers in the MRP	0.421	0.850
Total static heat load	1.879	4.622
Dynamic heat load: Nuclear heating at moderators	6.696	17.248
Total static + dynamic heat load	8.575	21.870

Table 6.11. Calculated heat load of one sampling line. Reproduced from [19]. © IOP Publishing Ltd. CC BY 3.0.

Component	Supply line to the sapphire window (W)	Return line from the sapphire window (W)
Supply pipe, vacuum break, and supports	1.97	1.95
Control valve	0.51	0.17
Sapphire window support	0.81	
Total	3.33	2.12

estimated to be 3.8 kW for a 1 MW proton beam operation [10]. When the beam is turned on, a kW-order heat load is suddenly applied to the hydrogen loop filled with incompressible fluid and the warmed hydrogen propagates at the flow rate. The density change (expansion) brings about an enormous pressure rise in the loop. The J-PARC CMS prepared the accumulator for a volume control and the heater for a thermal compensation to mitigate the pressure fluctuation caused below an allowable pressure of 0.1 MPa. The accumulator has a bellows filled with helium, which behaves as compressible fluid. The pressure fluctuation is absorbed by its spontaneous expansion and contraction, whose maximum volume change is 6.84 l [36, 37]. The functions of the heater are to minimize the volume affected by the temperature rise caused by the nuclear heating and to maintain the same heat load from the CMS to the helium refrigerator. Figures 6.33 and 6.34 show the dynamic behaviors of the J-PARC CMS loop for a 285 kW and 532 kW proton beam operations, respectively [10]. The pressure increases as soon as the proton beam is turned on. The temperature at the inlet of the heater (T07) starts to increases 20 s after the 286 kW proton beam's injection and is eventually increased by 0.65. The heater's power is reduced at a stretch by that corresponding to the nuclear heating

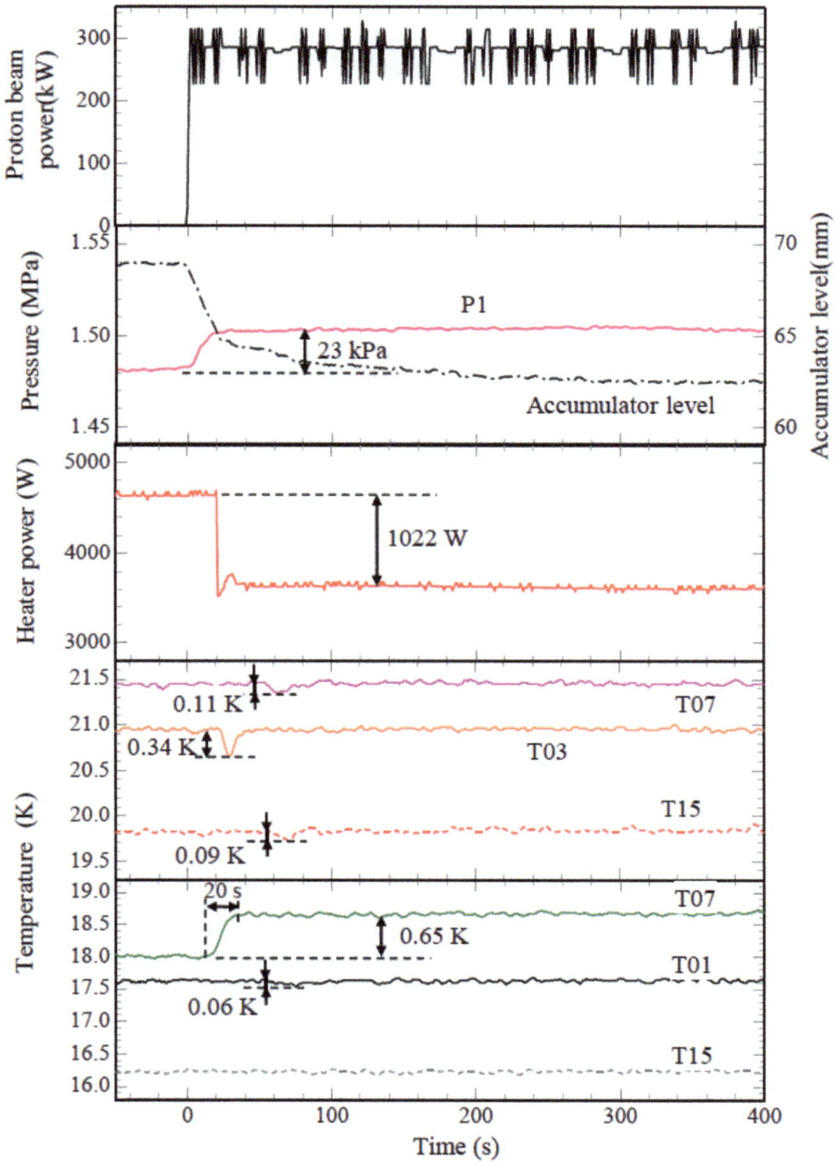

Figure 6.33. Dynamic behaviors of the J-PARC CMS loop for a 285 kW proton beam operation. Reprinted from [10], with the permission of AIP Publishing.

and is then adjusted by a PID control to maintain the outlet temperature of the heater (T03) at 20.95 K. It can be seen from this figure that the pressure is continuously increasing until the change of the temperature distribution has settled down and the pressure rise is mitigated by the contraction of the accumulator. The feed hydrogen temperature (T01) can also be kept at 17.6 K within 0.1 K. Tatsumoto *et al* [37] analyzed the pressure rise using a simplified mode based on

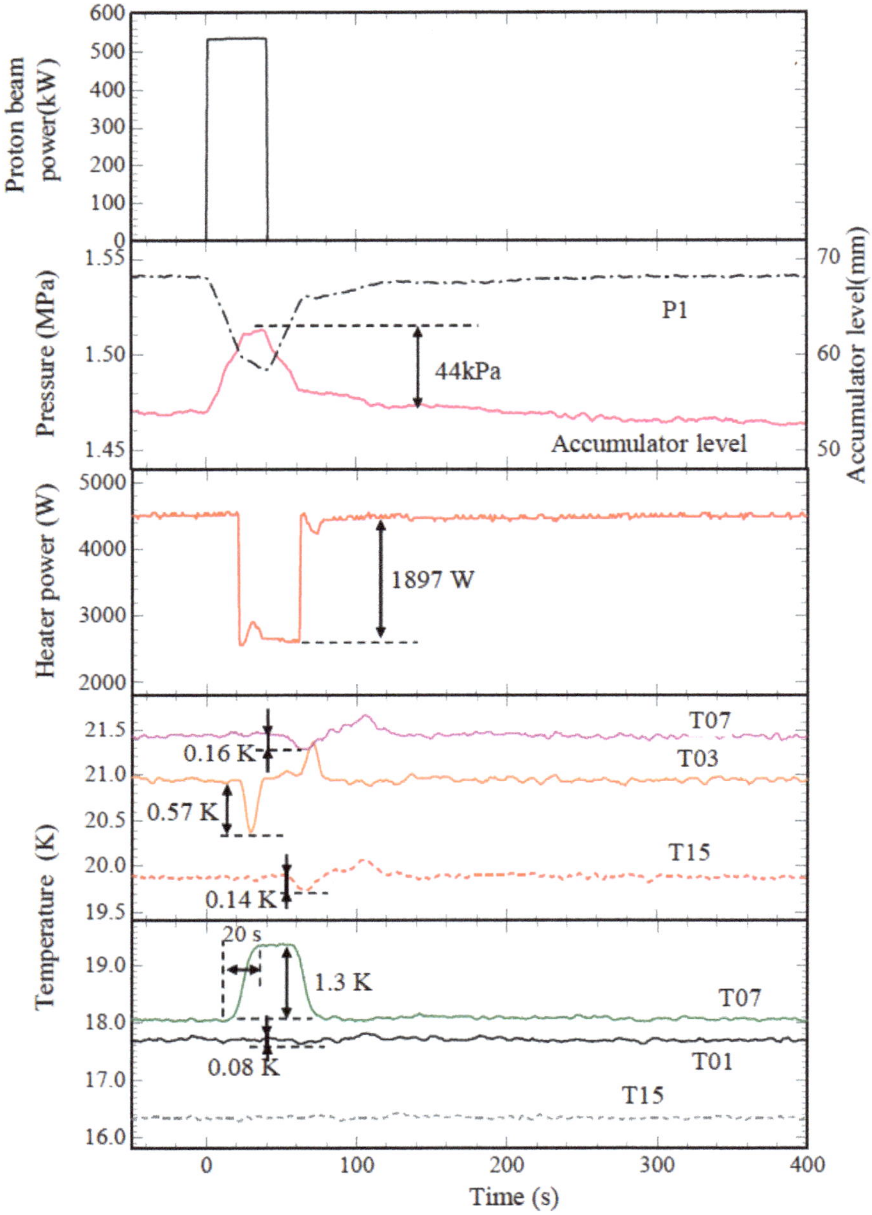

Figure 6.34. Dynamic behaviors of the J-PARC CMS loop for a 532 kW proton beam operation. Reprinted from [10], with the permission of AIP Publishing.

mass conservation. If it were not for the accumulator and heater, the pressure would increase by 1.9 MPa for the 1 MW proton beam operation. The predicted pressure rises were in a good agreement with the measured values. The accumulator and heater would make it possible to mitigate the pressure rise below 100 kPa for the 1 MW proton beam operation (figure 6.35).

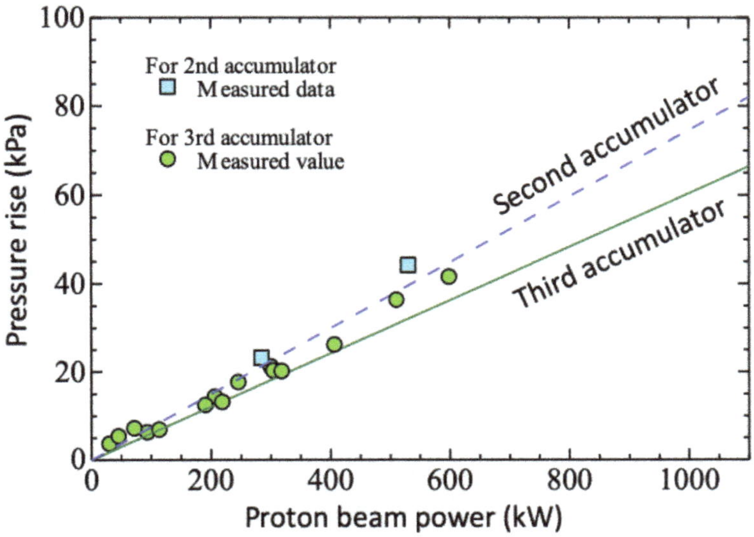

Figure 6.35. Effect of the proton beam power on the pressure rise. Reprinted [37], Copyright (2015), with permission from Elsevier.

6.6.3.2 ESS CMS pressure rise analysis caused by the beam injection

Similar to the J-PARC, the ESS CMS also forms a closed loop filled with liquid hydrogen where the total hydrogen mass is conserved. A slight density change would bring about an enormous pressure rise. Unlike the J-PARC and SNS, the ESS CMS has selected subcooled liquid hydrogen that has a clear vapor–liquid interface. The vapor phase of the PCB tank will work to absorb the pressure fluctuation. The thermal compensate will be implemented by adjusting the feed helium flow rate from the helium refrigerator to the CMS like the SNS CMS.

The pressure rise of the ESS CMS is analyzed, applying the same method as the J-PARC CMS mentioned above [36, 37]. The CMS loop can be divided into four sections based on the temperature distribution, as shown in figure 6.36 [25]: the supply line from the HX-1 to the moderators ($V_1 = 209.1$ l), the return line ($V_2 = 140.8$ l), liquid phase in the PCB (V_3), and vapor phase in the PCB (V_4). The total PCB volume ($= V_3 + V_4$) is 68.9 l. The average temperatures of the supply and return lines are 17.4 K and 17.6 K, respectively, while the proton beams are off. The temperature rise at the moderators is estimated to be 1.56 K at the mass flow of 0.5 kg s^{-1} for the heat load of 6.7 kW. In this analysis, the heat input applied at the moderators is assumed to be completely removed by the HX-1 The average vapor temperatures calculated in the next section are used.

6.6.3.3 Steady-state PCB operational condition

The PCB has the function not only to mitigate the transient pressure fluctuation when the proton beam is on or off but also to adjust the operational pressure to a set point. In this section, the steady-state PCB operational condition is studied and optimized. The details have been reported in [25]. As shown in figure 6.14, there are

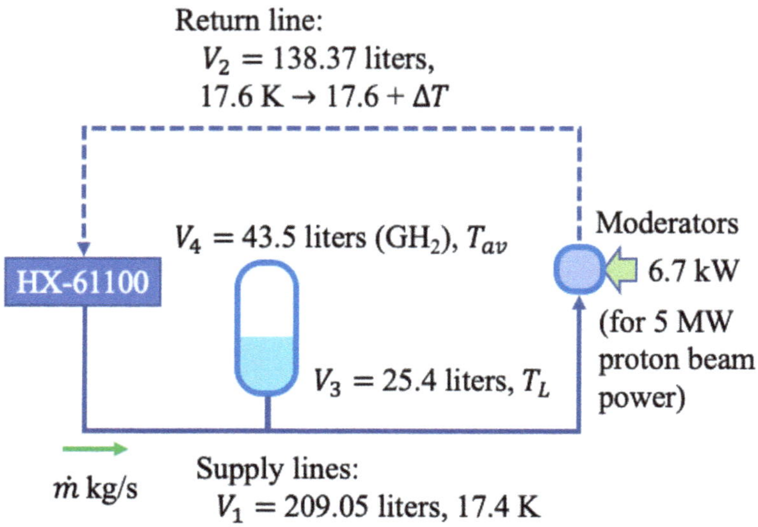

Figure 6.36. Simplified model to estimate the CMS pressure rise by nuclear heating. Reprinted/adapted by permission from Zhejiang University Press: [25]

four heaters wrapped on the PCB between each temperature sensor in its own sheathed tube that work as a level sensor. The evaporated gas is generated by the heater and produces internal pressure. The excessive gas is released by CV-62029 to the suction of the hydrogen pump via the condensation heat exchanger (HX-61200) to maintain the pressure at a set point of 1.1 MPa. The CV-62029 position should be controlled around 10 to 20% by the PID control. Meanwhile, the subcooled LH_2 with the temperature of 17.3 K that corresponds to the evaporated flow rate would flow into the PCB. Therefore, the liquid level should be always kept constant. The average vapor temperature, T_{av}, should be affected by the evaporated flow rate and the radiation heat load. The effects of the change of T_{av} on the pressure change at the pressure of 1.1 MPa are analyzed using the simplified model mentioned above. It can be seen from figure 6.37 that the pressure change is strongly affected by the change of T_{av}. The tendency gets smaller for higher T_{av}. This means there is less flow rate to vaporize the liquid hydrogen by the heater.

The average vapor temperature, T_{av}, is calculated by a CFD analysis, using ANSYS FLUENT, when a heat input is applied to the PCB by the heater. The analysis model of the vapor phase is shown in figure 6.38. An inlet boundary condition is applied to the liquid surface where the LH_2 temperature is kept at T_{sat} (= 31.9 K) at a saturated pressure of 1.1 MPa. The evaporated flow rates are calculated using the latent heat of 1.96×10^5 J kg^{-1} and the heater power, Q_w. A pressure boundary condition is applied to the top of the PCB. A wall condition is applied to the PCB wall where a radiation heat load is given because of the absence of a super insulation on the outside of the PCB above the expected liquid surface. As shown in figure 6.39, For $Q_w < 150$ W, the average vapor temperature is strongly affected by the evaporated flow rate. With an increase in Q_w, it approaches T_{sat}. The CV-62029 positions for each Q_w are also calculated under the pressure drop of 60 kPa. For $Q_w >$

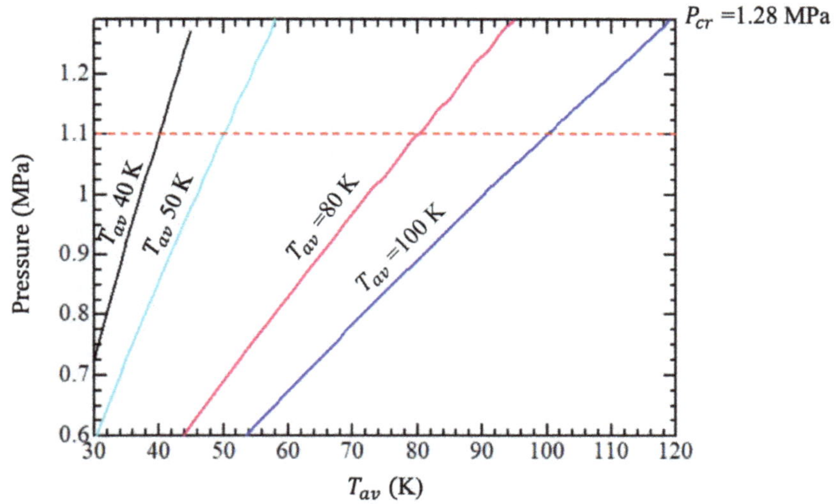

Figure 6.37. Effect of T_{av} on the pressure change at the nominal condition. Reprinted/adapted by permission from Zhejiang University Press: [25]

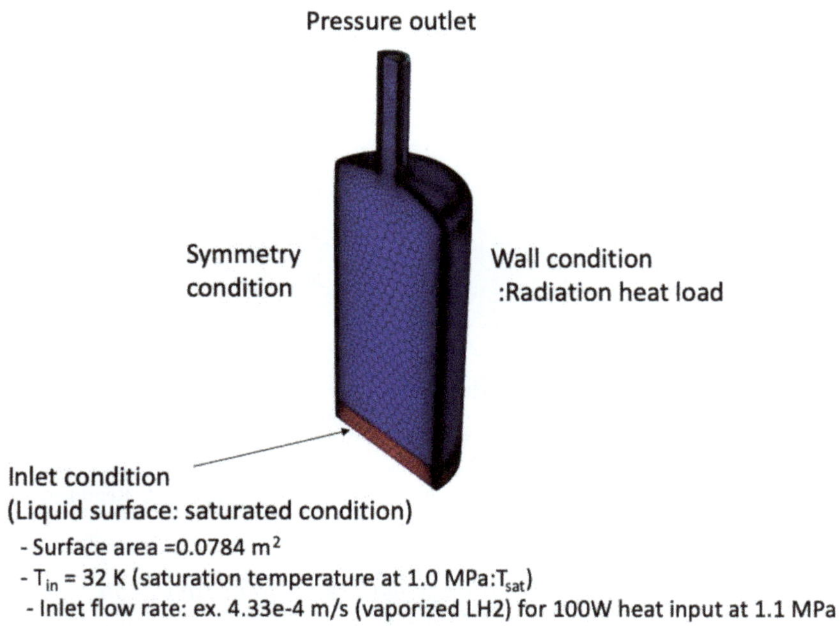

Figure 6.38. CFD analysis model to estimate the average vapor temperature, T_{av}. Reprinted/adapted by permission from Zhejiang University Press: [25]

50 W, which corresponds to the evaporated flow rate, m_{ev}, of 0.33 g s^{-1}, CV-62029 can be operated at around the position of more than 10%. Meanwhile the subcooled LH_2 with a temperature of 17 K would be refilled at the same flow rate of m_{ev} and the liquid level would be kept at constant.

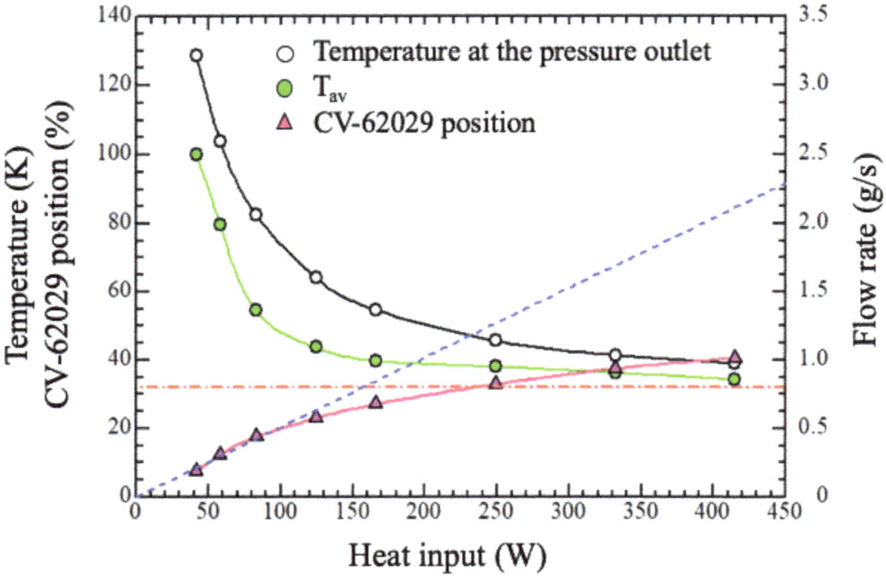

Figure 6.39. Calculated average vapor temperature, T_{av}, affected by the evaporation flow rate. Reprinted/adapted by permission from Zhejiang University Press: [25]

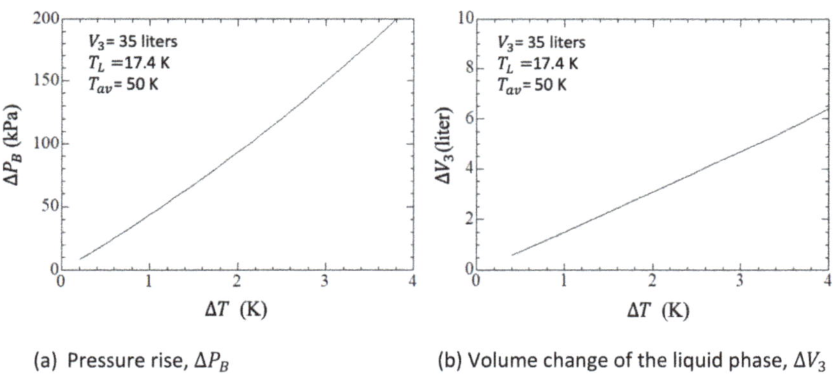

Figure 6.40. Effects of the return temperature fluctuation, ΔT, on the pressure rise, ΔP_B and the volume change of the liquid phase in the PCB, ΔV_3, for the two-moderator arrangement. Reprinted/adapted by permission from Zhejiang University Press: [25]

6.6.3.4 Pressure fluctuation mitigation when the proton beam is on

The effects of the return temperature fluctuation, ΔT, on the pressure rise, ΔP_B and the volume change of the liquid phase in the PCB, ΔV_3, e.g., for $V_3 = 35$ l, $T_L = 17.4$ K and $T_{av} = 50$ K for the two-moderator configuration are shown in figure 6.40. The pressure rise and the volume change are almost proportionally to ΔT. This is effective in increasing the flow rate to mitigate the pressure fluctuation.

Figure 6.41 shows the effect of the liquid volume of the PCB, V_3, and its temperature, T_L, on ΔP_B and ΔV_3 for a 5 MW proton beam operation. The average vapor temperature, T_{av}, is 50 K which corresponds to $Q_w = 58$ W, as shown

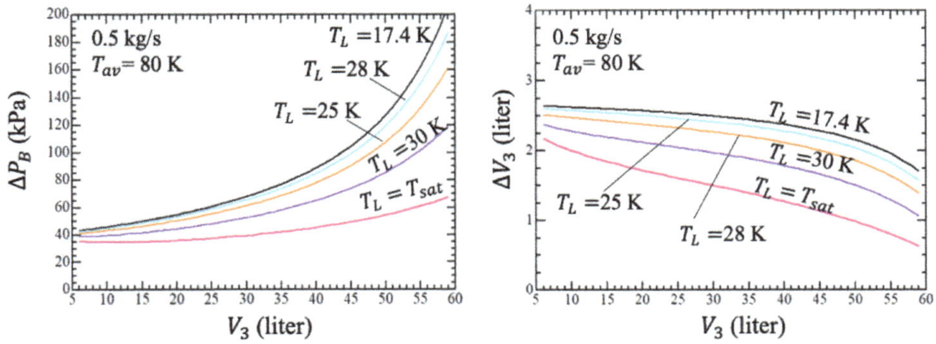

(a) Pressure rise, ΔP_B (b) Volume change of the liquid phase, ΔV_3

Figure 6.41. Effects of the liquid phase volume in the PCB, V_3, on the pressure rise, and the volume change of the liquid phase in the PCB for the two-moderator arrangement. Reprinted/adapted by permission from Zhejiang University Press: [25]

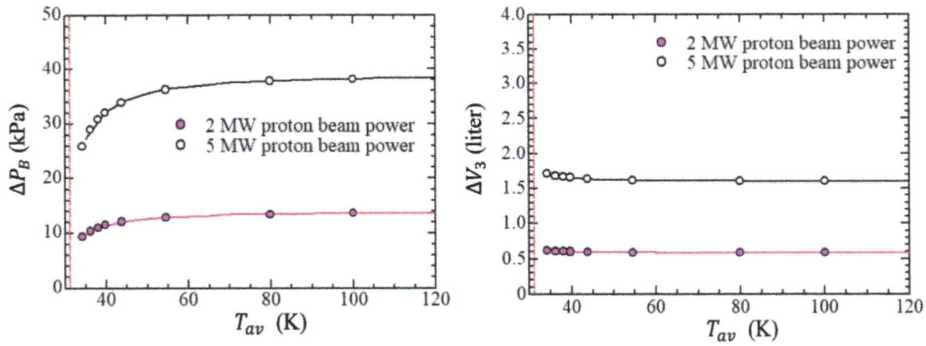

(a) Pressure rise, ΔP_B (b) Volume change of the liquid phase, ΔV_3

Figure 6.42. Effect of T_{av} on ΔP_B and ΔV_3 for $V_3 = 25$ l, $T_L = T_{sat}$ and rate $\dot{m}= 0.5$ kg s^{-1}. Reprinted/adapted by permission from Zhejiang University Press: [25]

in figure 6.39. The return temperature is increased by 1.6 K for the flow rate $\dot{m}$ of 0.5 kg s^{-1}. The pressure rise caused by the nuclear heating can be mitigated for larger liquid volume of V_3. In other words, it is effective in mitigating the pressure fluctuation to reduce the ratio of the volume with the temperature rise to that without the temperature change. Furthermore, the liquid phase temperature in the PCB should be maintained at the saturated temperature, T_{sat}. Based on the analysis results, the volume of the liquid phase was determined to be 25 l and the liquid temperature in the PCB is set to be T_{sat}.

Figure 6.42 shows the effect of T_{av} on ΔP_B and ΔV_3 for $V_3 = 25$ l, $T_L = T_{sat}$ and rate $\dot{m} = 0.5$ kg s^{-1}. It seems that the average vapor temperature has little influence on ΔP_B and ΔV_3 for $T_{av} > 55$K, although the pressure rise steeply gets smaller for $T_{av} < 55$K. There seems to be advantages of setting T_{av} to more than 55 K in light of reducing the required heater power to produce the evaporated GH_2. In order to

Table 6.12. PCB optimum operational conditions. Reprinted by permission from Zhejiang University Press: [25]

Q_W(W)	Q_L(W)	m_{ev} (g s^{-1})	T_{av} (K)	V_3 (l)	T_L (K)
60	64.9	0.31	80	25	T_{sat}

minimize Q_W, the vapor average temperature is determined to be 80 K where $Q_W = 60$ W and $m_{ev} = 0.31$ g s^{-1} in light of achieving the stable CV-62029 PID control. The position of CV-62029 is 12.5%, as shown in figure 6.39. The volume change of the liquid phase in the PCB increases by 1.6 l and the pressure rise can be mitigated to 38 kPa for the 5 MW proton beam power. The subcooled liquid hydrogen should be refilled at the flow rate of 0.31 g s^{-1} and has to be heated up to T_{sat} by a heater located around the bottom. The heater power, Q_L, of 64.9 W is required because the sensible heat is 2.12×10^5 J kg^{-1}.

After the proton beam is turned on, the liquid level is increased at 1.03 mm s^{-1} and the pressure also increases at 1.87 kPa s^{-1}. The pressure is eventually increased by 37.4 kPa. The PCB operational conditions have been optimized based on the analytical results and are summarized in table 6.12 [25].

The pressure fluctuation would be mitigated by the PCB under the conditions displayed in table 6.10 and it is important to avoid any feed hydrogen temperature fluctuation downstream of the HX-61100 such as the assumption of the pressure fluctuation. It is, therefore, necessary to study the method of the thermal compensation via the HX-61100 without any change of the feed hydrogen temperature by adjusting the feed helium flow rate from the TMCP when the proton beam is injected or stopped.

6.7 Safety

The CMS shall follow the Pressure Equipment Directive (PED) 2014/68/EU for pressure equipment. All commercial parts are specified and certified for PN25, which is above the design pressure of the CMS.

6.7.1 PRDs of the process line

PRDs shall be provided in hydrogen piping systems to relieve pressures that can exceed the maximum allowable working pressure. PRDs shall be included in liquid hydrogen piping to protect sections of piping where liquid or cold vapor can become trapped between valves or closures.

Figure 6.43 shows the CMS flow diagram and the locations of PRDs. The most fragile components are the cold moderators in the Twister, which have to be protected from overpressure. They define the system's design pressure. Every PRD is connected to the vent line to release gas to the ambient outside the building. Furthermore, every vacuum pump set exhaust is connected to another vent line parallel to the HVL. The cold pipes and vessels contain liquid hydrogen with a great potential of overpressure. The CMS has five separate vacuum spaces, A to E independently.

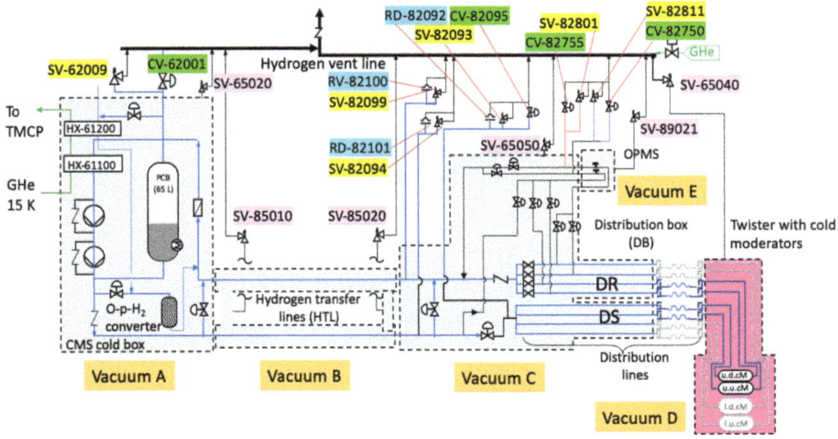

Figure 6.43. CMS flow diagram and the locations of PRDs. Reprinted from IPC Science and Technology Press, Guildford, Surrey, England, ©1978

Air ingress and its condensation on the cold surfaces usually results in significant heat loads of several kW m^{-2}, and would bring about a rapid and huge pressure rise of the process line. The sizes of the safety relief valves were determined under the condition of the loss of insulation vacuum failure mode of one vacuum space caused by air ingress according to the EN-ISO 21013-3 design code guidelines [17].

Additionally, redundant burst disks (RD-82101, RD-82100, and RD-82092) were prepared for every process volume that has to be protected. The set pressure $P_s = 17$ bara is the same as the design pressure. The diameters of these burst disks are calculated according to EN ISO 21013-3:2016.

Figure 6.44 shows common published values of 38 kW m^{-2} as maximum heat flux to equipment that is not covered with MLI [38]. Pipes and vessels that are protected by MLI will experience a maximum heat flux of about 6 kW m^{-2}, which is six times smaller than that for wrapping MLI. For every vacuum section, the surfaces have been calculated and multiplied with a constant maximum heat flux. This is a conservative approach because this heat transfer is a transient process and the temperature range of the CMS is higher than the data for a liquid helium system.

The set pressure, P_s, of the safety relief valve is 14.5 bar, which is higher than critical pressure of 12.9 bar. The discharge mass flow rate, W, is given using a total heat load, Q, and specific volume,ν, according to EN-ISO 21013-3 design code guidelines [17],

$$W = 3.6\left(\frac{Q}{L'}\right) \tag{6.24}$$

where

$$L' = \nu\left[\frac{\partial h}{\partial \nu}\right]_P \tag{6.25}$$

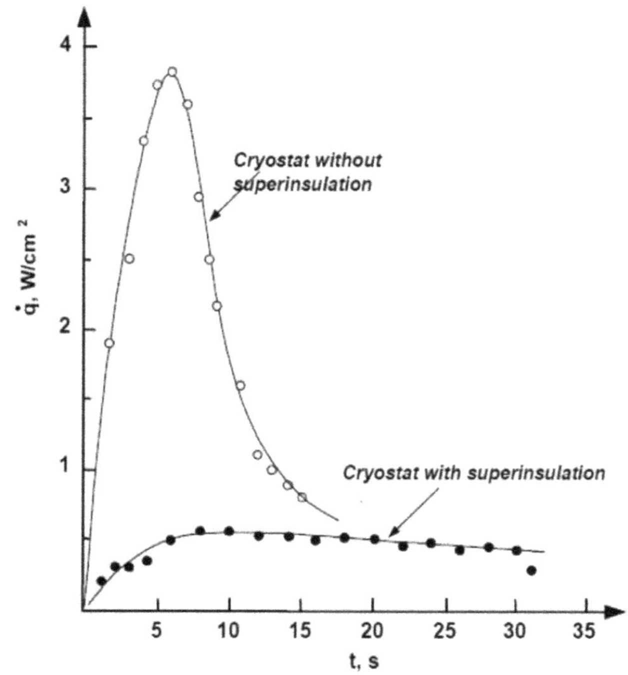

Figure 6.44. Transient heat flux for loss of insulation vacuum, air ingress, and condensation on equipment at liquid helium temperature of 4.5 K with or without MLI [38]. Reprinted from IPC Science and Technology Press, Guildford, Surrey, England, ©1978

L' (kJ kg^{-1}) is evaluated at the relieving pressure and temperature, T. The values of T and L' are determined to be associated with the maximum value found for ψ.

$$\psi = \frac{\sqrt{\nu}}{L'} \tag{6.26}$$

The calculated discharge mass flow rates and the total heat loads are summarized in table 6.13. The most severe design case for SV-62009, SV-82093, and SV-82099 is a vacuum loss to the Twister vacuum (vacuum D) accompanied with air ingress and condensation. Since no MLI is feasible for this region, the heat flux can be as high as 148 kW. For the calculation of SV-82094, the vacuum A failure case is selected because it is worse than the vacuum B failure case.

The minimum required PRD flow area is given by the following equation:

$$A_V = \frac{W}{0.2993 C K_{dr} \sqrt{\frac{P_i}{\nu_i}}} \tag{6.27}$$

where A_V is the PRD flow area (mm^2), K_{dr} is the PRD coefficient of discharge, P_i is the inlet pressure of the PRD, ν_i is the specific volume at the inlet of the PRD, and

$$C = 3.948\sqrt{\gamma\left(\frac{2}{\gamma+1}\right)^{\frac{(\gamma+1)}{(\gamma-1)}}} \tag{6.28}$$

Table 6.13. Calculated discharge mass flow rates and the total heat loads for air ingress into the vacuum space.

Vacuum space	Location	Volume (m^3)	Cold surface area (m^2)	Q (kW)	W (kg h^{-1})
Vacuum A	CMS CBx	4.3	3.9 (0.6 without MLI)	47.0	648
Vacuum B	HTL (MS and MR)	0.82	7.1	42.6	621
Vacuum C	DB, DS, and DR	5.9	22	132	1069
Vacuum D	Moderators	0.18	0.0 (3.9 without MLI)	148	1076
Vacuum E	OPMS sampling	0.031	0.34/0.41	2.0/2.5	28/35

where γ is the ratio of the specific heat capacity at constant pressure to that at constant volume.

6.7.2 Failure analysis for liquid hydrogen leaking into a vacuum envelope

A liquid hydrogen leakage through a crack is considered as the most severe failure scenario in order to determine the adequate size of a PRD on the vacuum space (VSD; Vacuum Safety Device) so that it works as a safety barrier without any hydrogen leakage into a room. The details have been reported in [18].

The vacuum loss caused by the hydrogen leak results in applying a huge heat load to the process line. When the process pressure reaches a set pressure of a safety relief device, the hydrogen is released to air via the HVL. Meanwhile, the leaked hydrogen flows through the vacuum envelope and is released via the VSD, which is also connected to the HVL. The failure analysis was carried out, dividing into the following three steps:

(1) A pressure rise analysis in the process and vacuum envelope.
   - Liquid hydrogen leak through a crack causes a vacuum pressure rise and the process pressure rise by increasing the heat load [18].

(2) Analysis of transient temperature reduction and pressure drop along the vacuum envelope, using a modified version of the one-dimensional transient thermal transport code described in section 6.6.1.5 [27]
   - Leaked hydrogen moves toward the VSD and is released at a mass flow rate. In this analysis, the release flow rate is calculated under the condition that the vacuum pressure is kept at the set pressure of 0.15 MPa.

(3) Analysis of transient temperature reduction and pressure drop along the HVL, which has been already described in in section 6.6.1.5 [27].
   - As an inlet boundary condition of the release flow rate and the release hydrogen temperature given by the second step, the pressure drop and vent line wall temperature reduction are calculated.

6.7.2.1 Pressure rise analysis—step 1

6.7.2.1.1 Postulated maximum crack size

At the J-PARC, a pressure rise analysis was performed for a cryogenic hydrogen leaking into a vacuum envelope of a coaxial transfer line with five layers: hydrogen supply, vacuum layer, hydrogen return, vacuum layer, and helium barrier [39]. As a postulated maximum crack size, the crack size of $Dt/2$, which is the same as the failure case of a primary coolant pipe in a reactor according to the US Nuclear Regulatory Commission (NRC) [40], was assumed (D: the pipe diameter and t: the thickness).

In general, when a water pipe bursts due to frozen water, a crack propagates along the pipe's length. In this analysis, a rhombic shape crack is applied as shown in figure 6.45, considering the actual burst pipe shape [18]. Half the cross-sectional area of the process pipe is postulated as a maximum crack size. A pipe perimeter is used as the length of the rhombic shape crack. The width is calculated from the postulated area and the length. For the analysis, a hydraulic diameter, d_H, is used and is larger than that for the J-PARC.

Figure 6.46 shows an analysis model that estimates pressure changes caused by the hydrogen leaked to the vacuum envelope through a crack. The whole process volume is divided into two sections: a liquid phase (0.344 m^3) and a gaseous phase (0.05 m^3), which correspond to the volume of gaseous phase in the PCB tank. The flow diameter of the pressure relief valve (PRV) was selected according to EN ISO 21013-3:2006, as mentioned in table 6.14. The required discharge mass flow rate through the VSD is estimated to avoid exceeding the design pressure of 1.5 bara. The required size of the VSD is determined based on this analysis result (table 6.15).

The leaked flow rate through the crack is calculated by the equation of an orifice flow meter.

$$\dot{m} = \frac{C_d}{\sqrt{1-\beta^4}} \varepsilon \frac{\pi d_H}{4} \sqrt{2\Delta P \rho_1} \tag{6.29}$$

where $\dot{m}$ is the mass flow rate, ρ is the fluid density, C_d is the coefficient of discharge, β is the diameter ratio of orifice diameter to pipe diameter, d_H is the hydraulic

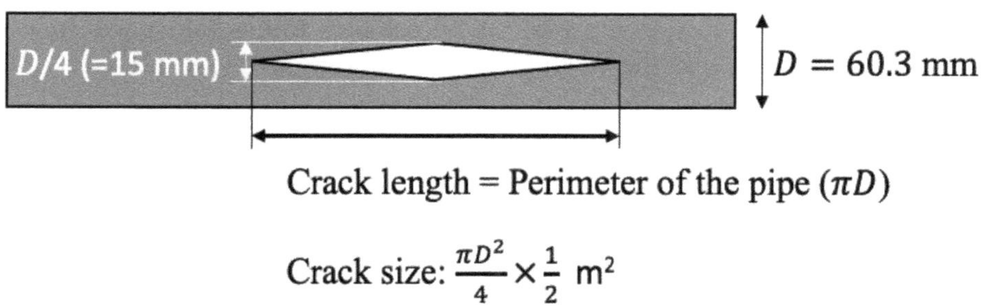

Figure 6.45. Postulated crack shape and size in this analysis. Reproduced from [18].

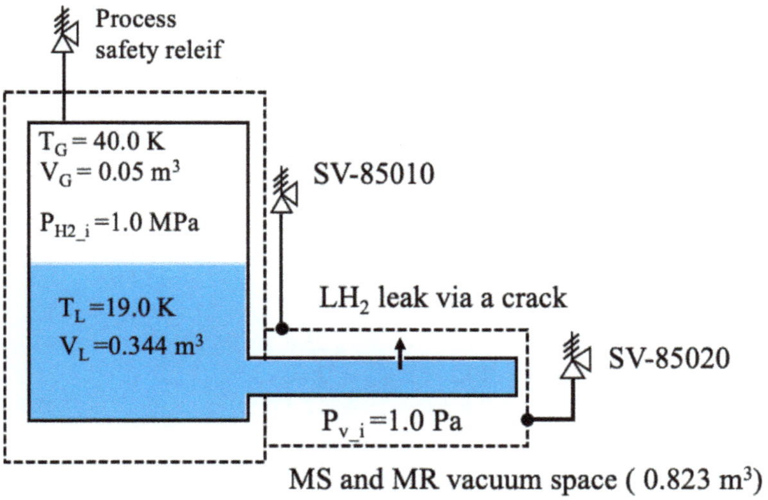

Figure 6.46. Analytical model for the pressure rise analysis. Reproduced from [18].

Table 6.14. PRD specifications.

Location	PRD	*Ps* (bar)	Protected section
CMS CBx	SV-62009	14.3	Full main loop, expect NV-62016 to CV-82008 and CV-82012 (if closed)
DB, DS,DR, and moderators	SV-82093	14.5	Cold moderators, if CV-82008 is closed
HTL (MS)	SV-82094	14.5	MS from NV-62016 to CV-82008 and CV-82012 (if closed)
HTL (MR)	SV-82099	14.5	Full main loop, expect NV-62016 to CV-82008 and CV-82012 (if closed)
OPMS	SV-82801	16.0	Sampling line for the upper moderators
OPMS	SV-82811	16.0	Sampling line for the future lower moderators

diameter which corresponds to the orifice diameter, and ΔP is the differential pressure. The subscript of 1 denotes the upstream part.

6.7.2.1.2 Analytical procedure

Detail of the analytical procedure is described in [18]. Initially, the liquid hydrogen in the process line is set at an average liquid temperature of 19 K and the operational pressure of 1.0 MPa. The temperature in the gaseous phase is set to 40 K. The pressure of the vacuum envelope is set to 10^{-4} Pa.

The leaked liquid hydrogen would expand uniformly across the entire vacuum space in an isenthalpic process. The leaked flow rate of liquid hydrogen is calculated. The heat flux of 6 kW m^{-2} is always applied to the process line as shown in

Table 6.15. Design choices for the pressure safety valves and the rupture disks.

PRD location	PRD	Min. calculated d_0 (mm)	Design choices d_0 (mm)	Set/burst pressure (barg)	Vacuum failure case
CMS CBx	SV-62009	20.1	23.0	13.3	D
DB and moderators	SV-82093	20.5	23.0	13.5	
	RD-82092	20.1	26.0	16.0	
HTL supply	SV-82094	12.8	15.0	13.5	A
	RD-82101	11.4	26.0	16.0	
DB and HTL return	SV-82099	19.5	23.0	13.5	D
	RD-82100	19.2	26.0	16.0	
OPMS (upper moderators)	SV-82801	2.6	12.5	15	E
OPMS (lower moderators)	SV-82811	2.9	12.5	15	

figure 6.44 [38], regardless of the vacuum pressure. All the heat load is applied to the liquid phase and is consumed in the temperature rise up to its saturated temperature. At saturated condition, it is consumed in the vaporization. If the pressure increases above the critical pressure, then the hydrogen is treated as a single phase. Averaged enthalpies and densities are obtained by a mass balance and energy balance. The temperature and the pressure are given by GASPAK [5] using the density and enthalpy as a parameter.

The leaked liquid hydrogen is evaporated on the vacuum envelope where film boiling is dominating. The film boiling heat flux of 1×10^5 W m^{-2} at a wall superheat of 300 K [41] is applied. The natural convection heat transfer between the vacuum envelope and surrounding air maintained at 300 K is calculated by equation (6.18). Once the vacuum pressure increases to 0.15 MPa, the hydrogen is released, maintaining the vacuum pressure at 0.15 MPa. Averaged enthalpy is calculated by the energy balance at the pressure of 0.15 MPa. The average temperature and densities are given by GASPAK using the average enthalpy and the pressure. The released flow rate is calculated by the mass balance. This calculation explicitly repeats with 1 ms time steps until the process hydrogen temperature increases up to 45 K, where 90% of hydrogen has been released.

6.7.2.2 Transient phenomena in the vacuum envelope—step 2

The details of the analysis procedures have been described in [18]. The heat transfer and pressure drop of the two-phase forced flow are considered. Single-phase forced-flow heat transfer is calculated by the Dittus–Boelter equation. The natural convection heat transfer from ambient air is considered by using equation (6.18). The effect of the condensation of moisture in air is ignored.

An enthalpy equation is solved by the finite volume method. The convection term is discretized by applying a central differencing scheme. Time integration is explicitly performed with a time step of 50 ms. The initial temperature of the vacuum envelope is set at 300 K. The inlet temperature is set to saturated temperature at 0.1 MPa due to the isenthalpic expansion process.

The pressure drop through the VSD is calculated by a convergent nozzle model. A VSD with a diameter of 100 mm is selected,

$$\dot{m} = A\sqrt{2\frac{\gamma}{\gamma-1}\rho_1 P_1\left(\left(\frac{P_2}{P_1}\right)^{\frac{2}{\gamma}} - \left(\frac{P_2}{P_1}\right)^{\frac{\gamma+1}{\gamma}}\right)} \tag{6.30}$$

Subscripts 1 and 2 donate vacuum space and backpressure region, respectively.

6.7.2.3 Transient phenomena in the HVL—step 3

The released hydrogen from the VSD close to the DB goes through the main HVL with an outer diameter of 168.3 mm and a thickness of 3.0 mm, meanwhile that from the VSD on the CMS CBx is merged to the main HVL through the sub-HVL with a length of 6.5 m and thickness of 7.0 mm.

The initial temperature and pressure are set to 300 K and 0.1 MPa over the entire HVL. The released flow rate and the temperature at the VSD, which are given by the second analysis, are applied as the inlet condition. Downstream of the merging point (at x= 31 m), the cold hydrogen passes through the HVL at the flow rate of $\dot{m}$.

The volume of a grid is $2.56 \times 10^{-3}\text{m}^3$. The MHVL and SHVL are divided into 288 and 52 grids, respectively. An enthalpy equation is solved by the finite volume method. The convection term is discretized by applying a central differencing scheme. Time integration is explicitly performed with a time step of 50 ms. The details of the analysis procedures have been described in section 6.6.1.5.

6.7.2.4 HTL failure

6.7.2.4.1 Analytical model

The details have been reported in [18]. The HTLs have a coaxial pipe structure with a length of 37. 6 m. The outer diameter of the process line, D, is 60.3 mm and its thickness, t, is 2.0 mm. The vacuum envelope has an outer diameter of 139.7 mm and a thickness of 2.0 mm. The vacuum envelopes of the supply and return HTLs (MS, MR) are connected at both ends to form one overall vacuum space, whose volume is 0.823 m^3, and are physically isolated from the CMS CBx and the DB as shown in figure 6.43. The pressure rise analysis was conducted assuming that the hydrogen leak occurs at the middle of the HTL.

When the process pressure increases to the set pressure of 1.43 MPa, the hydrogen is released from the PRV (SV-62009, SV-82094, or SV-82099). Meanwhile, the leaked hydrogen flows through the HTL vacuum envelope to both ends, where the vacuum safety relief devices (VSDs) (SV-85010 and SV-85020) are located. The hydrogen released from SV-85020, which is placed in the A2T access room, goes through the MHVL with the length of 36 m, while the outlet of SV-85010 is connected to the sub-hydrogen vent line (SVHL) with a length of 6.5 m and the SVHL is merged to the MHVL at $L = 33$ m.

The flow diameter of the PRV (SV-62009) was selected to be 20.1 mm at a set pressure of 1.43 MPa, as shown in table 6.12. The required discharge mass flow rate through the VSDs (SV-85010 and SV-85020) is estimated by this simulation not to exceed the design pressure of 0.15 MPa. The pressure drop through the VSD is calculated by equation (6.30).

The maximum crack size and hydraulic diameter, d_H, are determined to be $\pi D^2/2$ and 15 mm, as mentioned in 6.7.2.1.1.

6.7.2.4.2 Analytical results

For the postulated maximum crack size ($d_H = 15$ mm), the process pressure continuously decreases without working the PRV (SV-62009), as shown in figure 6.47. The liquid hydrogen temperature rises due to the vacuum loss and reaches the saturated temperature at $t = 15$ s. After that, the liquid temperature is decreased under the saturated condition, with a decrease in the pressure. Meanwhile, the vacuum pressure increases to 0.15 MPa at $t = 2.5$ s. and cold hydrogen is released via the VSD. The maximum flow rate is 0.7 kg s^{-1} at around $t = 10$ s. Most of the hydrogen is released not from the PRV but the VSD within 40 s. Figure 6.48 shows the effect of the crack size, d_H, on the maximum release hydrogen flow rate from the VSD and the PSV. For higher d_H, all the hydrogen is released from the VSD and the release flow rate decreases with decreasing in d_H. For $d_H < 8$ mm, the liquid hydrogen is mainly released from the PSV and the maximum flow rate is 0.43 kg s^{-1}.

Subsequently, a temperature propagation and pressure drop are analyzed using the time-dependence of the released flow rate described in figure 6.47(b) as an inlet boundary condition. As shown in figure 6.49, the two-phase region spreads to 7.5 m away from the crack and the hydrogen temperature drops down to 91 K at the location of the VSD. The lowest wall temperature is 60 K at the crack location at $t = 40$ s.

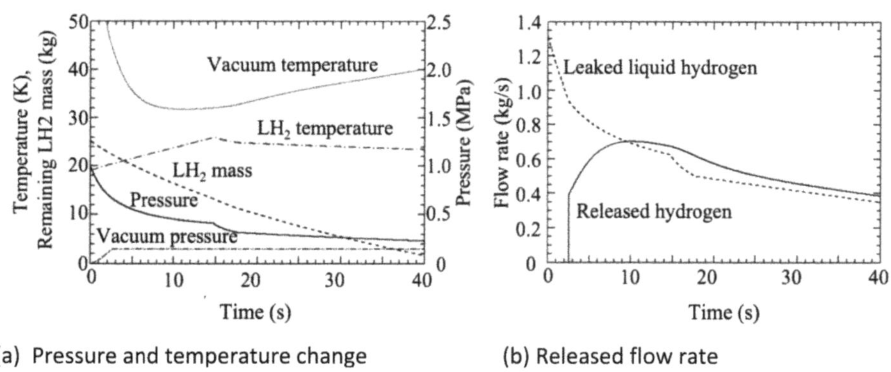

Figure 6.47. Analytical results for the postulated maximum crack ($d_H = 15$ mm). Reproduced from [18].

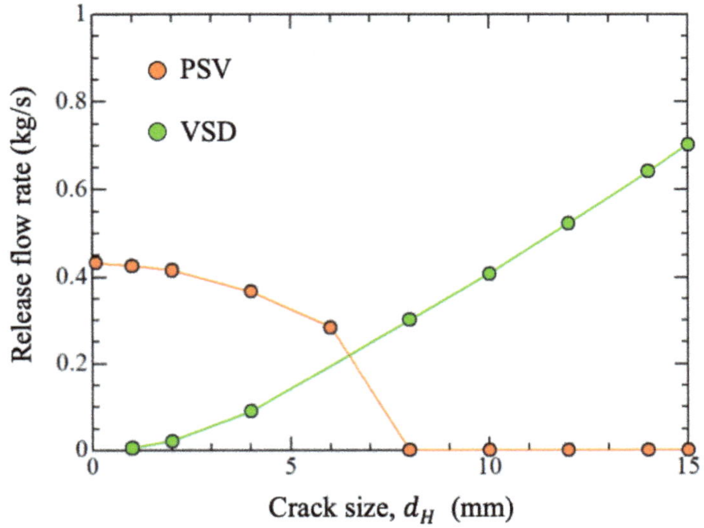

Figure 6.48. Effect of the crack size on the release hydrogen flow rate. Reproduced from [18]. © IOP Publishing Ltd. CC BY 3.0.

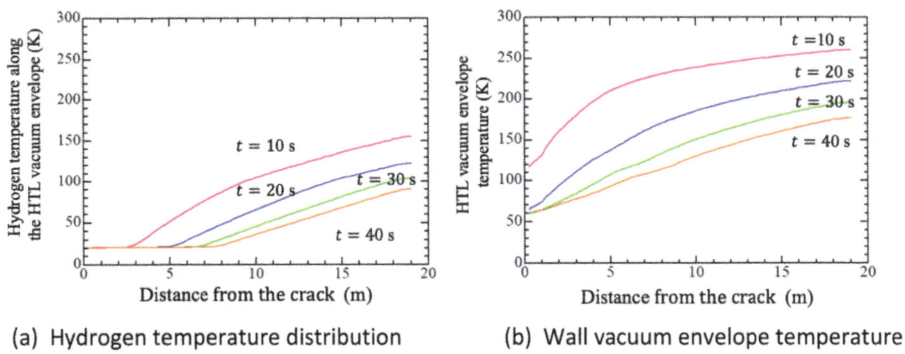

Figure 6.49. Temperature change along the HTL vacuum envelope during liquid hydrogen leak. Reproduced from [18]. © IOP Publishing Ltd. CC BY 3.0.

Figure 6.50 shows the effect of the VSD size on the pressure drop via the VSD, ΔP_{VSD}, which is calculated by equation (6.30) at the maximum release flow rate of 0.35 kg s^{-1} at 170 K, which is released from one side of the VSDs. Based on the analytical result, the diameter of the VSD is determined to be 100 mm where the pressure drop is 4.8 kPa at 350 g s^{-1}. Figure 6.51 shows the time variations in the pressure drops of the vacuum envelope, ΔP_{VE}, and the VSD, ΔP_{VSD}. The maximum pressure drop,ΔP_{VE}, is 15.4 kPa at around t= 11 s for a short duration of 5 s. The pressure drops released from SV-85010 and SV-85020 ($\Delta P_{\mathrm{VE}} + \Delta P_{\mathrm{VSD}} + \Delta P_{\mathrm{MHVL}}$ and $\Delta P_{\mathrm{VE}} + \Delta P_{\mathrm{VSD}} + \Delta P_{\mathrm{SHVL}}$) should be equal at the merging point (x= 31 m). The distributed flow rates in the vacuum envelope are estimated to be 50.6% and 49.4%, respectively. It can be considered that the postulated half flow rate used in this analysis would be reasonable.

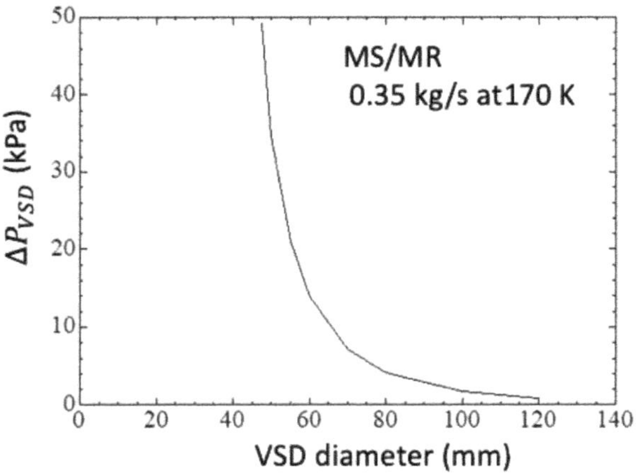

Figure 6.50. Effect of the VSD size (SV-85010 or SV-85020) on the pressure drops through the VSD, ΔP_{VSD}. Reproduced from [18]. © IOP Publishing Ltd. CC BY 3.0.

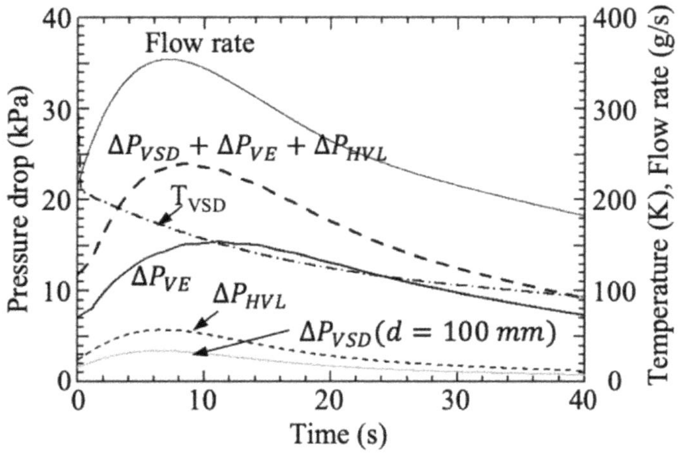

Figure 6.51. Time variations in the pressure drops for the leak at postulated maximum crack size. Reproduced from [18]. © IOP Publishing Ltd. CC BY 3.0.

As the final step, the temperature distributions along the MHVL for each time step were calculated using the released flow rate and the hydrogen temperature at the VSD, T_{VRD}, given by the step 2 calculation. There is a guide pipe at the roof penetration whose temperature should be maintained above 253 K, as mentioned in section 6.5.6. However, the wall temperature drop is higher than the lowest allowable HVL wall temperature of 175 K [27] to maintain the guide pipe above the design temperature. The pressure drop, ΔP_{HVL}, increases up to 5.8 kPa when the release flow is at 0.35 kg s^{-1}. For the hydrogen leak failure, the maximum total pressure drop, $\Delta P_{\mathrm{VE}} + \Delta P_{\mathrm{VSD}} + \Delta P_{\mathrm{HVL}}$, is no more than 24 kPa for a few seconds. The HVL size is sufficient to limit the back pressure below the design pressure.

6.7.2.5 Distribution line (DS/DR) failure

6.7.2.5.1 Analytical model and procedure

A failure case for liquid hydrogen leak to the DS or DR vacuum envelope is also analyzed in the same manner as that for the HTL mentioned in section 6.7.2.4. The four DLs (DS and DR) have a coaxial pipe structure with a length of 15 m. The outer diameter of the process line is 38.4 mm and its thickness is 2.0 mm. The DL vacuum envelope has an outer diameter of 104 mm and a thickness of 2.0 mm, and is connected to the DB at one end. The total volume of the vacuum space is 5.91 m^3. It is physically isolated from the vacuum envelopes of the HTL and the OPMS, as shown in figure 6.43. In the analysis, a liquid hydrogen leak into the DS or DR vacuum envelope occurs far away from the DB. If CV-82008 is closed, then liquid hydrogen between CV-82008 and NV-82016 is released from SV-82093 not to exceed the design pressure. The leaked hydrogen flows to the VSD (SV-65050) in the DB through the DS or DR vacuum envelope. The flow diameters of the PRVs were summarized in table 6.14. The required discharge mass flow rate through SV-65050 is estimated in the same way as mentioned in section 6.7.2.4 and the required VSD size was determined.

For the maximum crack size, half the cross-sectional area of the process pipe (5.79×10^{-4} m^2) is postulated and a rhombic shape crack with a length of 133.2 mm and width of 8.6 mm are used. The hydraulic diameter, d_H, is 8.6 mm.

6.7.2.5.2 Analytical results

Pressure behaviors in the process line and the distribution vacuum envelope for the postulated maximum crack size are shown in figure 6.52. The process pressure continuously decreases, meanwhile the vacuum pressure rapidly increases and the VSD is activated at t= 8.3 s. The maximum leak flow rate appears at around t= 22 s and is 0.252 kg s^{-1}.

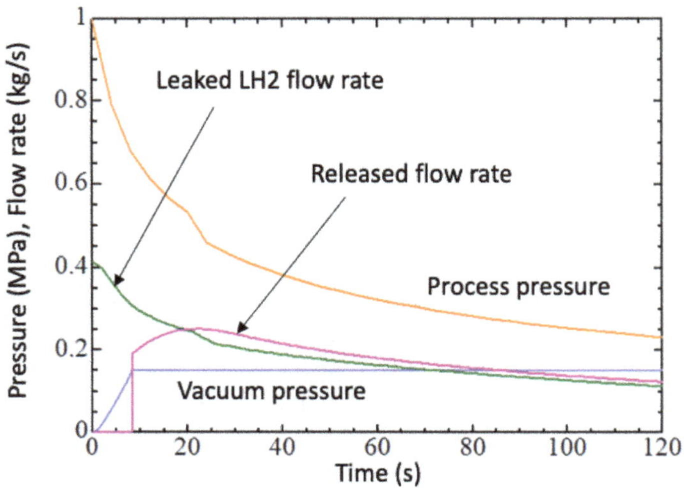

Figure 6.52. Time variations in the pressure drops for the leak at postulated maximum crack size.

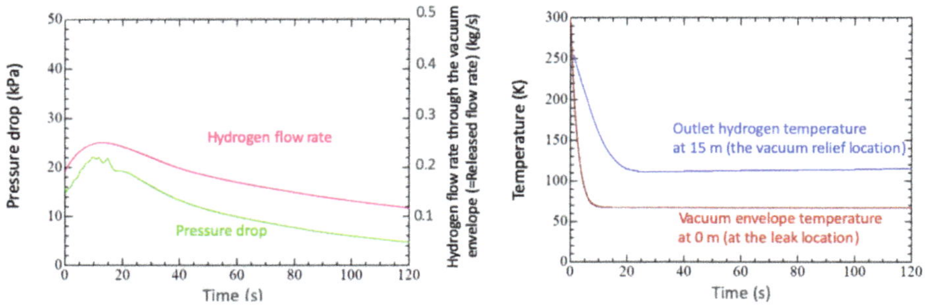

(a) Pressure drop over the DS/DR vacuum envelope.

(b) Hydrogen and wall temperatures.

Figure 6.53. Pressure drop along the vacuum envelope of the DL and the hydrogen and wall temperature drop at the positions of the VSD and crack.

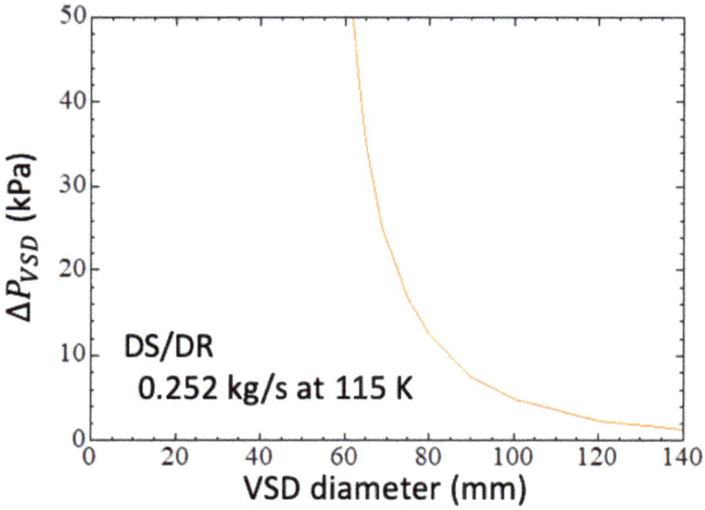

Figure 6.54. Effect of the VSD size (SV-65050) on the pressure drops through the VSD, ΔP_{VSD}.

As the second step, the transient hydrogen temperature distribution along the distribution vacuum envelop is analyzed, applying the released flow profile as the inlet condition, as shown in figure 6.52, according to section 6.7.2.2. As shown in figure 6.53, the wall temperature of the vacuum envelope at the location of the VSR is decreased down to 115 K. The maximum pressure drop appears around t= 10 s and is 22 kPa. The pressure drop of more than 20 kPa is continued for 10 s. The VSD size is determined to be 100 mm, as well as SV-85010 and SV-85020. The pressure drop is 1.65 kPa at 252 g s^{-1}, as shown in figure 6.54.

As the third step, the transient temperature reduction of the HVL during the releasing the hydrogen was analyzed and the pressure drop was estimated. The maximum pressure drop is 2.4 kPa and the lowest wall temperature at the roof penetration is calculated to be 276 K, which is much higher than the allowable lowest temperature of 175 K, as mentioned in section 6.5.6.

6.7.2.6 CMS CBx failure

A liquid hydrogen leak from a rhombic shape crack to the CMS CBx was considered in order to determine the VSD size of the CMS CBx. The process pipe diameter is 60.5 mm, which is the same as that of the HTL. Therefore, the hydraulic diameter, d_H, corresponding to the postulated maximum crack size is 15 mm. The internal cryostat void volume is approximately 4.3 m^3. The diameter of the cryostat vessel is 1.8 m (4 mm in thickness). Its inner surface area is 13.8 m^2. The process lines with the outer diameter of 60.3 mm are covered with 20 layers of MLI and its surface is 14.5 m^2.

The VSD (SV-65020) is placed on the top plate of the CMS CBx. The effect of the pressure drop caused by the leaked hydrogen through the CMS CBx can be negligible, unlike the case of the HTL and DL. Therefore, the step 2 procedure mentioned in section 6.7.2.2 was skipped. The required discharge mass flow rate through SV-65009 and average hydrogen temperature in the vacuum space were estimated in the step 1 procedure mentioned in section 6.7.2.1 and a pressure drop through the VSD was calculated.

As shown in figure 6.55, for the postulated maximum crack size of 15 mm, the process pressure continuously decreases and no hydrogen is released from the PSV, as well as the previous cases. Meanwhile the VSD works at $t=$ 2.7 s and the maximum leak flow rate of 0.7 kg s^{-1} appears at around $t=$ 10 s where the average hydrogen temperature through the vacuum space is 157 K.

Figures 6.56 and 6.57 show the effect of the VSD size on the pressure drop through the VSD at 0.7 kg s^{-1} and 157 K and the pressure drop through the HVL from the CMS CBx, whose length is 11.5 m and is 24.5 m shorter than that of the total MHVL from the DB. The maximum pressure drop through the vent line is 10.4 kPa at most. Even if the VSD size of 100 mm is selected as the same size as the others, the total pressure drop from the VSD to the exit of the vent line is 30.6 kPa. The VSD with 100 mm cross section is sufficient to limit the CMS CBx overpressure

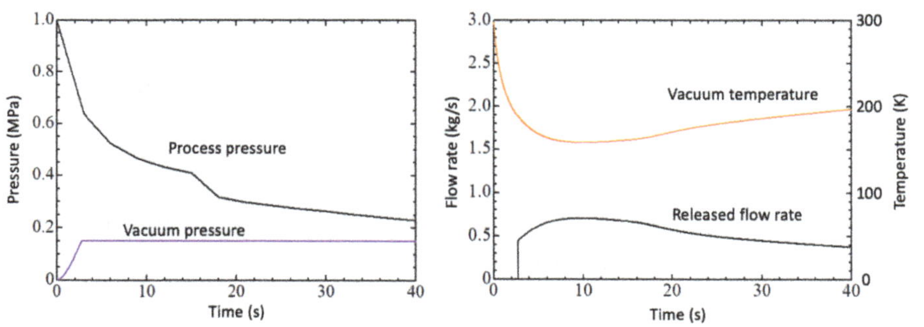

(a) Pressures in the process line and vacuum (b) Released flow rate and vacuum

Figure 6.55. Analytical results for the postulated maximum crack (d_H = 15 mm).

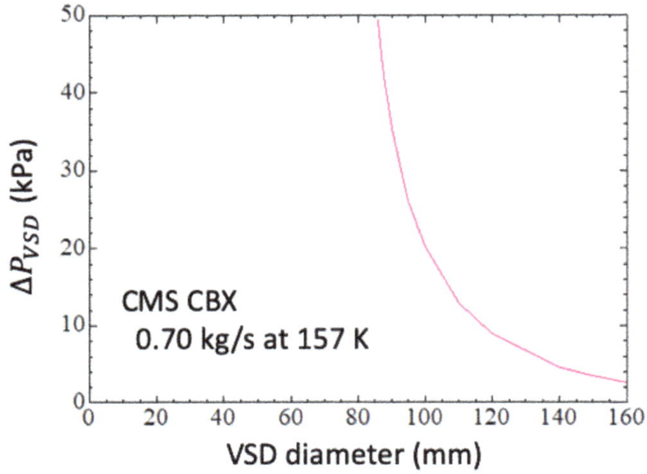

Figure 6.56. Effect of the VSD size on the pressure drop through the VSD (SV-65020) at 0.7 kg s^{-1} and 157 K.

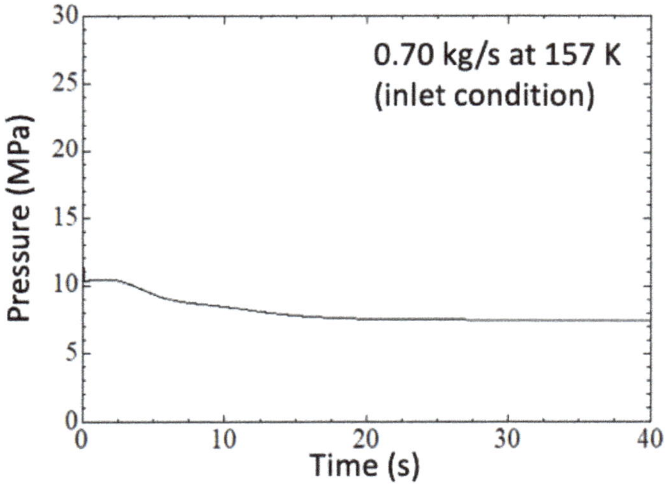

Figure 6.57. Pressure drop through the HVL from SV-65020 at 0.7 kg s^{-1} and 157 K as the inlet condition.

below the design pressure and release the leaked hydrogen safely in the case of a hydrogen leak failure.

6.7.2.7 Sapphire window failure

As an imaginable failure scenario for the OPMS, a liquid hydrogen leak caused by the sapphire window failure is considered. The details have been reported in [19].

6.7.2.7.1 Analytical model and procedure

Figure 6.58 shows the simulation model where the volumes of liquid and gaseous hydrogen are 0.344 m^3 at the average temperature of 19 K and 0.05 m^3 at 45 K, respectively. The OPMS vacuum chamber with a volume of 0.0895 m^3 is physically

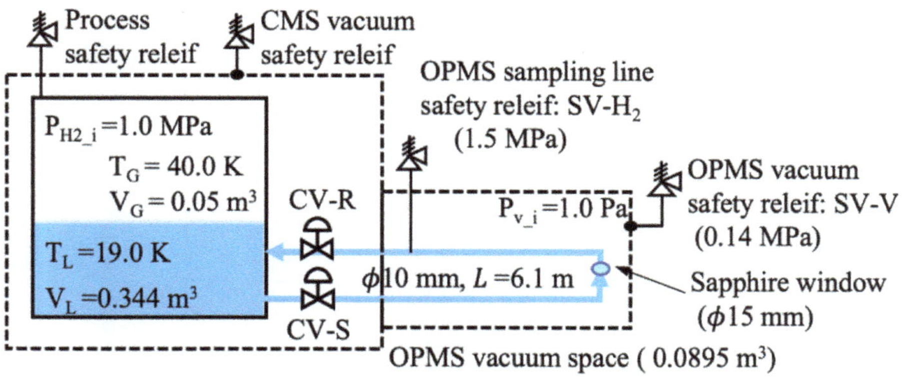

Figure 6.58. Simulation model for sapphire window failure. Reproduced from [19]. © IOP Publishing Ltd. CC BY 3.0.

isolated from the DB and is kept at 300 K and 1.0 Pa initially. If the sapphire window breaks, then the liquid hydrogen will flow into the OPMS vacuum chamber. Liquid hydrogen would be continuously supplied from the CMS process line through the sampling line with the inner diameter of 10 mm until the control valves CV-S and CV-R close. The leak flow rate should be driven by a pressure drop between the vacuum space and the process line through both of the sampling lines with a control valve ($Kv = 1.1$ and 100% position). A crack size of the same size as the sapphire window (ϕ 15 mm) is postulated in this analysis. The leaked liquid hydrogen is partially evaporated due to a film boiling heat transfer [34] on the vacuum chamber. The released flow rate is calculated, assuming hydrogen is continuously released maintaining the vacuum pressure at 0.14 MPa. The released hydrogen from the vacuum safety device (SV-V) is calculated by mass balance.

The OPMS vacuum loss would result in increased heat load only to the sampling line. The additional heat load from the OPMS vacuum loss is estimated to be 2.7 kW [19] and would not affect the hydrogen temperature in the process line. Therefore, a static heat load of 1.88 kW described in table 6.9 is continuously applied to the main process line.

6.7.2.7.2 Analytical results

Figure 6.59 shows the behaviors of the process line and the vacuum chamber, and the released flow rate through the SV-V (SV-89021). The maximum leak flow rate of 315 g s^{-1} appears just after breaking the sapphire window at $t = 0$ s. The vacuum pressure rapidly increases and the SV-V is activated at 1.1 s. The process pressure then decreases to 0.928 MPa. The released flow rate increases to the maximum flow rate of 197 g s^{-1} at $t = 13$ s. If the failure action control logic where the CV-S and CV-R are automatically shut off is activated within 1 s by detecting the OPMS vacuum degradation, then the process pressure degradation can be limited by 0.17 MPa. As well as the CMS CBx analysis, the step 2 simulation can be skipped because the VSD is placed close to the small OPMS vacuum chamber, as shown in figure 6.19. The pressure drop through the HVL is calculated to be 1.84 kPa at the maximum flow rate of 197 g s^{-1} and the temperature is 300 K. The pressure drop via the check valve

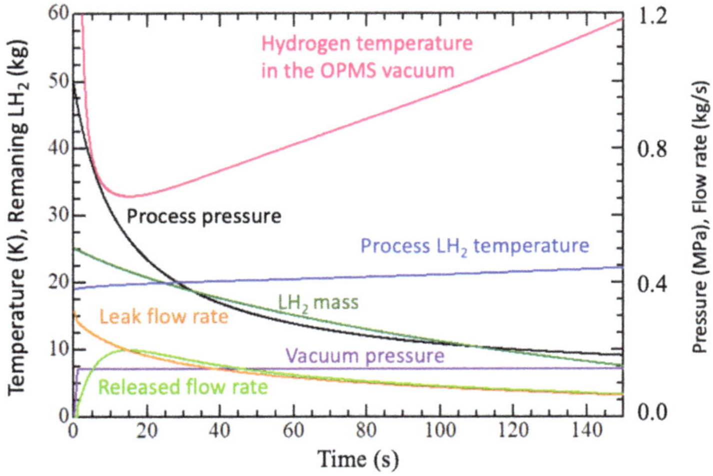

Figure 6.59. Sapphire failure analysis result. Reproduced from [19]. © IOP Publishing Ltd. CC BY 3.0.

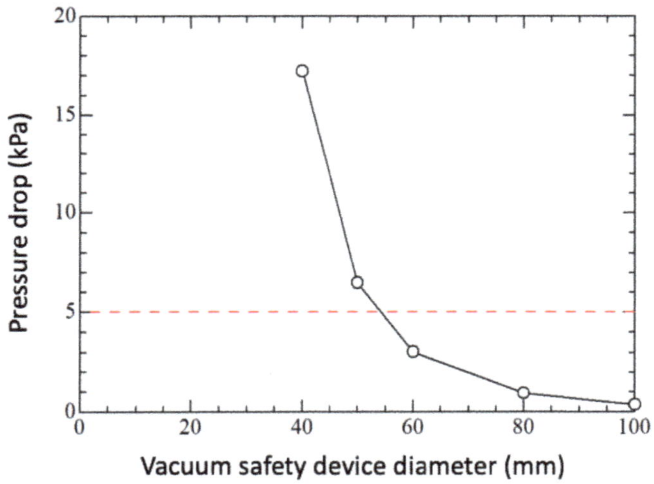

Figure 6.60. Effect of the VSD size on the pressure drop through the VSD (SV-89021) at 0.195 kg s^{-1} and 40 K. Reproduced from [19]. © IOP Publishing Ltd. CC BY 3.0.

($Kv = 760$) is 0.88 kPa. The actual pressure drop should be lower than the value at 300 K because the colder hydrogen has higher density and lower viscosity. The required cross-section area, A, of the VSD is estimated at the maximum flow rate of 195 g s^{-1} and the temperature of 40 K as shown in figure 6.60. A diameter of more than 60 mm is needed to reduce the pressure drop below 5 kPa. The diameter of the SV-V has been determined to be 100 mm as the same as the others.

6.7.2.8 Twister vacuum failure

A liquid hydrogen leak from a rhombic shape crack on the moderator pipe was considered in order to determine the size of SV-65040 for the moderator vacuum

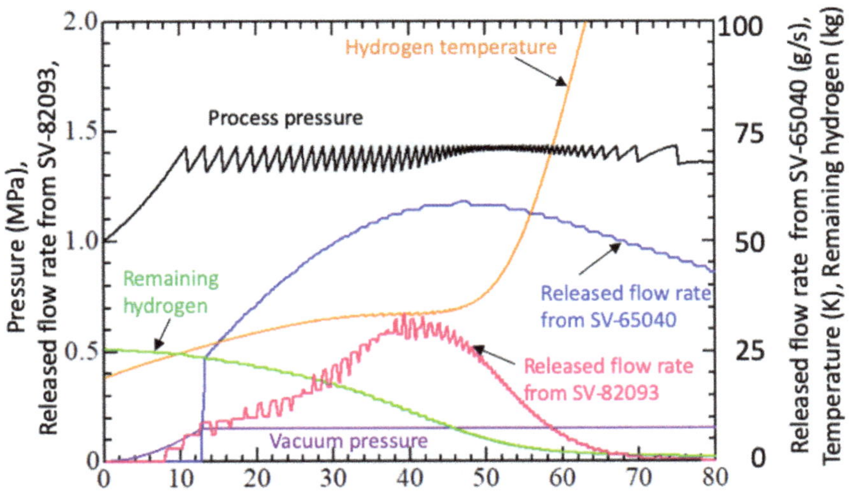

Figure 6.61. Effect of the relief size on the pressure drop.

space. The process pipe has a diameter of 25.0 mm and a thickness of 2.0 mm, and is not covered with MLI due to high radiation field. The hydraulic diameter of the crack, d_H, corresponding to the postulated maximum crack size is 4.4 mm. Supply and return pipes for one moderator in the twister are mounted in their own rectangular vacuum envelope with 68.7 × 37.6 mm. The rectangular vacuum length from the moderator to the bayonet joint on the top of the Twister is about 5.7 m. Each rectangular vacuum volume is 7.81×10^{-3} m^3. The hydraulic diameter of the rectangular vacuum envelop is 17.3 mm. There is a vacuum pipe with a diameter of 114.3 mm and a length of 16 m from the top of the twister to SV-65040 and a vacuum pump unit because the twister is in a high radiation field. The total volume of the vacuum space is 0.316 m^3.

Figure 6.61 shows the pressure change analysis for the postulated maximum crack size with $d_H = 4.4$ mm. Unlike the other cases, the process pressure increases and liquid hydrogen is released from the PSV, SV-82093, (with a set pressure of 1.43 MPa) on the process line. The maximum release flow rate is 0.6 kg s^{-1}. Most of the liquid hydrogen has been released $t=$ 50 s where the temperature increases above the critical temperature. On the other hand, the released flow rate via the VSD (SV-65040) rises to maximum (59 g s^{-1}) at around $t=$ 45 s.

For this case, hydrogen is released via not only SV-65040 (VSD) but also SV-82093 (PSV). The pressure drop driven by the released flow rate from the SV-82093 should be considered because it is much larger. Figure 6.62 shows the pressure drops through the moderator vacuum envelope and the HVL. For the VSD with the diameter of 100 mm, like the other VSDs, the pressure drop is calculated to be 0.3 kPa at 59 g s^{-1} and 300 K. This is negligibly small. The total pressure drop can be kept within 50 kPa, although the largest pressure drop is formed in the moderator vacuum envelope.

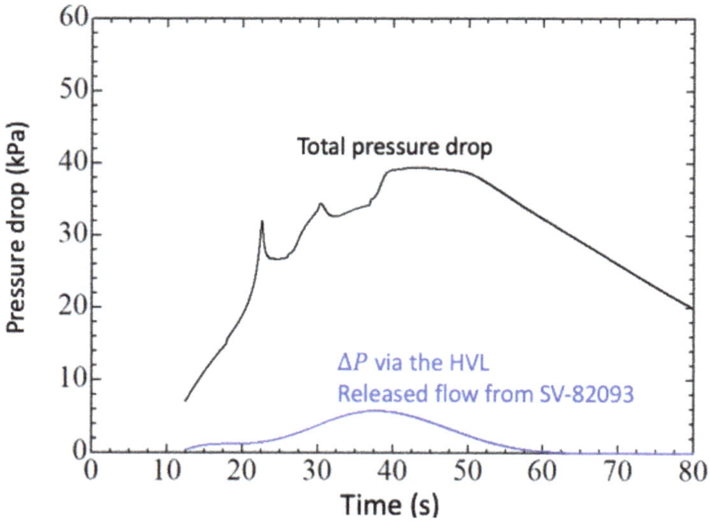

Figure 6.62. Pressure drop via the moderator vacuum envelope and the HVL.

6.7.3 Instrument air failure

There are two control release valves CV-62001 and CV-82095. If the pressure rise is relatively slow, then one of the control valves, CV-62001, can release hydrogen without actuating the PSV. The CV-62001 is placed on the top of the CMS CBx in the hydrogen room, while the CV-82095 belongs to the DB in the A2T access room. Each room has its own instrument air system. When an instrument air failure happens, hydrogen is released and the pressure is depressurized because of a fail-open function. The instrument air failure analysis was conducted using the analytical model mentioned in section 6.7.2.1.

6.7.3.1 CV-62001

The main function of CV-62001 is to prevent the whole CMS from exceeding the design pressure. This control release valve is located on the top of the PCB and is in parallel to SV-62009. An instrument air failure analysis during the cryogenic nominal operation (1.0 MPa and 19 K) was carried out. The simulation results of the CMS behaviors when CV-62001 ($Kv = 11$) is opened to 100% for the operating time of 1 s at t= 1 s are shown in figure 6.63. Only gaseous hydrogen in the vapor phase of the PCB, whose temperature is 80 K (as described in table 6.10), would be released because the operational pressure is lower than critical pressure. The pressure can be decreased down to atmospheric pressure within 3 s. The maximum release flow rate is then 0.319 g s^{-1}. The released hydrogen temperature is 78.2 downstream of CV-62001 due to the isenthalpic expansion process. Furthermore, the set pressure of the PID controller for CV-62001 should be set below not only the set pressure of 1.43 MPa but also the critical pressure of 1.29 MPa in order to be depressurized effectively and quickly.

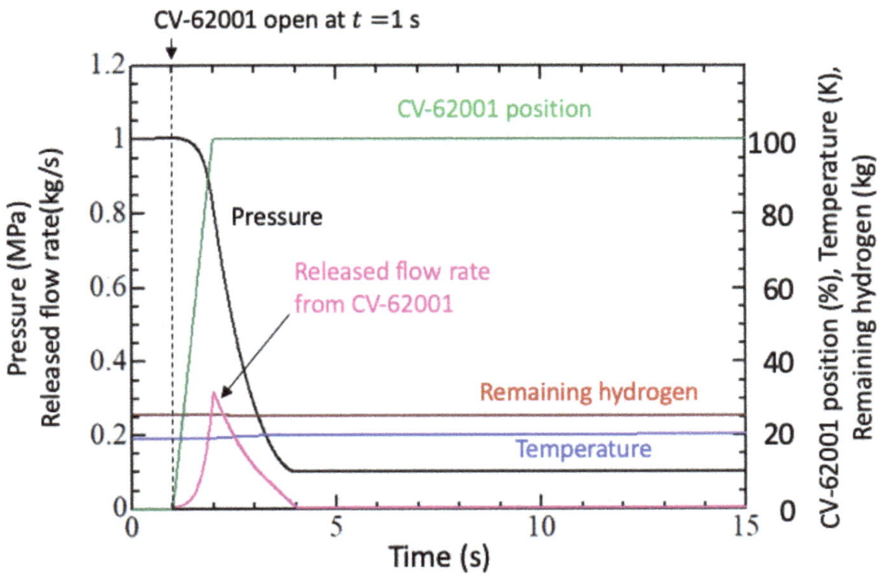

Figure 6.63. Simulation result of CMS behaviors when CV-62001 is opened to 100%.

6.7.3.2 CV-82095

The main function of CV-82095 is to prevent the moderators and their DLs (from CV-82008 to NV-82016) from exceeding the design pressure. If the pressure rise is relatively slow, then the CMS is depressurized by it before the PSV being operated. If a moderator vacuum failure happens, the CMS control system will detect the vacuum pressure degradation and will immediately close CV-82008 in order to isolate the moderators from the CMS main loop and simultaneously open CV-82095 to release the liquid hydrogen. CV-82095 is connected to the main transfer line, which is filled with liquid hydrogen, unlike CV-62001. When CV-82095 is opened during the nominal operation, liquid hydrogen would be released until a liquid-gas interface appears at the branch for CV-82095.

Figure 6.64 shows a simulation result when CV-82095 is opened to 100% for the valve operation time of 2 s. It takes more than 50 s until the pressure of 10 bar is decreased down to atmosphere pressure. The depressurization time seems to be longer than that for CV-62001. However, most of liquid hydrogen can be released by CV-82095. If you want to release a lot of liquid hydrogen from the CMS loop quickly, you should use CV-82095. On the other hand, if you want to depressurize the CMS to atmospheric pressure as soon as possible, you should definitely use CV-62001.

These are all active measures to prevent overpressure, which are not acceptable for reliable system safety. Passive equipment without need of personnel actions or control is required. Spring-loaded safety valves are recognized passive PRDs by PED.

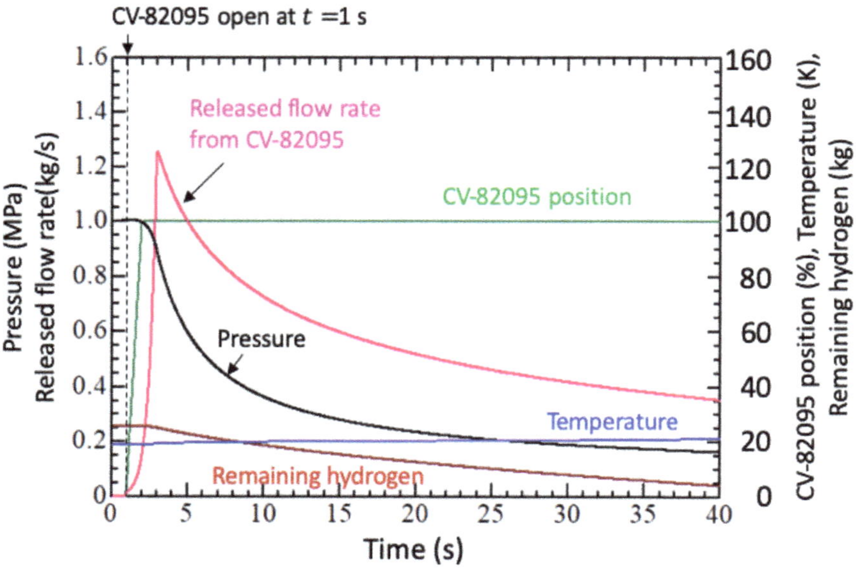

Figure 6.64. Simulation result of CMS behaviors when CV-82095 is opened to 100%.

Based on the maximum allowable pressure of PS = 16 bar.g (= 100%: the PRD characteristic, the pressure level in % according to PED), the set pressures of the relief devices have to be chosen in a way that the pressure at the cold moderators do not exceed the limit of 16 bar.g.

6.7.4 Atmosphere explosible and explosion protection

Hydrogen has wide flammable limits in air of 4 vol.% to 75 vol.%. There is a potential to form flammable and explosive atmospheres with the air if the hydrogen is leaked into a room. Directive 2014/34/EU is related to equipment and protective systems intended for use in potentially explosive atmospheres. An area classification and risk assessment shall be carried out with a conservative approach. The hazardous area is defined by SRVFS 2004:7 like Zone 0 to Zone 2. The CMS can be categorized into Zone 2 where an explosive gas atmosphere is not likely to occur in normal operation but if it does occur it will persist for a short period only. The process pipes that are welded and covered with a vacuum envelope, which can be considered as a safety barrier, does not fall into the category of atmosphere explosible (ATEX). However, an area around a mechanical connection such as a valve, Swagelok fitting, or flange is regarded as the definition of Zone 2 because there is a possibility of a hydrogen leak, although it is considerably low. The circumstance of 1 m around the mechanical connections in the Target building is defined as ATEX Zone 2. Furthermore, hydrogen leak sensors are placed in the room above the mechanical connections in order to immediately detect a hydrogen leak, trigger an alarm, remove of any potential ignition source, and release of hydrogen to the outside.

- All the CMS equipment is grounded.
- The ATEX zone area boundary will be marked on the floor.
- Electrical equipment within the restricted zone defined here shall be certified for ATEX.
- Access to the ATEX zone area is limited to the case where hydrogen concentration is below a permitted level.

On the other hand, for the GH2 filling station, the circumstance of 1.5m around the Gas Management Panel (GMP) is defined as ATEX Zone 2 and that of 0.5 m radius of the GH2 bundle connection is defined as ATEX Zone 1. There is no electrical equipment in ATEX Zone 1 for the filling station.

6.8 CMS CBx fabrication

The CMS CBx fabrication started in 2018 and was completed in October 2020 by the ESS in-kind partner: Forschungszentrum Jülich GmbH (FZJ), as shown in figure 6.65. The CMS CBx is separated into two vacuum sections with a diameter of 1.8 m and a total height of 1.6 m. They are suspended from the top flange to realize a compact cryostat stand and to make them easy to disassemble without cutting any process piping. Figure 6.66 shows how to disassemble the vacuum sections. Chain

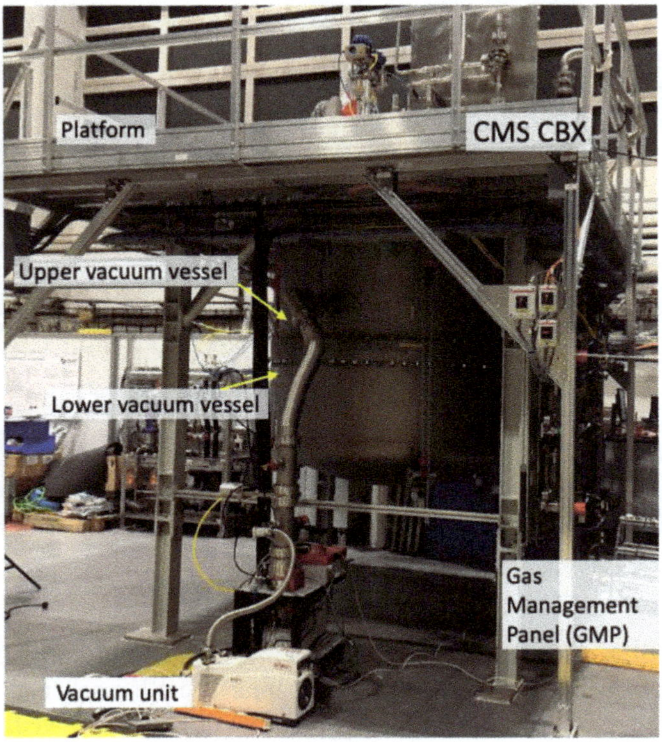

Figure 6.65. CMS CBX fabrication at FZJ. Reprinted by permission from Zhejiang University Press: [42]

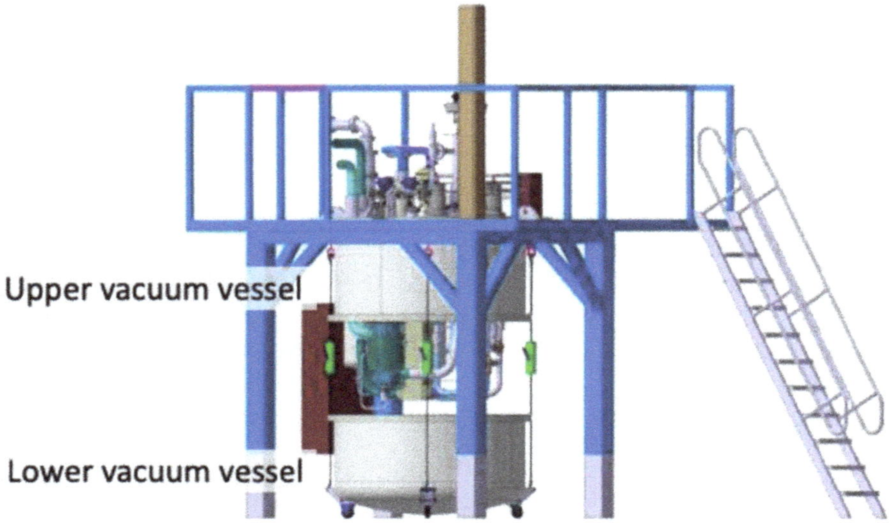

Figure 6.66. Disassembly of the two vacuum vessel sections of the CMS CBx.

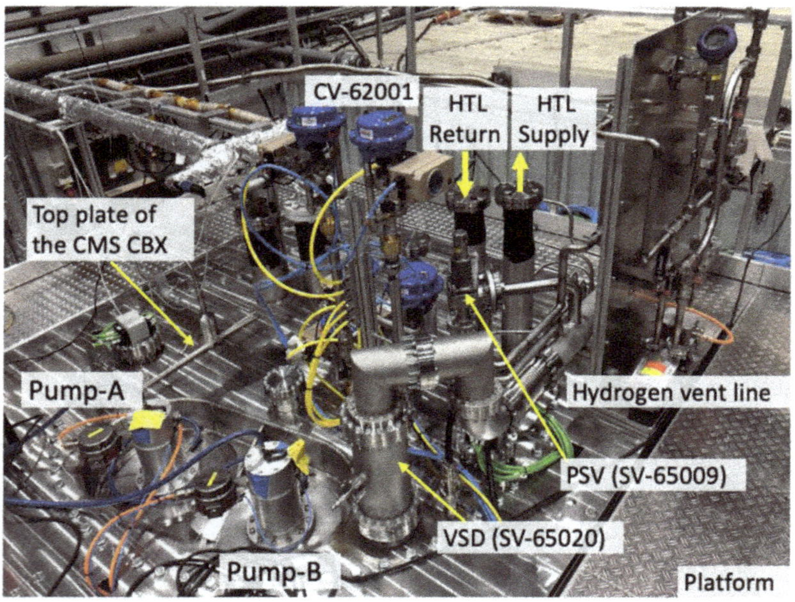

Figure 6.67. Top plate of the CMS CBX. Reprinted by permission from Zhejiang University Press: [42]

hoists are used to drop the sections. Rollers are welded and allow the piece to be moved to a free space outside of the platform area. The other rollers are bolted to the flange of the upper section, and can also be lifted down and moved.

Figure 6.67 shows the top flange of the CMS CBx. The HVL, to which all the release pipes are connected, is routed from the vacuum safety device (SV-65020) on the top flange. A GMP to supply GHe, GH_2 and GN_2 is mounted on one side of the CMS CBx. An electrical cabinet and a water-cooling system for the hydrogen pumps are temporarily placed close to it for the cryogenic test. Subsequently, the cryogenic test of the CMS CBx was performed using liquid nitrogen (LN_2) at the FZJ before delivery to the ESS site.

6.9 Cryogenic test of the CMS CBx

6.9.1 Development of a mixing system

Figure 6.68 shows the setup of the CMS CBx cryogenic test at FZJ [42]. A GN_2–LN_2 mixing system with a design pressure of 0.15 MPa was developed to adjust the cool-down temperature and maintain the CMS at a desired temperature, instead of the TMCP. There is a LN_2 dewar with a volume of 20 l on a scale. The liquid level was measured by the weight change. The LN_2 was supplied via PV-02 before the LN_2 run out. The feed of GN_2 splits into two lines. One leads to the bottom of the LN_2 cryostat to produce cold GN_2. The other is to supply ambient GN_2. The cold GN_2 is mixed with the ambient GN_2 and the feed temperature of TE-01 is adjusted.

6.9.2 Cool-down test

The cool-down operation of the cryogenic test consists of two phases: Phase I and Phase II. In Phase I, the CMS filled with GN_2 at 1 MPa and 296 K was cooled down to 100 K by the mixing system. One of the pumps ran at around 4000 rpm. Figure 6.69 shows the Phase I cool-down test result. The feed flow rate of the mixing system was around 10 g s^{-1} and the feed temperature was decreased at around 0.48 K min^{-1}. If temperature differences at the both ends of the heat exchanger

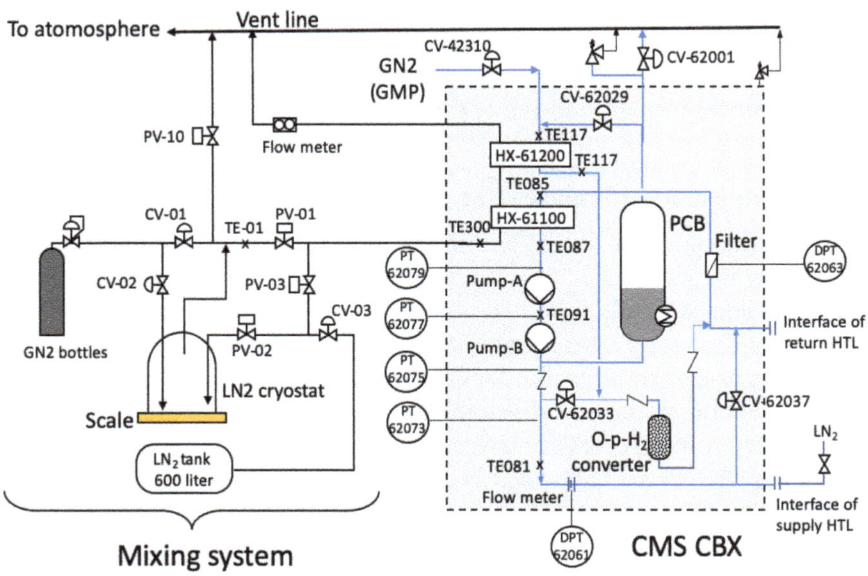

Figure 6.68. Overview of the developed mixing system for the cryogenic test using LN_2. Reprinted by permission from Zhejiang University Press: [42]

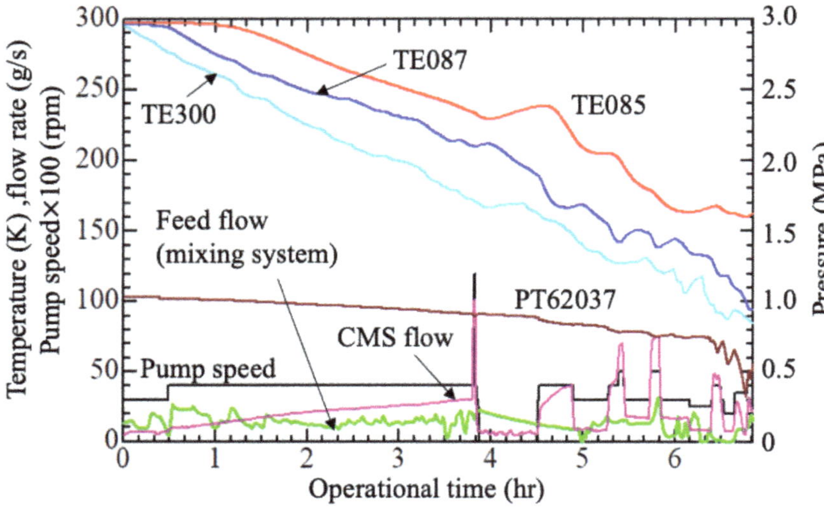

Figure 6.69. Cool-down operation at the cryogenic test. Reprinted by permission from Zhejiang University Press: [42]

(HX-61100) exceeded 35 K, then the decreasing speed of the feed temperature was tentatively stopped. No GN_2 was additionally supplied during the Phase I cool-down operation. Although the average nitrogen density remained the same in the CMS loop, the GN_2 density passing through the pumps locally was higher. The CMS circulation flow rate was increased with a decrease in temperature. It took 6.8 h to complete the cool-down of the CMS to 100 K. Finally, the pressure was decreased to 0.4 MPa. In Phase II, the CMS was depressurized and LN_2 was directly fed to the CMS at the interface of the supply HTL until a liquid level appeared above the highest liquid level sensor location. The evaporated GN_2 was released via CV-62001. On the other hand, the mixing system continuously supplies LN_2 to the HX-61100 in order to maintain the HX-61100 at around 80 K.

6.9.3 Hydrogen pump performance test

Two ball-bearing type centrifugal pumps are mounted on the top plate of the CMS CBx, as shown in figure 6.68. There is a vertical pipe with a length of 1.3 m in the pump suction side, where LN_2 should not be occupied during the pump stopping. In order to verify the pump's performance, as shown in figure 6.10, a discharge flow rate, $\dot{m}$, was measured by an orifice flow meter and a pump head, ΔP, was estimated by a difference between the two absolute pressure transmitters calibrated in advance. It was necessary to pressurize the CMS and to get a subcool condition so that the pump was capable of sucking the LN_2 just when the pump started. Figure 6.70 shows the measured pump performance under the conditions of (a) liquid nitrogen (LH_2) at 78.8 K and 0.51 MPa, (b) cold GN_2 at 123 K and 0.49 MPa, and (c) GN_2 at 297 K and 0.5 MPa. The pump head decreases monotonically with an increase in the flow rate. With an increase in the pump speed, both ΔP and $\dot{m}$ are higher.

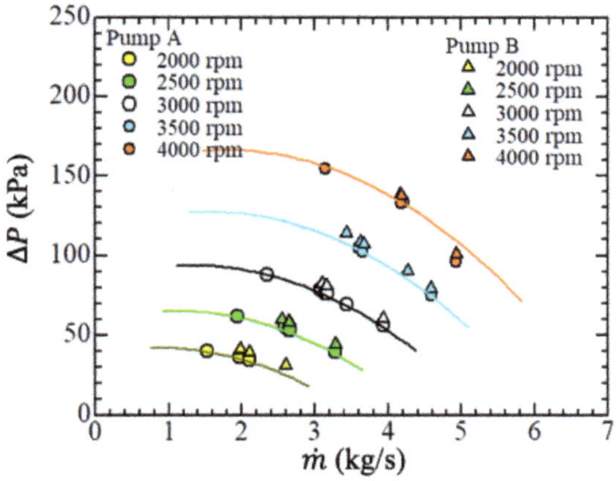

(a) LH_2 at 78.8 K and 0.51 MPa

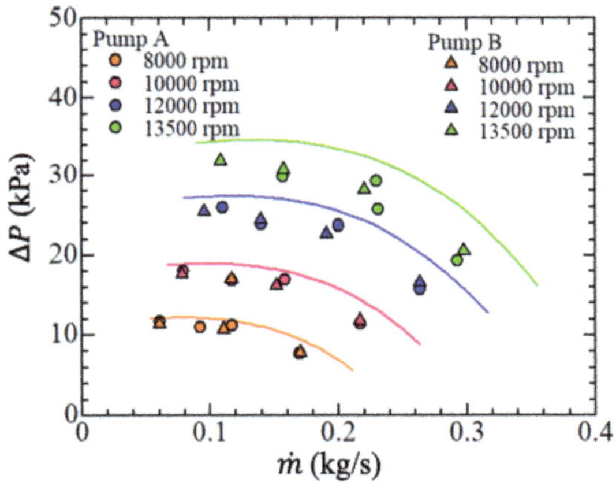

(b) GH_2 at 123 K and 0.49 MPa

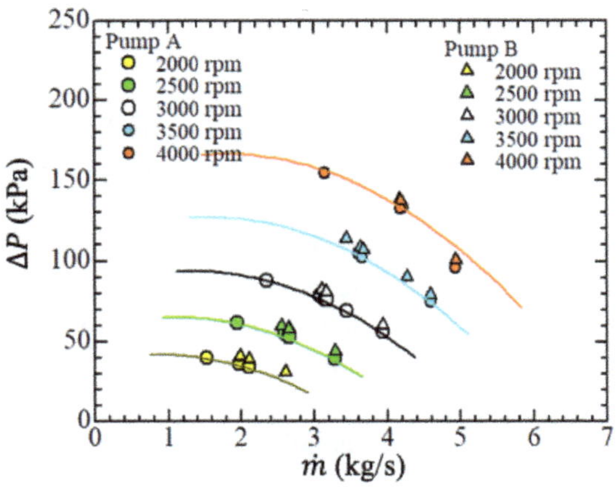

(C) GH_2 at 297 K and 0.5 MPa

Figure 6.70. Hydrogen pump performance test in LN2 and GN2. Reprinted by permission from Zhejiang University Press: [42] (a) LH_2 at 78.8 K and 0.51 MPa. (b) GH_2 at 123 K and 0.49 MPa. (C) GH_2 at 297 K and 0.5 MPa.

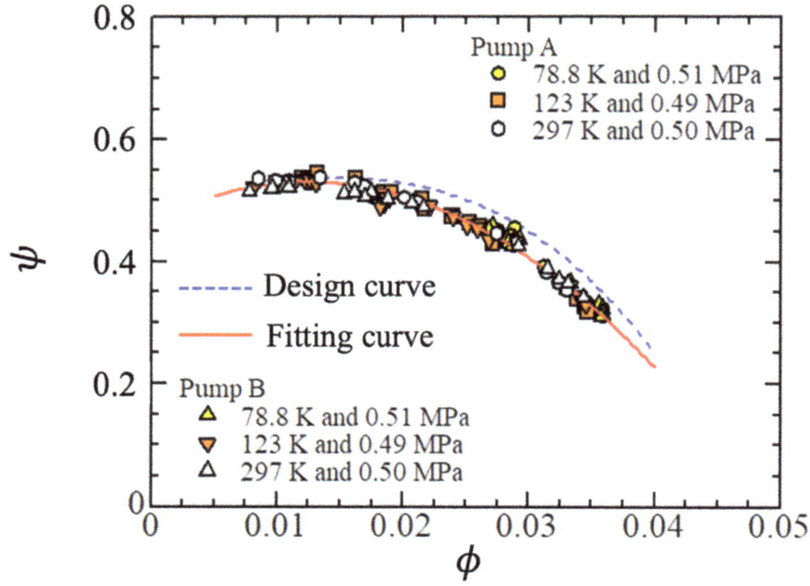

Figure 6.71. Dimensionless pump performance curve. Reprinted by permission from Zhejiang University Press: [42]

It is known that the performance curve can be arranged using dimensionless expressions of a head coefficient, ψ, discharge coefficient, ϕ, and the wheel speed, u_s, as mentioned in section 6.5.1. The non-dimensional expressions of the measured pump characteristics are shown in figure 6.71. It was verified that all the measured pump performances were on the same curve within 10%., independent of the temperature and revolution speed. A new fitting curve is obtained based on the experimental data. It would be helpful and useful to study a CMS operational procedure and the optimum parameters.

6.9.4 Pressure drop

The pressure drops over the CMS CBx were measured under the condition of (a) LN_2 at 79 K and 0.5 MPa and (b) GN2 at 130 K and 0.5 MPa. The bypass valve position (CV-62037) was maintained at 100%. As shown in figure 6.72, the flow rates were changed by the pump's revolution speed. In comparison, the predicted pressure drops under the same conditions of the cryogenic test The measured pressure drops in LN_2 and GN_2 agreed with the calculated values within 15%. It was verified that the CMS CBx has been fabricated as designed.

6.9.5 Pressure control at the nominal condition

The PCB is prepared to maintain the pressure at a set point. Six temperature sensors are located along the height of the PCB to identify the liquid level, which correspond to 0 (TE-100), 3.14 (TE-102), 10.2 (TE-104), 15.7 (TE-106), 21.2 (TE-108), and 25.9 l (TE-110) as shown in figure 6.14. There are four heaters on the side of the PCB and

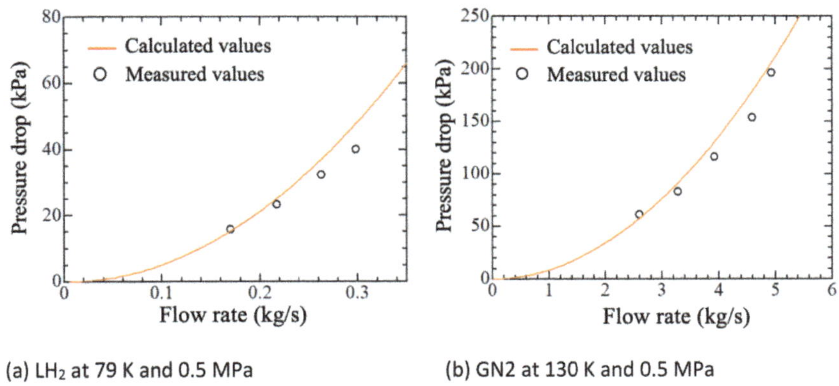

Figure 6.72. Measured pressure drops. Reprinted by permission from Zhejiang University Press: [42]

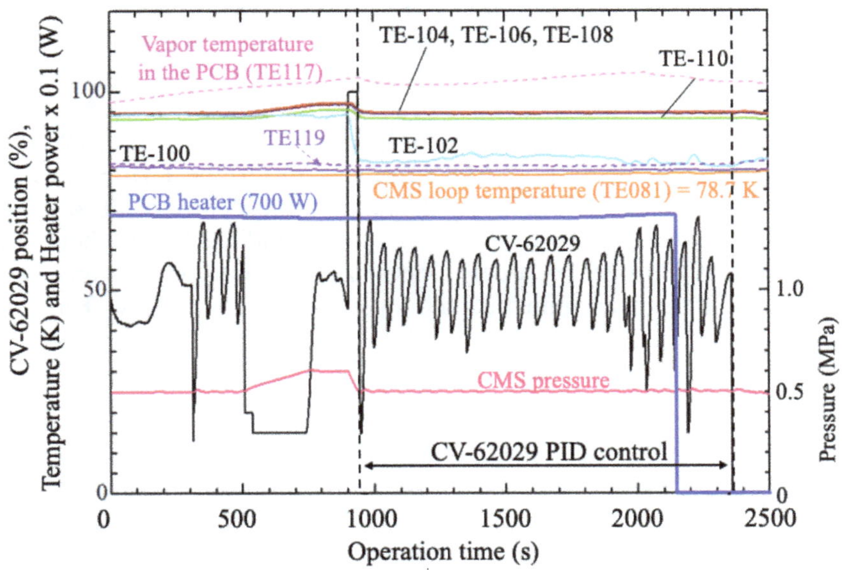

Figure 6.73. Pressure control test result. Reprinted by permission from Zhejiang University Press: [42]

a release valve (CV-62029) on the top of it. The heater produces evaporated gas and increases the pressure. The excess gas is not released by CV-62001 but by CV-62029 in order to adjust the pressure to the set point and is condensed via the HX-61200. Figure 6.73 shows a pressure control functional test result of the PCB. The temperature of LN_2 in the main loop was maintained at 78.7 K at 0.5 MPa. The two pumps were operated at the speed of 3000 rpm and the liquid level of LN_2 in the PCB was maintained above the highest location of the level sensors. The total heater power of 700 W was applied and the LN_2 temperature was maintained at its saturated temperature. Otherwise, most of the heat load would be consumed to make the LN2 temperature increase. The vapor temperature was kept around 102 K

(TE-117) by a circulation flow through the CV-62029. The released evaporated GN_2 was able to be condensed and cooled down to 81 K (TE-119) through the HX-61200. lN_2 was refilled into the PCB and the temperature around the bottom (TE-100 and TE-102) was kept close to the loop temperature. It was verified through the cryogenic test that the CV-62029 PID control and the HX-61200 works well and that the pressure can be adjusted to the set point of 0.5 MPa as designed, although the PID control parameters of CV62029 should be studied.

6.10 CMS installation status at the ESS site

The cryogenic test had been completed at the FZJ and the CMS CBx was delivered to the ESS site on November 2021. The CMS was lifted to the hydrogen room, which is located at the fourth floor of the ESS Target building, as shown in figure 6.74. Around the same time, fabrications of the 38 m long HTLs and distribution valve box were completed. The installation of the CMS started in January 2022, and was completed by the end of May 2022, except for the final parts from the DLs to the moderators in the connection cell area. The installation status as of May 1, 2022 is shown in figure 6.75. All of the installation works (the CMS CBx, the HTL, the Distribution box, and the GH_2 feed line) have been completed, with the exception of the moderators, the DLs from the DB to the Twister, and a GMP for the GH_2 filling station, which has been completed in February 2024. As the first step, the commissioning plans to be continuously conducted using helium for 10 months, instead of hydrogen, in order to optimize the operational parameters of the CMS and TMCP, and accumulate their operational experiences. The first-time hydrogen operation of the TMCP-CMS without the moderators will be carried out in summer 2023 in order to study the mitigation of the pressure and temperature

Figure 6.74. CMS CBX delivery to the H_2 room in the target building by a crane. Reprinted by permission from Zhejiang University Press: [42]

(a) CMS CBX in the H_2 room

(b) Distribution box, HTL and HVL in A2T access room

(c) GH_2 filling station

Figure 6.75. CMS installation status. Reprinted by permission from Zhejiang University Press: [42]

fluctuation. The moderators will have been installed at the end of 2023 and the commissioning of the whole CMS will be carried out.

6.11 Suggestions and lessons learned

- Build in as much time as possible for prototype tests. We found such tests to be extremely valuable for items such as the hydrogen pumps.
- Use helium or other nonflammable gas for the first commissioning of the hydrogen system. This allows the discovery of issues and the development of detailed operating procedures prior to the use of hydrogen.
- We had issues with the thermal contact of the heater to the PCB. These issues were related to differential thermal contraction, even though we installed

indium foil between the heater and the buffer wall. We were eventually able to solve this by wrapping copper wool around the outside of the heater to make up the gap that formed when the heater expanded as it heated up.

- Whenever possible, use industrial class rather than laboratory class components for items such as flow meters, vacuum pumps, and power supplies. These will be more robust and should be more easily integrated into industrial control systems.
- Have an expert in ATEX who is also aware of local requirements and closely involved from the beginning in all aspects of design, manufacture, and installation. This expert should participate in all reviews. This will reduce the need for any rework, driven by ATEX requirements, at the end of the project.

References

[1] Andersen K, Bertelsen M, Zanini L, Klinkby E B, Schönfeldt T, Bentley P M and Saroun J 2018 Optimization of moderators and beam extraction at the ESS *J. Appl. Crystallogr.* **51** 264–81

[2] Bessler Y, Henkes C, Hanusch F, Schumacher P, Natour G, Butzek M, Klaus M, Lyngh D and Kickulies M 2017 *IOP Conf. Ser: Mater. Sci. Eng.* **171** 012131

[3] Iverson E B *et al* 2003 *Proc. ICANS-XVI* (Neuss Germany) p 707

[4] Barron R F 1985 *Cryogenic Systems* 2nd edn (Oxford: Oxford University Press) pp 47–50

[5] Arp V, McCarty R D and Fox J R Gaspak Version 3.30, USA Copyright by Cryodata, Inc.

[6] Robert D and McCarty 1975 *Hydrogen Technological Survey—Thermophysical Properties,* (Aerospace Safety Reseawrch and Data Institute NASA Lewis Research Center) p 518

[7] AGabriel T and HainesThomas JMcManamy J 15 May 2003 Overview of the Spallation Neutron Source (SNS) with emphasis on target systems *J. Nucl. Mater.* **318** 1–13

[8] Oyama Y 2003 'Present status of J-PARC high intensity proton accelerator project in Japan,' in *The Proc. of ICANSXVI* ed G Mank and H Conrad (Jülich: Forschungszentrum Jülich GmbH) pp 7–11

[9] Kornegay F C, Jones K, Haines J, Myles D 2011 *Spallation Neutron Source Final Safety Assessment Document For Neutron Facilities* (Oak Ridge, TN: OAK Ridge National Laboratory) pp 3–35

[10] Tatsumoto H, Ohtsu K, Aso T, Kawakami Y and Teshigawara M 2014 Operational characteristics of the J-PARC cryogenic hydrogen system for a spallation neutron source *Adv. Cryog. Eng. AIP Conf. Proc.* **1573** 66–73

[11] Tatsumoto H, Aso T, Ohtsu K, Sakurayama H and Kawakami Y Performance test of a centrifugal supercritical hydrogen pump *Proc. 22nd Int. Cryogenic Engineering Conf.* pp 377–82

[12] Tatsumoto H, Aso T, Otsu K and Kawakami Y 2015 Development of a partitionable accumulator with pressure tolerance for the cryogenic hydrogen system at J-PARC *Phys. Procedia* **67** 123–8

[13] Tatsumoto H, Aso T, Kato T, Ohtsu K, Maekawa F and Futakawa M 2010 Design of a high power heater for the cryogenic hydrogen system at J-PARC *Cryogenics* **51** 315–20

[14] Tatsumoto H, Aso T, Ohtsu K, Hasegawa S and Maekawa F 2009 Performance test of a helium refrigerator for the cryogenic hydrogen system in J-PARC *Proc. 22nd Int. Cryogenic Engineering Conf.* pp 711–6

[15] Tatsumoto H, Lyngh D, Bessler Y, Klaus M, Hanusch F, Arnold P and Quack H 2020 Design status of the ESS cryogenic moderator system *IOP Conf. Ser: Mater. Sci. Eng.* **755** 012101

[16] Arnold P, Hess W, Jurns J, Su X, Wang X L and Weisend J G 2015 ESS Cryogenic System Process Design *IOP Conf. Ser: Mater. Sci. Eng.* **101** 012011

[17] *EN-ISO 21013-3:2016 Cryogenic Vessels-Pressure—Relief Accessories for Cryogenic Service —Part3: Sizing and Capacity Determination* (ISO 21013-3 2016)

[18] Tatsumoto *et al* 2022 Failure analysis of leaks due to cracks in hydrogen transfer lines of ESS cryogenic moderator *IOP Conf. Ser.: Mater. Sci. Eng., Adv. Cryog. Eng.* **1240** 012115

[19] Tatsumoto *et al* 2022 Design of an *in situ* measurement system for ortho-para liquid hydrogen fractions at ESS *IOP Conf. Ser.: Mater. Sci. Eng., Adv. Cryog. Eng.* **1240** 012117

[20] *CGA G-5.5, Hydrogen Vent Systems* 3rd edn 2014 (Compressed Gas Association)

[21] Tatsumoto H, Ohtsu K, Aso T and Kawakami Y 2015 Pressure and temperature fluctuation simulation of J-PARC cryogenic hydrogen system *IOP Conf. Ser.: Mater. Sci. Eng., Adv. Cryog. Eng.* **101** 012109

[22] Iverson E B and Carpenter J M 2003 *Proc. Int. Collaboration. Adv. Neutron Source (ICANS-XVI)* pp 707–18

[23] Sakamoto *et al* 2012 Technical design report of spallation neutron source facility in J-PARC *Japan Atomic Energt Agenct Report (JAEA-Technology)* **2011-035** pp 129–31

[24] Klaus M, Schwab A, Haberstroh C, Eckhardt K, Beßler Y, Baggemann J and Cronert T 2019 Ortho-Parahydrogen Mixer, Catalysts, Measurement Devices and their Application *IOP Conf. Ser.: Mater. Sci. Eng.* **502** 012161

[25] Tatsumoto H, Arnold P, Segerup M, Tereszkowski P, Beßler Y and Lyngh D 2023 Study of pressure and temperature fluctuations applied to the ESS cryogenic moderator system *Proce. of the 28th Int. Cryogenic Engineering Conf. and Int. Cryogenic Materials Conference 2022. ICEC28-ICMC 2022 (Advanced Topics in Science and Technology in China, vol 70)* (Cham: Springer) pp 133–40

[26] Tatsumoto H, Shirai Y, Shiotsu M, Naruo Y, Kobayashi H and Inatani Y 2015 Heat transfer characteristics of a horizontal wire in pools of liquid and supercritical hydrogen *J. Supercond. Nov. Magn.* **28** 1185–8

[27] Tatsumoto *et al* 2022 Design of a hydrogen vent line for ESS cryogenic moderator system *IOP Conf. Ser.: Mater. Sci. Eng., Adv. Cryog. Eng.* **1240** 012116

[28] Colebrook C F 1938 Turbulent flow in pipes with particular reference to the transition region between the smooth and rough pipe laws *J. Inst. Civ. Eng* **11** 133

[29] Ergun S 1952 Fluid flow through packed columns *Chem. Eng. Prog.* **48** 89–94

[30] Tatsumoto H, Aso T, Ohtsu K, Kato T and Futakawa M 2010 Development of a simulation code for a cool-down process of the cryogenic hydrogen system *Adv. Cryog. Eng.* **55** 1154–61

[31] Van Sciver S W 1986 *Helium Cryogenics* (New York: Plenum) p 251

[32] Giarratano P J and Smith R V 1966 *Adv. Cryog. Eng.* **11** 492

[33] McAdams W H 1954 *Heat Transmission* (New York: McGraw-Hill Book Co)

[34] Frost W 1975 *Heat Transfer at Low Tempeatures* (New York: Plenum) p 235

[35] Van Sciver S W 1986 *Helium Cryogenics* (New York: Plenum) pp 328–33

[36] Tatsumoto H, Aso T, Ohtsu K, Uehara T, Sakurayama H, Kawakami Y, Kato T and Futakawa M 2012 Design of a compact type cryogenic accumulator to mitigate a pressure fluctuation caused by a sudden kW-order heat load *Adv. Cryog. Eng* **57** 368–75

[37] Tatsumoto H, Aso T, Ohtsu K and Kawakami Y 2015 Development of a partitionable accumulator with pressure tolerance for the cryogenic hydrogen system at J-PARC *Phys. Proc.* **67** 123–8

[38] Lehmann W and Zahn G 1978 Safety aspects for LHe cryostats and LHe containers *Proc. ICEC Lond.* **7** 569–79

[39] Tatsumoto H, Aso T, Hasegawa S, Ushijima I, Kato T, Ohtsu K and Ikeda Y 2006 Pressure rise analysis when hydrogen leak from a cracked pipe in the cryogenic hydrogen system in J-PARC *Adv. Cryog. Eng.* **51** 753–60

[40] US Nuclear Regulatory Commission, Standard Review Plan NUREG-75/87, Section 3.6.1

[41] Brentari E G, Giarratano P J and Smith R V 1965 *Techinical Note 317* (National Bureau of Standard) p 4

[42] Tatsumoto H, Beßler Y, Rosenthal E, Arnold P, Kickulies M, Boros M, Horvath A, Segerup M, Tereszkowski P and Lyngh D Fabrication, cryogenic test and installation of the ESS cryogenic moderator system *Proc. 28th Int. Cryogenic Engineering Conf. and Int. Cryogenic Materials Conf. 2022. ICEC28-ICMC 2022 (Advanced Topics in Science and Technology in China* **vol 70** (Cham: Springer) pp 203–10

IOP Publishing

Cryogenic Technologies at the European Spallation Source
A big science case study
J.G. Weisend II

Chapter 7

The test and instruments cryoplant and auxiliary systems

P Arnold

In this chapter the requirements and early design choices for the test and instruments cryoplant (TICP) are addressed, along with procurement, design, installation, and commissioning experiences. The name 'test and instruments cryoplant' indicates the cryoplant's main tasks, which is first to provide cooling to the elliptical cryomodule test stand and second to provide liquid helium to the neutron instruments and their sample environments [1].

Besides these core cryogenic requirements, the TICP work unit also comprises a local and later side-wide helium recovery system, purification and storage units, liquid helium and nitrogen filling, and other commonly used piping and other auxiliary systems.

7.1 Introduction

The first of the three plants, the TICP was installed, commissioned, and acceptance tested in 2018 by ESS and Air Liquide Advanced Technologies (ALAT). The plant consists not only of a standard compressor system and coldbox but also of a process vacuum system for 2 K operation, internal and external helium purifiers, liquid helium tank, filling and boil-off station, and a helium recovery system. It is heavily integrated in the overall cryogenic installations at ESS [2]. See also chapter 2.

The helium purification is used for all ESS cryoplants, the warm helium storage tanks, and the helium recovery system. The helium recovery system is interconnected with a warm helium collector line from the Neutron Instrument Suites; with the TICP filling and mobile dewar boil-off station; with the TICP and accelerator cryoplant (ACCP) large LHe storage tanks; with all cryoplants adsorber and purge connections; and with the safety relief valve discharge protecting the subatmospheric circuits of the test stand and linear typo: accelerator (LINAC).

doi:10.1088/978-0-7503-3223-1ch7

Figure 7.1 shows a much-simplified block diagram of the major TICP components as it was used in the procurement process.

7.2 TICP procurement preparation

7.2.1 Heat load requirements and core design choices

As described in [1, 2], the main cryogenic performance requirements of the TICP compressor and coldbox system originate from the needs of the elliptical cryomodule (CM) test stand (TS2). Based on a technical note on the expected heat load of a single elliptical CM and another internal technical note on the expected heat load of the cryogenic multitransfer line and valve box, the heat load requirements have been derived as outlined in tables 7.1–7.5, also documented in [3]. In chapter 2, section 2.2.5 of this book, the safety factor philosophy and formula used are described. For sizing and specifying the TICP, the same approach has been followed.

"The 'installed' figures in tables 7.1 and 7.3 take into account the safety factors of tables 7.2 and 7.4, respectively."

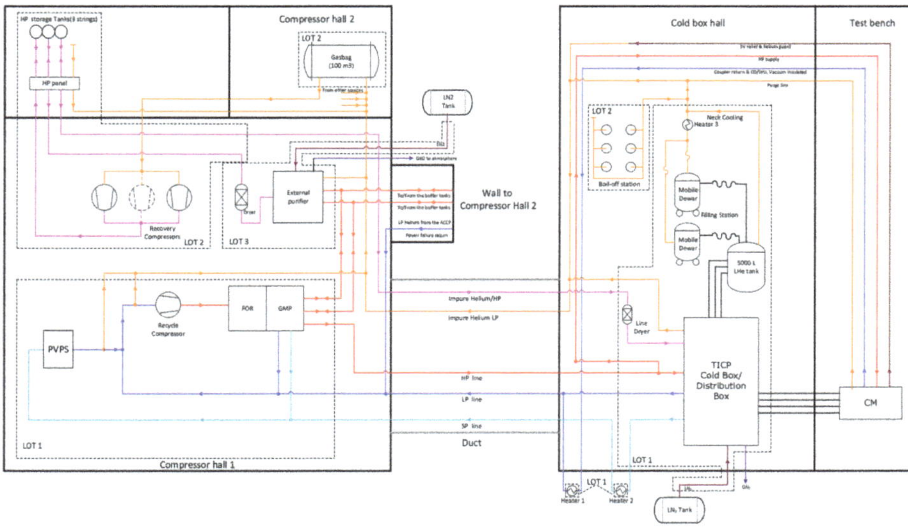

Figure 7.1. Simplified TICP block diagram. (Courtesy of ESS)

Table 7.1. One high beta CM heat load, [W].

	2 K		5 K	50 K
	Static	Dynamic	Static g s^{-1}	Static
Predicted one HBCM	13.3	24.4	0.09	46.5
Installed one HBCM	**26.8**	**49.1**	**0.19**	**93.6**

Table 7.2. CM safety factors used for the test stand.

	F_0	F_{ud}	F_{us}
2 K	1.15	1.75	1.75
5 K	1.15	1.75	1.75
50 K	1.15		1.75

Table 7.3. Test stand CDS heat load, [W].

	2 K		5 K	50 K
	Static	Dynamic	Static g s^{-1}	Static
Predicted one TS-CDS	4.1	0.0	0.00	170.0
Installed one HBCM	**7.1**	**0.0**	**0.00**	**293.3**

Table 7.4. CDS safety factors used for the test stand.

	F_0	F_{ud}	F_{us}
2 K	1.15		1.5
50 K	1.15		1.5

Table 7.5. Requirements for the test and instruments cryoplant.

Supply of 4.5 K 3 bara helium	4.0	g s^{-1}	Constant level
Return of 4 K, 30 mbar helium	3.8	g s^{-1}	
Return of 300 K, 1.25 bara helium	0.2	g s^{-1}	
Shield load	387	W	@~40 K

In chapter 3, the configuration of the cryoplant, CDS, and CM has been shown and described for the ACCP, which is essentially the same approach for the TICP. The cryoplant provides a flow of sub-cooled supercritical helium at 3.0 bara and 4.5 K at its interface to the CDS for the 2 K circuit cooling. Within the estimated limits of the CDS heat load on the 4.5 K supply, we assumed that the 2 K heat exchanger in the CM is sufficiently efficient to cool the supply flow to 2.2 K. This defines the feed enthalpy to the cavity cooling at 5.024 J g^{-1}, i.e., at the outlet of the 2 K heat exchanger. The return enthalpy, the cold side inlet to the 2 K heat exchanger, is defined as saturated vapor at 2 K, i.e., 25.041 J g^{-1}. With these two enthalpies and the heat loads from table 7.1 (static + dynamic) we can derive the required mass flow. The required coupler cooling flow needs to be added to this flow.

For the thermal shield load, the figures from CM (table 7.1) and CDS (table 7.3) have to be added. The resulting load specification figures for the cryoplant supplier is then listed in table 7.5.

Applying rough factors for converting heat load to liquefaction load assuming exergetic equivalence (1 g s^{-1} ≈ 100 W @ 4.5 K; 1 W @ 40 K ≈ 0.07 W @ 4.5 K, 1 g s^{-1} ≈ 30 l h^{-1}) the load requirements in table 7.5 translate to a helium liquefier size of ~127 l h^{-1} constant level liquefaction, which is well in the standard range of the typical plant suppliers.

Different from the ACCP (chapter 3), the returning vapor to the cryoplant is not used in heat exchangers because cold subatmospheric compression is not feasible for such small flows. Instead, the returning vapor flow is heated and compressed in process vacuum pumps to the low cycle pressure. Hence, for the TICP, the supercritical 4.5 K helium flow translates to constant level liquefaction [4]. This also leads to the perhaps surprising result that the CDS heat load on the 2/4.5 K circuit is of almost no or only marginal (performance of 2 K heat exchanger) importance.

For the cryoplant supplier it is important to note that the specified 3.98 g s^{-1} supply flow has to be guaranteed at constant liquid helium level in the LHe dewar. Normally, for liquefiers, the level in the LHe dewar is rising during operation. A rising liquid level pushes helium vapor from the dewar to the coldbox, boosting the plant performance to up to 15% greater compared to constant level, as in our configuration. We could assume our plant to provide around 150 l h^{-1} liquid helium with rising level without internal purifier operation.

The shield temperatures have at first assumed higher (~70 K–80 K) by our in-kind partners for the CMs. However, for the cryoplant–CM system, the optimal temperature is not only defined thermodynamically but also where physically coupling out a cold flow is possible. For a 'standard'-kind of cryoplant like the TICP, the natural out-coupling position is upstream of the first expansion turbine, which happens to be at around 35 K–40 K. To align CM heat load specification with ACCP requirements in the same temperature range, a minimum supply temperature of 33 K and a maximum return temperature of 53 K have been chosen for the TICP.

In addition to the CM test stand, the TICP is supposed to supply liquid helium to external and internal clients, namely Lund University, Max IV, and the Neutron Instruments Science Activities Division. The external clients only came later, after the design process, on the list.

The sample environments team provided an estimated need of liquid helium, as shown in figure 7.2, based on a full suite of 22 instruments and an extrapolation of the consumption in comparable facilities with similar magnets, cryostats, and sample environment facilities such as ILL or ISIS.

The resulting 7500 l/month are to be considered net requirement or 'payload,' and need to be escalated for the liquefier's specification, as shown in table 7.6. It quickly became obvious that the cryoplant, designed for TS2 operation, could easily support the requested liquefaction needs.

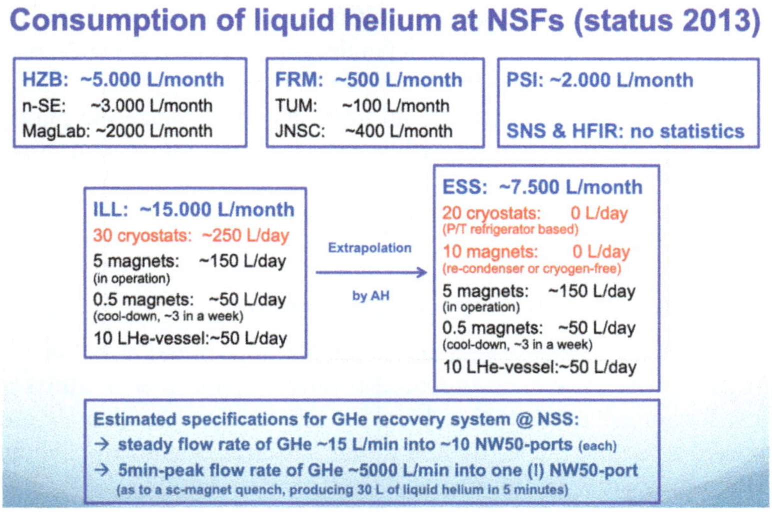

Figure 7.2. Helium liquefaction requirements of the science activities division. (Courtesy of ESS)

Table 7.6. Liquefier sizing assumptions.

TICP requirements from NSS/SAD		
Payload	7500	l/month
Include transfer losses from storage tank to mobile dewars (−30%)	10 714	l/month
Include losses of internal purifier operation (−20%)	13 393	l/month
Include safety factor 1,3 (higher Dewar heat loads, losses at client)	**17 411**	**l/month**

Vendors were asked in the tendering process to design for TS2 operation and at the same time guarantee minimum helium liquefaction rates with and without LN2 precooling. In addition to price and several quality criteria, the bidders were also rated by these figures. Guaranteed and later tested liquefaction rates are discussed in section 7.5 and in [5].

An additional sizing requirement was that the recycle (main) compressor of the TICP should be able to act as a recovery compressor for the evaporating helium from the LINAC in case of a power shutdown, assuming the required utilities to run the TICP compressor in that case are available. The compressor needs to be able to compress at least 50 g s^{-1} clean helium from 1 bara to 14 bara in order to store the helium inventory of the LINAC in the pure helium storage (PHS) tanks.

When operating to supply the CM test stand, the characteristic plant load is ideal for liquid nitrogen precooling. It was therefore decided to request a plant design using liquid nitrogen precooling for the test stand service. This reduces the plant size significantly, and cuts electrical power and cooling water consumption in half [4].

For the normal TICP liquefaction operation in an open loop, there is the choice to precool with liquid nitrogen or not, depending on the specific needs at the time of operation.

The cryoplant specification includes an incentive to optimize the plant for rising level liquefaction without precooling while operating the internal purifier.

A variable frequency drive for the recycle compressor helps to cope with the different load scenarios and reduces energy consumption.

7.2.2 Other subsystem requirements and design choices

To operate the TICP as helium liquefier, sufficient high-pressure buffer volume and liquid storage volume have been prepared to limit the number of starts and stops, and to allow for longer maintenance breaks. Case studies (see also chapter 2) have been performed to define the sizing and design requirements of the helium storage, gasbag, HP compressors, and purifiers.

For the gasbag sizing, a base load was considered including:

- Return of the helium for the neutron instruments sample environment, i.e., 7500 l/month as mentioned in table 7.5.
- Boil off from 20,000 l LHe dewar from ACCP (during standstill) of about 0.5% per day.
- Boil off from 5000 l LHe dewar from TICP (during standstill) of about 0.8% per day.
- Purge and warm-up flow from the internal purifier during half of its operation time of an estimated 300 h/month.

This sums up the base load to about 16 $m^3\ h^{-1}$ on average.

Additional to the base load are some special cases, such as the filling of mobile dewars from the LHe tank. We can assume a 200 l dewar filled in 30 min with a loss of 30% to the recovery system. Assuming that the typical operating points of large gasbags are from 10% to ~50%–60% level, a gasbag size of 100 m^3 seemed adequate for the purpose because the compressor hall is big enough to allow for a large gasbag. That also provides some margin against helium loss in case of a power outage, and consequently tripping of systems that need some time to recover even in case of UPS and diesel generators for parts of the system.

For the recovery compressors that compress impure helium from the gasbag to the storage, we specified a minimum capacity of 75 $m^3\ h^{-1}$ in order to never be restricted by NSS operation, such as cool down of a magnet and simultaneous dewar filling. We provided space and utilities for two recovery compressors to provide for redundancy with the option to add a third compressor in the future if need be. One compressor in the order of 35–40 $m^3\ h^{-1}$ would run a couple of times per day and in the order of 1–1.5 h to empty the gasbag between the typical operation points, which seems a reasonable operation scheme.

Furthermore, a 4 ×4 high pressure (HP) matrix panel with automated valves has been specified to provide interconnection possibilities for pressure source lines from three separated HP bundle strings and one spare connection that is used to tie-in mobile gas bundles on one side and four process lines—from HP compressors, to external purifier, to internal purifier, and back to gasbag for depressurization of the bundles. The HP bundles have a geometrical volume of 12 m^3 combined and can be pressurized up to 200 bar.

The duty specification of the stand-alone purifier was to purify an average flow of at least 5 g s^{-1} with up to 2000 ppm air impurities or 1.7 g s^{-1} average flow at up to 10,000 ppm air impurities, whereby the average flow was defined as the total purified helium mass divided by the cycle time, comprising all steps from operation, regeneration, purging, cool down, and to resuming operation [5].

The sizing specification came out of discussions with typical suppliers to optimize adsorber bed volume–mass flow ratio incorporating reasonable assumptions on the to be expected flows and impurity levels.

The PHS tanks were from the beginning thought to be a separate procurement from the cryoplants and their sizing is basically a result of providing a second fill for the helium inventory of the superconducting LINAC with the given space constraints on site.

7.2.3 TICP scope and split in procurement lots

In order to boost competition on different elements of the TICP scope, ESS took the decision to split the system up in three different lots so that not only the typical two European cryoplant suppliers with in-house turbo expander technology and decade long experience would be permitted to bid. The scope split is shown in figure 7.3 and consists essentially of:

Lot 1: The core refrigerator system including process vacuum pumps to achieve 2 K and including LHe dewar and filling station.

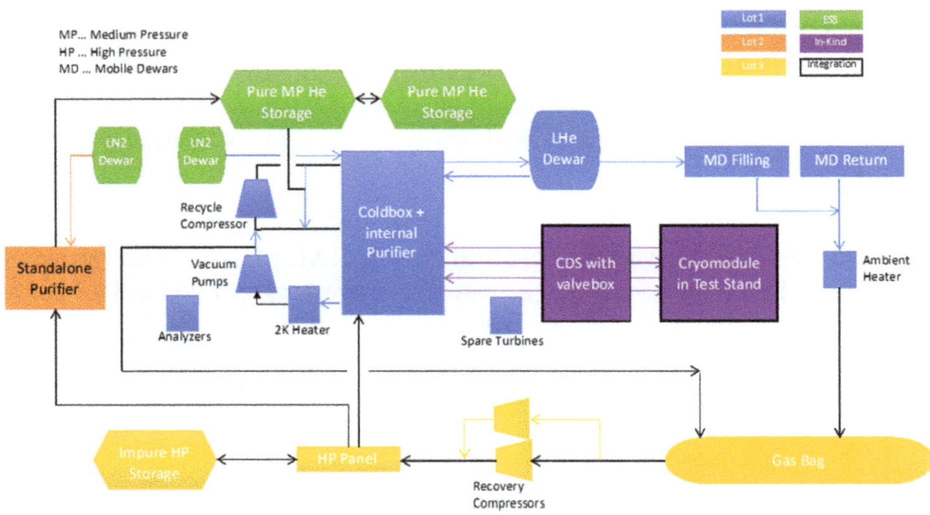

Figure 7.3. Scope split. (Courtesy of ESS)

Lot 2: The helium recovery system including gasbag, HP compressors, and storage.

Lot 3: An LN2 cooled stand-alone external purifier.

The rest of the system—particularly most interconnecting piping and clean storage tanks and also civil infrastructure, utilities, connections to the cryomodule test stand and the recovery system and the optical fiber network connections—were provided by ESS as part of separate procurements or in-kind deliveries.

The main reason to define precisely these lots is that they consist of adequately independent subsystems that can be designed and acceptance tested autonomously from each other. In an earlier iteration, the LHe dewar, mobile dewars, and filling station comprised a separate fourth lot; however, the refrigerator is so tightly connected (physically by transfer lines, logically by controls, functionally by heat loads) to the LHe dewar and filling station that handling with potentially two different suppliers would have been very challenging.

Another decision was to include the process vacuum pump station (PVPS) in Lot 1 with the main refrigerator, even though other laboratories have done this separately. We did not want to use only oil free pumps and then you have to take care to use the same compressor oil for PVPS and recycle compressor, which would be difficult to define before the suppliers are known.

Since bidders were permitted to offer price deductions for ordering a combination of two or three lots, it turned out that ALAT handed in the best offer in the combination of all three lots.

It is, of course, difficult to tell the outcome if there had been different suppliers, but ESS received a functional, integrated system for a good price. There have been difficulties with the exact piping interfaces, which will be discussed in section 7.4, but all in all this strategy proved to be quite successful.

7.3 TICP design choices

The TICP supplied by Air Liquide is essentially a highly standardized system with several extra components and features that are important for its use at ESS. An overview of the main components in the ESS controls system is displayed in figure 7.4.

7.3.1 TICP warm compression system

Air Liquide designed a cryoplant based on the HELIAL ML™ helium refrigerator with a Kaeser DSDX 305 SFC used as recycle compressor, operated at 14.5 bar and a maximum flow rate of 56 g s^{-1} [5]. The recycle compressor model has an internal frequency converter (hence the name SFC) which proved to be very practical for maintaining the low pressure (LP) of the plant stable even in transient operation, such as cryomodule pump down that may otherwise lead to rather violent perturbations.

Another feature is that the recycle compressor has been specified to operate with the same oil as the compressors for ACCP and TMCP, which is less hygroscopic than the normally used BREOX oil provided as per default by Air Liquide.

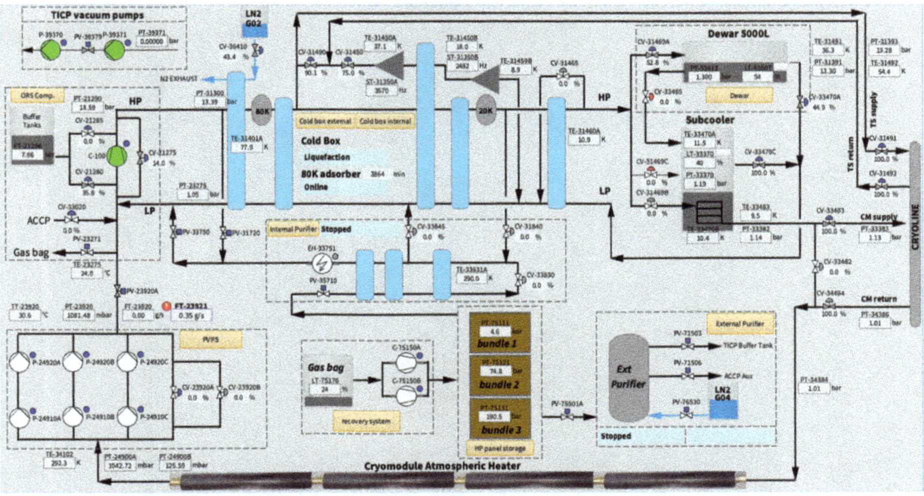

Figure 7.4. Overview of TICP in EPICS OPI. (Courtesy of ESS)

The oil removal system was furnished with a third oil coalescer to provide additional safety that is otherwise not a standard. This is shown on the backside of the GMP/ORS skid in figure 7.6.

The gas management system (figures 7.5 and 7.6) contains several additional valves compared to a standard liquefier, connecting to:

- An additional line to the PHS so that helium recovered from the LINAC can be stored in different PHS tanks.
- A line to the suction side of the PVPS to permit independent acceptance testing.
- A connection to the helium guard of the PVPS valves operating in subatmospheric conditions.

The vapor pressure of 2 K helium is 31.9 mbar. Taking into account return valves, transfer line, and ambient heater pressure losses, the PVPS was specified for an inlet pressure of 27 mbar and designed for an inlet pressure of 20 mbar at 4.0 g s^{-1} flow.

The PVPS consists of three parallel trains of a root blower–rotary vane pump–assembly from Leybold, see figure 7.7. The RUVAG blowers have variable frequency drives, permitting rotation speeds between 20 and 70 Hz, and use the single stage SOGEVAC rotary vane pumps as backing pumps. The blowers are not oil free but accept the same oil as the Kaeser recycle compressor.

7.3.2 TICP coldbox system

The TICP coldbox contains additional equipment compared to a standard helium liquefier coldbox. Most importantly, of course there is a transfer line terminal for the four-fold multi cryogenic transfer line 4.5 K, 3 bar supply, VLP return, thermal shield supply and return to the test stand. In addition, there is a 4.5 K phase

Figure 7.5. ORS and GMP front. (Courtesy of ESS)

Figure 7.6. ORS and GMP back. (Courtesy of ESS)

Figure 7.7. PVPS. (Courtesy of ESS)

separator vessel with immersed copper coil to subcool the 4.5 K helium to the test stand. The thermal shield circuit has supply and return valves, as well as a bypass valve and built in test heater.

During execution of the project there have been discussions about the optimal location of the 80 K air adsorber where it was decided to place it directly downstream of the LN2 precooling stage, assuming that the plant operates predominantly with LN2 precooling switched ON. This means that the air adsorber will not be effective when there is no LN2 precooling, which was considered to be the case rather rarely and certainly not with TS2 operation including PVPS. Had the air adsorber otherwise been placed just upstream, the first expansion turbine stage of the plant would likely have to suffer often from clogging of the first heat exchangers where impurities would freeze out.

The coldbox contains also an internal freeze-out purifier that is currently not used as frequently because the plant mostly serves the test stand and the liquefaction/refrigeration capacity drop due to purifier operation is substantial. Upstream of the freeze-out purifier there is a line drier for the impure helium feed. The assembly is shown in figure 7.8.

The returning VLP helium from the test stand passes the coldbox only through the transfer line terminal and return valve before it is warmed up in a specially designed ambient heater. The specific design requirements are to provide very reliable warming with minimal failure possibilities and minimal pressure drop. Therefore, a passive heater with large heat exchange area was preferred. This ambient heater consists of four welded finned tube segments with an inner swirl, each of the four segments turned 45° with respect to the previous one to ensure proper turbulence in the pipe and avoid a cold inner jet.

Part of the coldbox system is the 5500 l liquid helium dewar. The dewar has, apart from the three coldbox connections: two-phase supply, vapor return and liquid return, two connections for decant lines including pneumatic. The decant lines,

Figure 7.8. TICP coldbox, 5500 l liquid helium dewar and line drier (from right to left). (Courtesy of ESS)

normally used to fill mobile dewars, can also be used the other way around, i.e., to push liquid helium from the gas supplier's mobile dewars into the storage vessel. This is currently the principal method to top up the helium inventory of the ESS.

The filling station also contains a weight scale and temperature measurements in the vapor outlet used to switch off the filling, so the operator does not need to wait next to the station until the filling is done. Furthermore, there is a boil-off station with space for up to 20 mobile dewars, flow meters, and passive flash gas heater.

7.3.3 TICP central recovery system

As described in section 7.2.2 the ESS central helium recovery system was part of the delivery of the TICP. The recovery piping has been engineered and built successively by ESS to permit site-wide feeding into it, from all three neutron instrument halls, the workshop and laboratory area, and all the three cryoplants. The helium recovery line backbone also acts as 'guard helium' for the discharge of the safety valves of the LINAC cryodistribution, the cryomodules, and the test stand.

All helium is recovered in the 100 m^3 gasbag, see figure 7.9, that itself is protected by an active pneumatic release valve and a passive release valve that looks like an oil vessel, whereby the oil level has the equivalent function of a spring force, see figure 7.10.

The passive safety valve is designed for a flow of 104 g s^{-1} at a slight overpressure of up to 30 mbar(g), which is the expected maximal failure flow through a cryomodule safety valve that in turn is sized for failure flow of JT and cool-down valve in the cryomodule itself.

The gas from the gasbag is compressed in one or two of two Girondon Sauer HP compressors, each with a capacity of 43 m^3 h^{-1} helium compressed from 1 bar(a) to 200 bar(g), see figure 7.11.

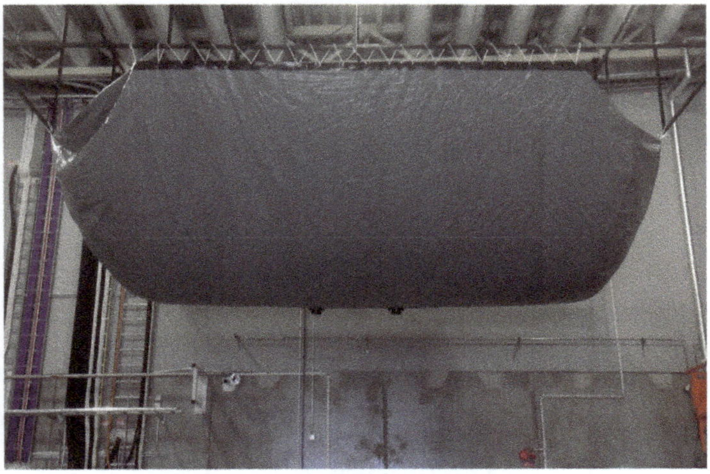

Figure 7.9. Gasbag. (Courtesy of ESS)

Figure 7.10. Safety valve of the gasbag. Reproduced from [6]. © IOP Publishing Ltd. CC BY 3.0.

The downstream HP valve panel (figure 7.12) was executed as specified and distributes the gas to the storage bundles (figure 7.13) and/or purifiers.

In figure 7.14, the functional arrangement of the helium recovery system is displayed in a simplified flow sheet, illustrating the high level of interlinking with the other cryoplants and users.

An essential part of the recovery system is the liquid nitrogen cooled purifier, shown in figure 7.15, based on the ULTRAL™ design with seven cold adsorber beds and a standard upstream dryer [5].

Figure 7.11. HP recovery compressors. Reproduced from [6]. © IOP Publishing Ltd. CC BY 3.0.

Figure 7.12. HP valve panel. (Courtesy of ESS)

Figure 7.13. HP storage. Reproduced from [6]. © IOP Publishing Ltd. CC BY 3.0.

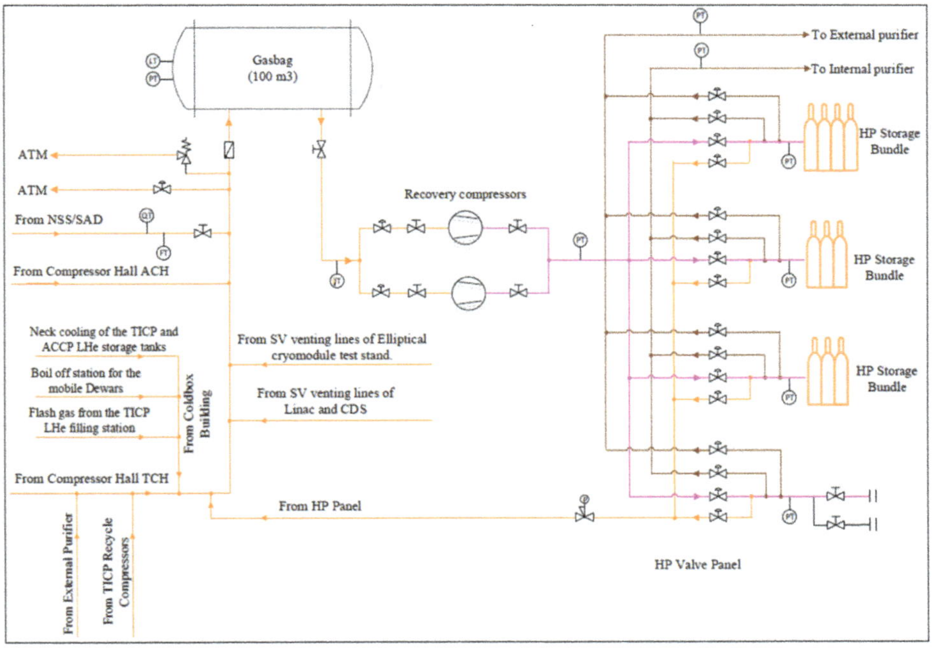

Figure 7.14. Simplified flow sheet of the TICP helium recovery system. Reproduced from [6]. © IOP Publishing Ltd. CC BY 3.0.

Figure 7.15. External purifier including drier. (Courtesy of ESS)

Additionally, Air Liquide delivered a variety of impurity detectors for different levels of impurity. In the recovery system there are nitrogen detectors and dewpoint meters in the %-range. In the warm compression system of the closed loop circuit there is a Linde-built multicomponent detector for moisture and nitrogen in the range 0–200 ppm that is even furnished with a pyrolizer and analyzer for detection of oil vapors that; however, is not really in use due to reliability reasons. Furthermore, there is a Linde-built mobile multicomponent detector for ppm-range moisture and nitrogen that can be used elsewhere, particularly at the coldbox if needed.

7.4 TICP installation

The TICP's installation, commissioning, and testing were characterized by huge delays and too optimistic planning. Figure 7.16 illustrates the foreseen durations of the project up to delivery and up to completion (vertical axis), project achievements in reality (horizontal axis), and revised monthly planning.

The component delivery was a few months delayed but on the overall project duration broadly okay. The installation and commissioning phase, however, was totally underestimated. Three months for the on-site work appear to be ambitious even for a standard liquefier with few components close by. For a project like that at ESS, a green-field facility with a large number of separate skids at two different locations, non-trivial interfaces, several non-standard components, and a very detailed time-consuming acceptance testing specification, it should have been clear that three months are not realistic.

One of the biggest contributors to the delay was perhaps an insufficient interconnecting pipe support concept from ALATs sub-supplier before the start of

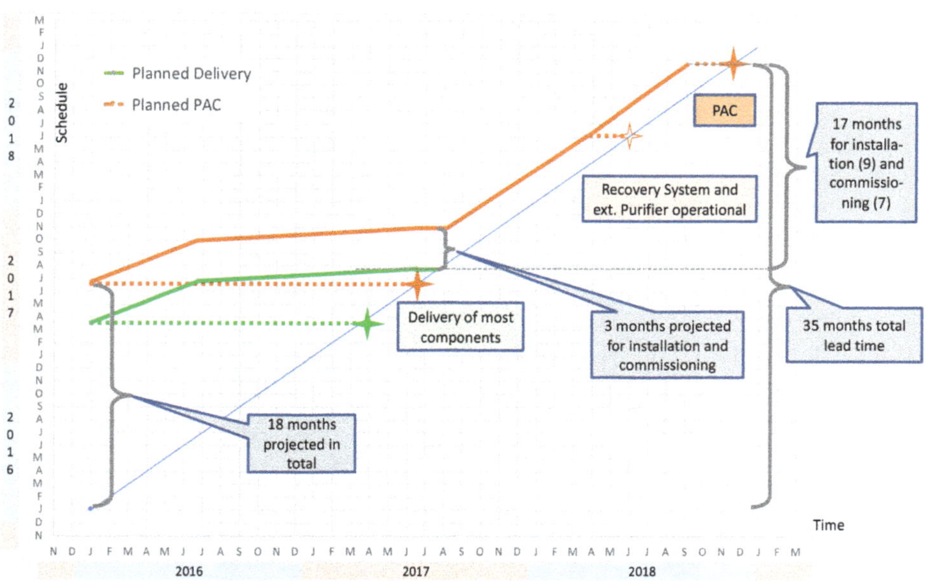

Figure 7.16. TICP project phase schedule vs. real time line. (Courtesy of ESS)

installation and a lacking stress calculation, particularly with respect to the interface points to the ESS supplied piping.

One crucial challenge was the correct support for the ambient heater for the returning subatmospheric flow from the CM test stand, also being used for the acceptance testing, shown in figures 7.17 and 7.18. The entire ambient heater is more than 18 m long and shrinks when in use, which needs to be carefully considered when supporting this pipe and its interconnecting pieces.

Another challenge of the ambient heater was liquefied air collecting on the drip tray and causing a lot of condensate to rain down from the drip tray. So, a secondary drip tray acting as additional fins that enhances thermal exchange with ambient air and provides a gap to the primary drip tray was installed. An even more effective measure was putting up a fan, connected to a perforated tube above the finned tube, which increases the heat transfer coefficient by providing forced flow over the fins.

Another source of difficulties was the pressure safety relief valve of the gasbag, which is basically an oil bucket that was connected to the gasbag on the upstream side and to the safety relief header on the downstream side with a flexible plastic hose. More details are provided in [6].

Several issues that showed up during start-up and commissioning required rectification by ALAT or their sub-supplier, such as a heavily vibrating LN2 filling line for the external purifier, wrong wiring of a superconducting level probe in the LHe tank, an uninsulated line for the vapor cold nitrogen discharge after the liquid

Figure 7.17. Finned tube. Reproduced from [5]. © IOP Publishing Ltd. CC BY 3.0.

Figure 7.18. Ambient heater for 4 K subatmospheric return flow from CM test stand, consisting of four welded finned tubes. Reproduced from [5]. © IOP Publishing Ltd. CC BY 3.0.

nitrogen phase separator upstream the coldbox, no functioning oil level switches at the PVPS, a broken thermal shield test heater, expansion turbine dummies (deployed at commissioning during system conditioning) that did not fit, several process leaks, and more. In addition, issues arising from ESS contributed to delays, e.g., ongoing cable tray installation with blocking scaffolding, power outages, involuntary interaction with the ACCP commissioning as detailed also in [6], and a pre-mature EPICS CSS version and archiver that ESS is responsible for.

7.5 TICP commissioning and acceptance testing

The general acceptance test strategy had already been defined at the procurement stage of the plant. Further details have been developed during preliminary and detail design by ALAT, including definition of the sensors to be used and measurement failure analysis [5].

Three main operation scenarios have been tested thoroughly.

7.5.1 Helium purification by means of the external purifier

Although specification and equipment design seem quite straightforward, correct acceptance testing is not as easy. We decided to test the 10,000 ppm case using dry nitrogen. Topping up clean helium gas bundles with nitrogen would not work due to insufficient homogeneity of the mixture within limited time. In order to achieve a more homogenous and time-constant gas mixture over a long time, i.e., for at least one entire purifier cycle, a simultaneous feed of helium and nitrogen had to be ensured. After several failed attempts to feed nitrogen at the high-pressure valve panel spare connection and to feed nitrogen upstream the gasbag, finally the best place was found directly upstream of the purifier itself very close to the inlet impurity measurement. Estimating the correct gas flow rates of helium and nitrogen based on pressure indicators or even weight scale proved to be too inaccurate or instable over time. Adjusting the flow manually as a function of the online impurity measurement gave the most reliable results.

The enforced acceptance testing of this stand-alone purifier, often a bit underestimated, led to several control logic adjustments and some limited hardware modifications based on suboptimal initial result. However, at the end we can report that the purifier performed at 180% of the guaranteed capacity, which is an excellent final result. Switching to regeneration is still triggered based on time but is now primarily triggered as a function of outlet impurity at the fifth out of seven adsorber beds (where the impurity measurement tap is located). The time constant is, within limits, adjustable and used now only as secondary regeneration trigger [5].

7.5.2 Helium liquefaction including coldbox internal purifier

As explained in [4] the liquefaction capacity of the TICP coldbox is secondary to its performance to supply refrigeration to the cryomodule test stand at ESS. During the procurement phase, the bidders were to give the maximum expected liquefaction rates, both with and without liquid nitrogen precooling and with a helium feed including 1% air impurities that their proposed cryoplant, designed for test stand operation, would provide. These values then had to be guaranteed and constitute the baseline of the acceptance tests in this scenario.

Further test conditions were a rising level in the liquid helium storage dewar and a constant level, over time average, of the pure helium gas buffer tank, i.e., the internal purifier performance needs to match the liquefier performance. To verify this apparently trivial condition is in practice quite time consuming, yet indispensable to enforce controller tuning for good plant performance. The helium flow cycle established to validate these requirements is illustrated in figure 7.19.

By connecting more PHS tanks as feeds for the internal purifier, a long test duration could be ensured. Furthermore, this setup permits precise helium mass balances to double-check the level measurement in the liquid helium storage tank.

Air Liquide guaranteed in their proposal a helium liquefaction rate of 137 $\mathrm{l\,h^{-1}}$ consuming 18 $\mathrm{g\,s^{-1}}$ liquid nitrogen and a rate of 40.7 $\mathrm{l\,h^{-1}}$ without liquid nitrogen precooling, but with 2% impurity. After discussions between ALAT and ESS during commissioning, it was decided to relax the impurity content from 2% to 1% in order to optimize the purifier for a more relevant use case, to be consistent with the ESS specified external purifier requirements, and to ease the acceptance testing itself. The nitrogen feeding was done in a similar way as for the external purifier tests. We implemented our lessons learned and used the well-proven setup with the same impurity measurement, which was specifically mounted on the line dryer upstream the coldbox internal purifier.

As with the external purifier, several control logic adjustments and substantial controller fine-tuning were necessary before an ultimately successful acceptance test could be conducted. The plant reached a liquefaction performance of 138.7 $\mathrm{l\,h^{-1}}$ with LN2 precooling at 1% N2 feed impurity, see figure 7.20, whereby it has to be noted that the internal phase separator level was decreasing during the test, meaning that the net liquefaction is slightly lower than shown in the trendline. Without LN2

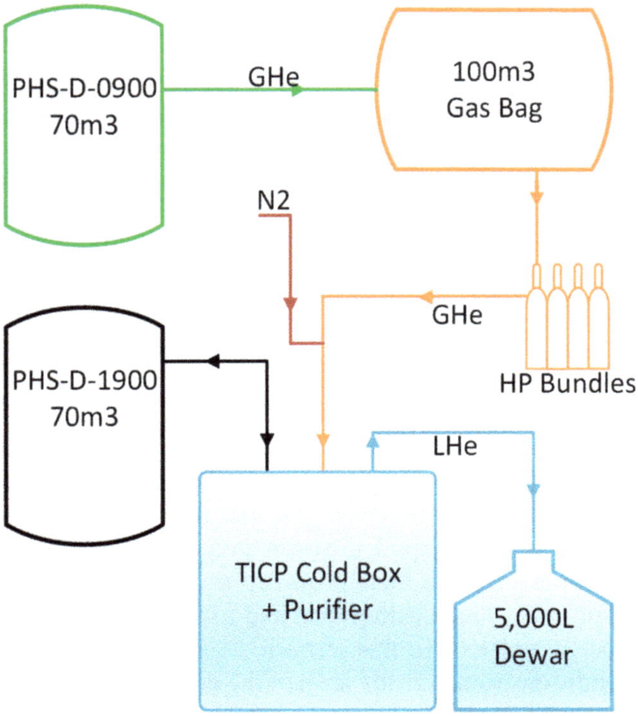

Figure 7.19. Test setup of the liquefaction and internal purifier operation. Reproduced from [5]. © IOP Publishing Ltd. CC BY 3.0.

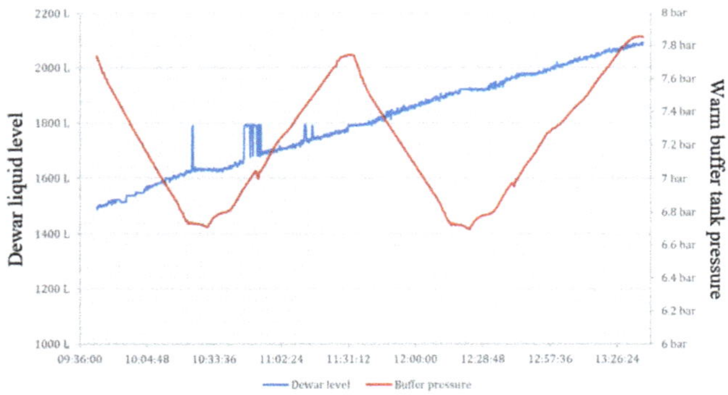

Figure 7.20. Liquefaction w/LN2 precooling and internal purifier operation. Reproduced from [5]. © IOP Publishing Ltd. CC BY 3.0.

precooling, the liquefaction rate was measured as $54.5\,\mathrm{l\,h^{-1}}$ at 1% N2 feed impurity. The LN2 precooled operation mode has passed the test with very little margin only and could possibly have failed had we tested with the original 2% feed impurity. In contrast to the external purifier, the internal purifier consumes refrigeration power from the expansion turbines, which reduces the liquefaction performance. In addition, the purified helium is discharged to the low-pressure side of the recycle compressor, i.e., it has to be compressed again. However, the liquefaction rate is more than sufficient for its purpose at ESS.

A point of concern was the apparently very high liquid nitrogen consumption, which could not be measured precisely during acceptance testing. Data collected during operation now indicate a more reasonable consumption rate that is furthermore being monitored by ESS operators.

7.5.3 CM test stand operation and PVPS

Test stand operation is the main design case of the plant and the performance guaranteed by ALAT matches the requested 4 g s^{-1} 'constant level liquefaction' with LN2 precooling (14 g s^{-1}) plus 387 W thermal shield load at <40 K supply temperature. Of the supplied 4 g s^{-1}, a returning still cold vapor flow of 3.8 g s^{-1} shall be warmed up and returned to the cycle at 27 mbar via a vacuum pump system, whereas 0.2 g s^{-1} cools the power couplers and returns at ambient pressure and temperature to the coldbox.

The thermal shield load was tested by a simple bypass with electrical test heater, continuously switched on during all performance testing with the bypass valve simulating the specified maximal pressure drop of 500 mbar. In hindsight it would have been beneficial to add a flow meter in the shield circuit, not only for the TICP acceptance testing but particularly for operation of the test stand where capturing the heat load on the thermal shields is of high interest yet difficult to obtain without proper flow measurement. A flow measurement in this circuit would have required sufficiently precise temperature measurements, which is also not a given.

In order to test the plants performance without extra equipment, such as a 2 K phase separator and an electrical heater, we decoupled the coldbox test and the PVPS test. This also helped to verify the different process margins of the liquefier and the PVPS.

The 3 bar, 4.5 K supply helium from the internal subcooler of the phase separator vessel was warmed-up in an ambient heater mounted on the wall of the coldbox hall, as shown in figures 7.16 and 7.17. In a first approach, we tried to control this helium flow such that the liquid helium level of the 5000 l dewar was kept constant in order to find out the maximum rate supported by the plant. Setting the correct control parameters turned out to be unfeasible due to the slow level change. Instead, we decided to set a stable flow, using the flow meter downstream of the PVPS, by a constant position of the supply valve and allow a slight dewar level rise, indicating some margin on the plant.

This test was successful and a 4.5 K supply mass flow of 4.1 g s^{-1} was achieved, simultaneously to the specified 387 W thermal shield load and a nearly constant liquid level in the LHe dewar.

Acceptance testing of the PVPS was done using a bypass valve in the gas management panel, which regulates the flow of high-pressure helium from the Kaeser compressor to the suction of the PVPS. The pressure upstream of the PVPS can be controlled very precisely by this small bypass valve and by variable speed control of the roots blowers. In the test we achieved a maximal flow of 5.4 g s^{-1} at 25 mbar suction pressure and ±0.1 mbar pressure stability. Furthermore, special care has been taken to ensure very low disturbances should a pump train be switched off or on to accommodate long term load adjustments by using the large bypass valve. Fast load fluctuations at the test stand shall be regulated by means of heaters in the cryomodules.

7.6 Operation experiences and continuous improvement

7.6.1 Preparation of cryoline and test stand operation

The commissioning and acceptance testing of the TICP were performed without the cryogenic distribution system and CM test stand in place. In order to be able to test the of the cryoline to the test stand heat load, some adjustments were necessary.

To measure a temperature difference between supply to and return from the cryoline, the helium (normally leaving the subcooler of the phase separator vessel at 4.5 K) is controlled to feed helium at a higher temperature. We also need to make sure that the flow downstream of the expansion valve in the valve box returns in purely gaseous form and not in the two-phase state. A stable feed temperature setting of 7 K was achieved by a combination of valve settings at the cold end and attenuation of the expander turbines in order to produce less refrigeration.

Another task was estimating the heat load on the thermal shield of the cryoline. In contrast to the main 4.5 K supply and subatmospheric return circuit, in the shield circuit there are neither flow meters nor sufficiently precise temperature measurements. The one really accurate measurement that we have is the electrical power of the shield test heater. The possibility of different shield bypass valve settings allowed us to do a parameter study in order to help deduce the shield mass flow from its bypass valve settings.

7.6.2 Modifications and fine-tuning

An issue due for optimization are the rather long cool-down and warm-up times of the TICP coldbox. During the bidding phase, ALAT estimated cool-down times of less than 3.5 h (with LN2 precooling) and warm-up times of less than 8 h, which is certainly possible as theoretical calculations show. During commissioning, however, the various cool-downs took about 8 h and the warm-ups about 24 h. We concentrated first on reducing the warm-up time due to the larger potential time reduction.

Four temperatures are regarded for defining the end of the warm-up: the cold end temperatures of the warmest and coldest heat exchangers, and the top and bottom temperatures of the phase separator.

An important role for the warm-up is played by the phase separator with its internal subcooler. Flow through the subcooler coil such as during CM acceptance testing means a high heat load for the cryoplant and an elegant way to bring heat to the coldest part of the coldbox. This means, however, that the phase separator has to be in use (mode enabled) for the warm-up, which is not naturally intuitive when the

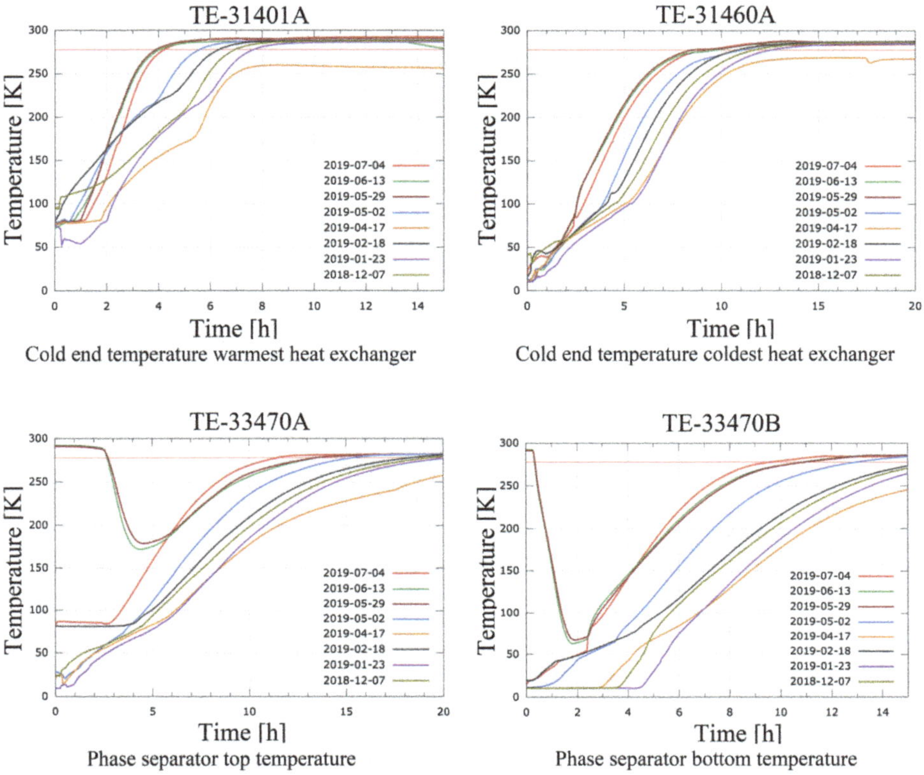

Figure 7.21. Essential coldbox temperatures for warm-up. Different colors indicate different test dates and the red-horizontal line indicates the warm-up end temperature definition of 278 K for all temperature measurement locations. Reproduced from [5]. © IOP Publishing Ltd. CC BY 3.0.

phase separator was not used during nominal operation when the plant was (for example) only used as a liquefier.

As the curves in figure 7.21 indicate, the warm-up times could successfully be shortened but there is still room for further improvement, particularly with respect to the phase separator. Its top temperature does currently limit the total warm-up time to about 11 h.

Additional small modifications of the TICP include an automated mechanism to empty impure HP bundles into the gasbag once the pressure is lower than accepted by either the internal or external purifier. ESS fine-tuned automation and parametrisation of a second unloading valve in the gas management panel that is used during helium recovery of the ACCP's inventory when ACCP and TICP LP circuits are connected. Furthermore, ESS integrated the possibility for remotely switching the gas analyzer on and off, and connecting this functionality to the operation of the recycle compressor to avoid situations of the gas analyzer running without helium feed, which can potentially damage the analyzer cell.

ESS has furthermore performed a number of improvements on the controls, such as updating the alarm table with easily readable additional information regarding possible cause and mitigation of alarms, integrating the system in the ESS alarm service, correcting and updating engineering units and trends, streamlining the look and feel of the OPI, and creating an overall controls' display. Control circuits have been added to facilitate cryomodule filling and filling of mobile dewars has been simplified with push buttons at the filling station.

7.7 Summary and lessons learned

Following smooth engineering, procurement, and manufacturing phases, a lot of issues on both ALAT's and ESS's side have led to substantial delays during the installation and commissioning phases. After successful acceptance testing, the TICP was handed from ALAT to ESS in December 2018. The plant fulfills its key performance parameters and even overperforms in certain areas, particularly the external purifier and the PVPS. Minor performance parameters such as LN2 consumption, warm-up and cool-down times did not meet the initial design estimations but are being fine tuned at ESS as the operating experience is building up amongst the cryogenic team.

Yet, there is a number of points that in retrospect could have been better.

7.7.1 Process design lessons learned

The plant specification was tailored to prompt a 'standard helium liquefier' from the typical suppliers with a few 'extras' to accommodate the special needs of ESS. The main goal of this strategy was to keep the capital investment reasonably low and use mostly well-proven standard equipment without the need of extensive engineering, and this played out rather well for ESS in general. However, some details should have been specified additionally for better usability at ESS.

- One point is that typical helium liquefiers use CTLS as rather cheap but also not very accurate temperature sensors. ESS specified only the 4.5 K, 3 bar supply temperature measurement at the interface to be of a very accurate Cernox temperature sensor measurement. It would have been better to also specify such accurate temperature measurement for the thermal shield supply and return flow because these measurements are used to characterize the shield load on the CMs in the test stand.
- At the same time, it would have been very useful to have a venturi type flow meter for the thermal shield flow.
- The special gasbag safety valve as in figure 7.20 has been described in section 7.3.3 with a safety incident presented in [6]. In retrospect it is clear that combining the gasbag safety valve discharge with the normal process safety valve discharge was a mistake that had to be rectified by ESS later.

7.7.2 Mechanical design lessons learned

- We have reasons to believe that the vacuum barrier design at the TICP coldbox interface to the multitransfer line for the test stand is suboptimal. The sleeve connecting vacuum pipe with the plate where the process line sleeves are connected is rather short and thick walled. The thermal shield lines are not

Figure 7.22. Gasbag before installation of tube for level balancing and regular inflation. (Courtesy of ESS)

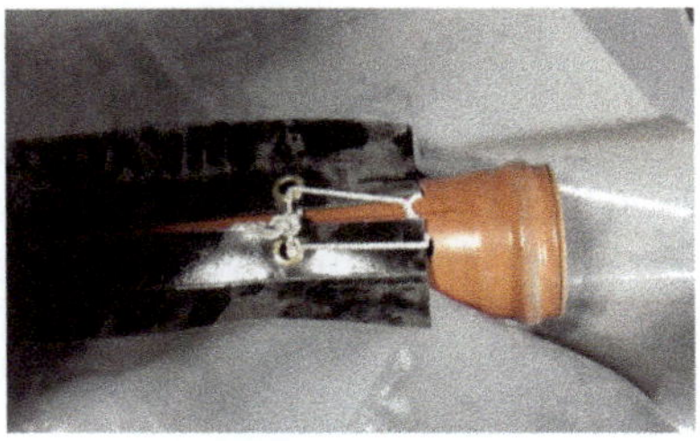

Figure 7.23. Gasbag, with installation of tube for level balancing and regular inflation. (Courtesy of ESS)

connected with shorter sleeves, meaning that this plate has a warmer temperature, bringing more heat to the transfer line interface. This would fit to our observations that there appears to be a much lower cooling power from the TICP when running the test stand compared to operation with its test equipment.

- The big liquid helium storage tank was specified and delivered with a filling connection and additional valve so that the tank can be filled from external mobile liquid helium dewars. However, the filling connection is placed on the backside of the helium tank, inaccessible for connection with a filling tube. External filling is now done using the same decant connections that are used for taking liquid helium out of the tank to mobile dewars. It turned out that the external connection with valve is not only useless but contributes to an increased static heat load of the LHe tank. For external filling, pure pressurized helium gas has to be diverted from the coldbox warm valve panel in order to pressurize the external mobile dewar to permit filling. This connection was retrofitted by ESS's technicians but could have been part of the initial delivery.
- Another improvement would be a weight plate on top of the 100 m^3 gasbag to ensure uniform expansion and accordingly level increase and decrease of the gasbag. This is rather difficult to retrofit after installation when all equipment is in place and access by mobile crane is hardly possible. As a provisional solution, a sand-filled tube has been placed on top of the gasbag in order to avoid irregular inflating of the gasbag, as shown in figures 7.22–7.24.

Figure 7.24. Gasbag after installation of tube for level balancing and regular inflation. (Courtesy of ESS)

- An important lesson learned is to more actively manage the implementation of mechanical interfaces, particularly with respect to piping supports. Defining interface points with allowed forces and moments is to be considered the start of a discussion and commonly agreed approach. A lot of calculations and work had to be redone, including by ESS, to end up with a technically acceptable solution.

7.7.3 Controls lessons learned

- At the time of the TICP controls implementation the ESS EPICS controls setup was still very rudimentary and guidance existed only in the form of written style guides. As described in detail in [7], an early development of a tool chain for the different suppliers is instrumental. Particularly for tasks like PLC-IOC communication but also aligned instrumentation representation and widgets, a library from ESS should have existed earlier instead of leaving this completely to the individual plant suppliers.
- Furthermore, it was beneficial if more controls literate people participated in commissioning and testing or, better, at an earlier stage of controls development to make sure that the controls are more transparent and consistent. Having parameter settings sometimes on controller faceplates, sometimes on parameter screens or state machine overview tables, sometimes both, sometimes exclusive, makes it more difficult to learn and not easy to retrofit in a later state.
- Having a clear alarm table including possibly cause and remedy not only on paper but popping up on the screen with the alarm is also a great help during operation.

7.8 Summary

The cryoplant is operated at least monthly to support the CM test stand and to liquefy helium for Lund University's different institutes, including MAX IV and at a small but increasing rate ESS's internal liquid helium users. The cryoplant and particularly its helium recovery and purification are tied closely into the rest of cryogenic system and is working satisfactory.

Still, a number of issues need to be addressed in the foreseeable future. It starts with software updates to better display sequences and facilitating manual intervention into certain steps to mitigate system hang-ups or speed up and hop over unnecessary steps but also covers required hardware changes, such as on a blocked filling line from the big dewar, constant leak search, and small upgrades for better plant handling.

References

[1] Weisend J G, Arnold P, Hees J F W, Jurns J M and Wang X L 2015 Cryogenics at the European spallation source *Phys. Proc.* **67** 27–34

[2] Garoby R, Vergara A, Danared H, Alonso I, Bargallo E, Cheymol B and Sordo F 2017 The European spallation source design *Phys. Scr.* **93** 014001

[3] Hees W, Arnold P, Fydrych J, Spoelstra H, Wang X L and Weisend J G 2015 The ESS cryomodule test stand *Phys. Proc.* **67** 791–5

[4] Arnold P, Hees W, Jurns J, Su X T, Wang X L and Weisend J G 2015 ESS cryogenic system process design *IOP Conf. Ser.: Mater. Sci. Eng.* **vol. 101** (IOP Publishing) p 012011

[5] Arnold P, Barbier A, Bigeard S, Goncalves R, Gourlet P, Nilsson P and Su X T 2020 The ESS test and instruments cryoplant—first test results and operation experiences *IOP Conf. Ser.: Mater. Sci. Eng.* **755** 012095

[6] Su X T *et al* 2020 Helium management of ESS cryoplants with common safety relief header and recovery system *IOP Conf. Ser.: Mater. Sci. Eng.* **vol. 755** (IOP Publishing) p 012104

[7] Arnold P, Boros M and Nilsson P 2021 ESS cryogenic controls design *EPJ Tech. Instrum.* **8.1** 8

Chapter 8

Elliptical cavity cryomodules and the ESS superconducting radio frequency collaboration

C Darve, P Bosland and P Pierini

This chapter describes the requirements, design, and construction of the European Spallation Source (ESS) elliptical-cavity cryomodules. Initial testing of the components and cryomodules is described. The production of these cryomodules is accomplished via the ESS Superconducting Radio Frequency (SRF) Collaboration. This collaboration involves ESS and numerous in-kind partners. The development and management of this collaboration is discussed in this chapter. Lessons learned, both technical and organizational, are also provided.

8.1 ESS SRF collaboration related to elliptical components

8.1.1 In-kind management for elliptical cavities and cryomodules

In the context of the ESS superconducting radio-frequency linear accelerator (SRF LINAC), the ESS SRF collaboration was developed to achieve the challenging SRF requirements of the 2013 SRF LINAC layout [1–9]. During the initial phase of the ESS establishment, European institutes and stakeholders for the SRF cryomodules collaboration have been identified thanks to Letters of Intent (LoI).

The development of the SRF components for the ESS proton accelerator was possible thanks to the ESS SRF collaboration team, which was composed of highly experienced European institutes, and using lessons learned from existing high-power proton beam and electron accelerators.

Due to the large number of elliptical-cavity cryomodules, and to comply with the in-kind process, several partner institutions were selected by ESS to complete the installation of the nine medium-beta cryomodules and 21 high-beta cryomodules in the cold LINAC [10–34]. The spoke configuration and collaboration are described in [35–48] and in chapter 9. The production of prototypes and series elliptical cryomodules was distributed between several institutions, as shown in table 8.1. The medium-beta and high-beta series cavities were manufactured and tested by

doi:10.1088/978-0-7503-3223-1ch8

Table 8.1. Distribution of the in-kind contributions for the elliptical cryomodules. (Courtesy of ESS)

Activities	In-kind contribution responsibility
Prototype medium and high-beta cryomodule	CEA Saclay and CNRS-ICJ Lab
Series medium-beta cavities production	INFN Milano
Series high-beta cavities production	STFC Daresbury
Cryomodule component production	CEA Saclay
Cryomodule assembly	CEA Saclay
RF power tests	ESS Lund and IPJ PAN
Support prototype testing of elliptical cryomodule	Uppsala and Lund Universities

INFN and STFC, respectively, before being transported to the French Alternative Energies and Atomic Energy Commission (CEA) for assembly into the cryomodules. Following this, the cryomodules were transported to Test Stand (TS2) in Lund to be tested at high power, in collaboration with the Institute of Nuclear Physics PAN Polish institute, before being installed in the ESS tunnel (see chapter 10).

Communication and coordination of each interface were key to the success of the project and required continuous effort [11]. In unison with the ESS local workforce, the elliptical SRF linear accelerator was prototyped and constructed based on a collaboration with European research institutes: CEA-IRFU in Saclay, IJCLab – Laboratoire de Physique des 2 Infinis Irène Joliot Curie (CNRS-ICJ), Istituto Nazionale di Fisica Nucleare , Laboratory for Accelerators and Applied Superconductivity (INFN-LASA), Science and Technology Facilities Council (STFC) Daresbury, Uppsala and Lund universities, as shown in table 8.1.

Beyond the signature of seven in-kind agreements, the collaboration technical and administrative proof of concept was achieved with the design, manufacture, and test of elliptical cryomodule technology demonstrators. A Medium-beta Elliptical Cavity Cryomodule Technology Demonstrator (M-ECCTD) and a High-beta Elliptical Cavity Cryomodule Technology Demonstrator (H-ECCTD) were designed, fabricated, and tested in CEA Saclay. The cryostat design was completed by CNRS-ICJ Lab and the cavity package was completed by CEA Saclay. The SRF component performance measurement and successful assembly resulted in the manufacture of every series cryomodule component, after numerous improvements at several steps of the component life cycle. For instance, two high-beta, six medium-beta cavities, and 10 1.1 MW power couplers (PCs) were designed and tested by CEA. Then, four cavity packages were assembled into the prototype, named M-ECCTD, in the equipped vacuum vessel.

LASA manufactured and tested a new design prototype bare cavity in close collaboration with industry. This cavity design was 'plug-compatible', i.e., all its interfaces were identical to the original cavity design and the two designs are interchangeable. To limit technical risk and to validate series cavity performance, the LASA cavity was added to the CEA cavities string, before being assembled into the M-ECCTD.

8.1.2 Towards an ESS collaborative model for the elliptical cryomodules

The organization of the ESS project and the management of the SRF cryomodule design, fabrication, and implementations has evolved as the ESS has matured from a green-field initiative. This section gives a short historical perspective of the in-kind partnership.

As mentioned earlier, the identification of SRF capabilities for building the ESS cold LINAC has been captured thanks to the European Institute LoI that were collected during the earlier phase of the ESS project.

A large in-kind partnership has been established with the support of five main European physics laboratories, sharing their expertise and infrastructures to produce the required ESS technology. That in-kind partnership was born from proactive collaborations in the field of particle accelerators. Additional lessons learned regarding in-kind contributions are given in chapter 12.

On the one hand, in a top-down approach, collaboration agreements were initiated by ESS in 2011 in the 'Cooperation Agreement in the Field of Neutron and Accelerator Sciences to the ESS Design Phase', which was based on utilizing the experience of CEA-IRFU and CNRS IJCLab in the design and construction of superconducting accelerating structure, and achieved the goal of kick-starting the ESS design update phase. The main purpose of the agreement was to enable an early start of the design, prototyping, and testing of key parts of the ESS accelerator before in-kind contracts and other contributions could be secured from the future member states. The agreement covered the design, prototyping, and testing of superconducting accelerating structures and a normal conducting low energy radio-frequency (RF) quadrupole accelerating structure for ESS.

On the other hand, in a bottom-up approach, two work packages (WPs) have been created in the accelerator project (ACCSYS) to manage the elliptical (WP05) and the spoke (WP04) cryomodule fabrication, respectively. WP05 covers the design, construction, assembly, and testing of the elliptical cavities and cryomodule prototypes (M-ECCTD and H-ECCTD); the design, construction, and assembly of the medium-beta and high-beta series cavities and cryomodules with all their components; and the testing of the first three cryomodules at the ESS in-kind partner and transport of all cryomodules to ESS in Lund.

When the cryomodules were fabricated, a new WP19 was created. This WP covers activities relating to superconducting cavities and cryomodules once they have been delivered from in-kind partners to ESS in Lund. This includes the SRF part of acceptance tests at Facility for Research Instrumentation and Accelerators (FREIA) for spoke cryomodule (chapter 11) and at TS2 in Lund for the elliptical cryomodules (chapter 10), as well as installation in the tunnel, commissioning, and future operation. Additional WPs were essential to operate the cryomodules, i.e., WP08/RF, WP10/TS2, WP11/cryogenics, including the cryogenic distribution system (CDS). Still, for the sake of this book's scope, we will limit our description to the work completed in the framework of the cryomodule fabrication. The project leaders of WP04 and WP05 were based in Saclay and ICJ-Lb, respectively, whereas the management was channeled via a unique ESS WP leader, charged as the CEA and CNRS IJCLab WP deputy leaders.

It is worth noting that the scientific and technical exchange developed during the fabrication of the spoke and elliptical cryomodules, which by the nature of SRF required competencies, has been an excellent vehicle to strengthen the collaboration between French institutes such as CEA-IRFU and CNRS IJCLab.

A demonstrator of each family of cryomodules was planned with our in-kind partners in order to prove the feasibility of the cold LINAC layout. Following the project requirements, the layout of the ESS accelerator was frozen in 2013, limiting the number of SRF cavities, while requiring them to operate at their ultimate accelerated gradients.

At the earlier stage of the ESS project, technology demonstrators were designed, manufactured, and tested for the elliptical cavities and cryomodules. These demonstrators validated the technologies to be implemented in the ESS SRF LINAC [2]. This prototyping phase has included the design, manufacture, and testing of eight elliptical cavities and PCs [4], which were assembled and tested in the cryomodules.

8.1.3 The SRF technology

The choice of SRF technology was a key element in the development of the ESS accelerator. The SRF LINAC is composed of three families of cavity strings, which operate at 2 K. Figure 8.1 shows that the 56 m long section of spoke cavities operates at 352.21 MHz, whereas the 77 m long section medium-beta and 179 m long section high-beta elliptical cavities operate at 704.42 MHz.

Cavities are housed in a cryomodule separated from the following cryomodule by a quadrupole doublet that is used for transverse beam control. This SRF LINAC accelerates the proton beam from 90 MeV, as input energy of the 13 spoke cryomodules, to 216 MeV at the inlet of the nine medium-beta elliptical cryomodules, and to full energy at 2.0 GeV outlet of the 21 high-beta elliptical cryomodules. Table 8.2 summarizes the distribution of SRF cryomodules per family that operate in the tunnel.

The ESS SRF LINAC has been designed to deliver a time-averaged proton beam power of 5 MW to the target at completion, with a nominal current of 62.5 mA.

Beam optic studies helped drive the design of the electromagnetic resonators (EMRs) in order to transfer energy from the RF sources to the beam. Hence, the SRF cavities and cryomodules are designed using the requirements defined in table 8.3.

Hence, the medium-beta and the high-beta cavities will reach accelerating gradients of 16.7 and 19.9 MV m^{-1}, respectively.

The ESS LINAC is fully segmented and cavities are grouped into cryomodules. The spoke cavities and elliptical cavities are gathered two-by-two and four-by-four in their cryomodules, respectively. Accordingly, the superconducting LINAC is composed of 26 double spoke cavities, 36 medium-beta elliptical cavities, and 84 high-beta elliptical cavities.

Thanks to the small difference in the elliptical cavities' length (56 mm), the medium and high-beta cryomodules are housed in a similar vacuum vessel. It is

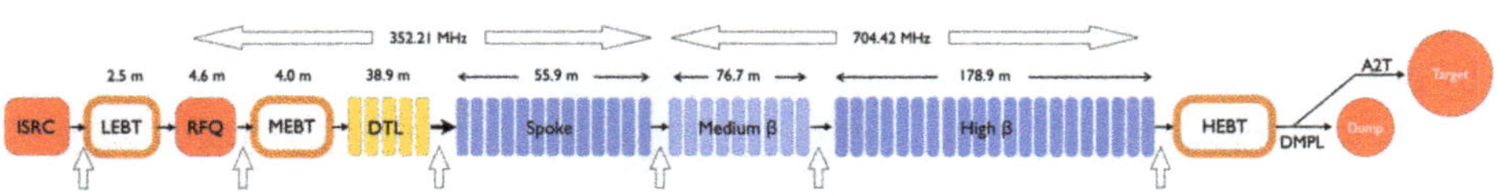

Figure 8.1. Block diagram of the ESS accelerator layout. Reproduced from [1]. © IOP Publishing Ltd. CC BY 3.0.

Table 8.2. Distribution of cryomodules to operate in the ESS tunnel. (Courtesy of ESS)

	Spoke	Medium-β	High-β
β	0.5	0.67	0.86
# CM	13	9	21
Cav./CM	2	4	4
# Cav.	26	36	84
CM L [m]	2.9	6.6	6.6
Sector L [m]	56	77	179

Table 8.3. Cavity EMR requirements. (Courtesy of ESS)

Requirements	Spoke	Medium	High
Frequency (MHz)	352.21	704.42	704.42
Geometric beta	0.50	0.67	0.86
Nominal Accelerating gradient (MV m^{-1})	9.0	16.7	19.9
Epk (MV m^{-1})	39	45	45
Iris diameter (mm)	56	94	120
RF peak power (kW)	335	1100	1100
Qext	$1.75–2.85 \times 10^5$	7.5×10^5	7.6×10^5
Min Q0 at nominal gradient	1.5×10^9	$> 5 \times 10^9$	$> 5 \times 10^9$

worth mentioning that it was made possible to replace two spoke cryomodules with one medium-beta or one high-beta cryomodule. To achieve a uniform lattice and increased reliability of the superconducting LINAC as a whole, the period length, i.e., the length of cryomodule plus a quadrupole doublet, medium, and high-beta in the elliptical section, is the same and is also equal to twice the spoke cryomodule length, which permits their replacement in the unfortunate event where the gradient in one structure is not achieved by the sequence of installed cavities.

Since the main components were manufactured by institutions located outside the Lund area, the integration and interfacing of each resulting component had to be carefully planned [12].

Figure 8.2 shows a 3D view of the elliptical cryomodules.

Finally, each cryomodule houses four cavities operating at 2 K and 704 MHz. LASA and STFC provided the medium-beta and high-beta elliptical cavities, respectively. The elliptical cavities were assembled with their fundamental PCs and cold-tuning systems (CTSs), before being inserted into cryomodules in CEA Saclay, using the experience learned from the European-XFEL project.

In addition to the M-ECCTD and the H-ECCTD, the first three medium-beta cryomodules series and the first three of the high-beta cryomodules series have been tested in CEA-IRFU prior to being shipped to ESS in Lund. Then, the performance measurements are repeated after the TS2 has been validated [14].

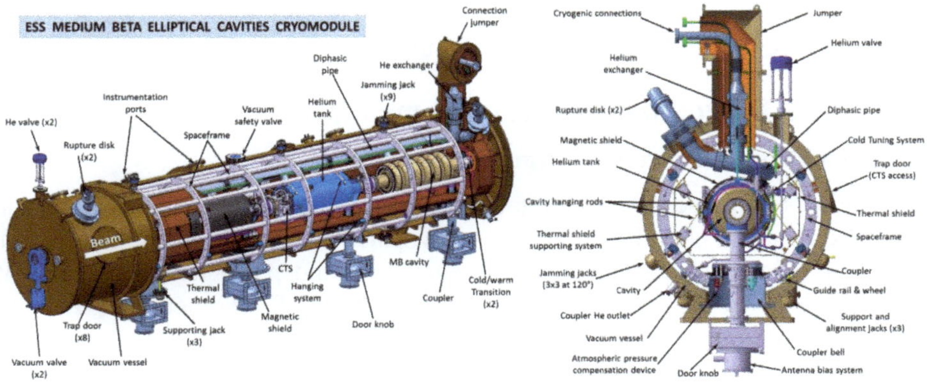

Figure 8.2. Design of the ESS elliptical cryomodule. Reproduced from [1]. © IOP Publishing Ltd. CC BY 3.0.

8.1.4 Manufacture cost estimate and schedule for elliptical cryomodules

The cost estimate for the elliptical cryomodules' WP (WP05) design and construction were developed in collaboration with our main in-kind partners CEA/Saclay. The detailed cost estimate and the different breakdown structures were included in the 2013 Cost Book. These estimates were benchmarked against earlier SRF accelerator projects, such as European-XFEL, SNS, and SPL. The prototyping phase for the elliptical cryomodule also held a competitive bid, involving Fermi National Accelerator Laboratory (FNAL) J-Lab, and CEA-IRFU.

Deviations from the cost book from 2013 followed a strict process entailing control change requests, and were tracked accordingly.

The Technical Annexes (TAs) bonding the ESS and the in-kind partners captured the schedule, cost, and, especially, the scope of the work. In the same way as the cost estimate, the schedule estimate was captured in the TAs, and monitored by the relevant WP leaders over the years. ESS planners supported the WP with the overall estimate. Any deviations and changes were tracked and captured in the project rebaselines.

Several other key scientific institutions have joined this partnership for the LINAC series manufacture and integration as ESS in-kind contributors, and by 2016 agreements of more than 100 million Euros had been signed between partner laboratories.

The in-kind partnership in charge of the elliptical cavities and cryomodules fabrication, and inferred from the 2011 collaboration agreements, had been augmented after 2014 by INFN-LASA in Milano (IT) and STFC in Daresbury (UK). The production of the cryomodules with elliptical cavities was distributed between several institutions. LASA and STFC provided the series medium-beta and high-beta elliptical cavities, respectively. In 2016, LASA proposed an updated RF design for the series medium-beta cavities, and gave momentum to the ESS SRF collaboration. The elliptical cavities were then assembled with their fundamental PCs and CTSs before being assembled into cryomodules in CEA Saclay, using the experience learned from the XFEL project. Finally, the cryomodules were transported to Lund to be tested at high power in collaboration with the IPJ Polish institute, before being installed in the ESS tunnel.

8.2 Elliptical cryomodules technical management

8.2.1 Elliptical cryomodule project management

The technical requirements and the statement of the work have been clearly captured in the TAs signed between ESS and each of the partner institutions. In addition to the explanation in chapter 2, synergies between the in-kind partners and the ESS WP leader have enabled the successful construction of the SRF cryomodules for the ESS operation. In addition, the ESS subproject leader and CEA in-kind management have met regularly during the accelerator monthly meetings and collaboration boards to address general and administrative management using a top-down approach.

Daily technical follow-up meetings between WP leaders and the technical responsibilities of both parties were key to the success of this large SRF cryomodule collaboration. Continuous and harmonious weekly meetings permitted ESS to track the details of the progress and addressed all possible issues, including at the interfaces. Continuous communication was possible thanks to the use of ESS tools adapted to prompt communication, e.g., Confluence, and final project documentation were captured in a product life-cycle management platform that was based on the Enovia product, called Collaborative Home ESS (CHESS).

Lessons learned and status reports of each partner's technical activities and integration were followed-up on a weekly basis to ensure an effective collaboration.

In the case of the elliptical cryomodule, the work breakdown structure (WBS), which is used to track the project progress, depends strongly on the product breakdown structure (PBS). Hence, this WBS is an Elliptical project management tool that is used to identify the status of the main components.

8.2.2 Technical requirements applied to elliptical cryomodules

Hundreds of requirements were identified in order to best integrate the SRF components in the ESS tunnel. The requirements were also used to define the geometrical and functional interfaces with the conventional facilities, the control system, and to define the operating modes for the ESS SRF LINAC [9]. The high-level requirements were defined in order to optimize the layout of the ESS superconducting LINAC [10].

The ESS accelerator requirements were defined at different levels.

The requirements for the SRF LINAC's cavities and cryomodules were quality defined using so-called disciplines. The relevant disciplines for the SRF LINAC were the EMR, radio-frequency system (RFS), cryogenics (CRYO), and vacuum (VAC).

The requirement PBS for the accelerator is shown in figure 8.3, and enabled the identification of [10]:

- Level 1: Overall ESS facility
- Level 2: System projects, e.g., the LINAC
- Level 3: LINAC sections, e.g., medium-beta LINAC (MBL)
- Level 4: Disciplines, e.g., EMR, RFS, CRYO, VAC etc.

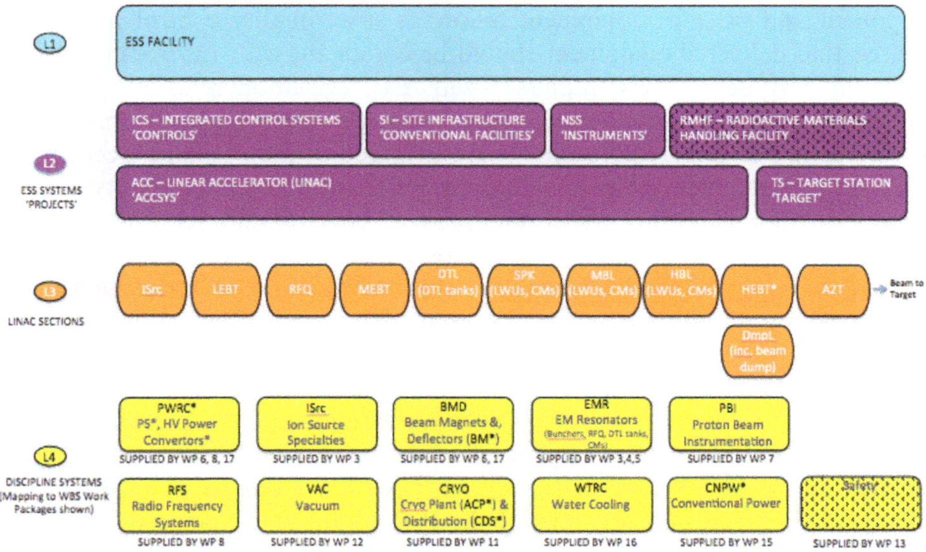

Figure 8.3. PBS for the definition of requirements. (Courtesy of ESS)

Each requirement was described in a comprehensible manner and was intended to identify:

- The part of the system that performs the action
- The action to be performed
- The part of the system acted upon
- Where the result of the action is sent/received
- When the action is performed and/or when the action is not performed
- The rationale—the reason or benefit achieved

Hundreds of requirements were identified in order to best integrate the SRF components in the ESS tunnel. Interface requirements between the different levels and the different disciplines were then developed.

The technical requirements of components and subcomponents formed the bonding agreement between ESS and each in-kind partner institution, and were reported in the TAs. Institutions were chosen based on their capacity to deliver the operating components.

8.2.3 Quality management and equipment compliance applied to elliptical cryomodules

Due to the interfaces and the mosaic complexity of equipment to be assembled at the ESS facility, strict quality processes have been implemented.

Comprehensible WBS have been used to guide the SRF project.

As the ESS project matured, the WP leaders have addressed technical issues in compliance with the ESS's quality management rules. The life cycle of each component and piece of equipment follows a strict quality control process. ESS required that delivered equipment should be compliant with European Directives. The main purpose of the compliance with EU legislation is to ensure that the equipment used at ESS is safe to operate, with regards to both personnel and the environment. The ESS requirements were captured in the in-kind agreements and TAs. The TAs required the parties to implement and maintain a quality assurance and safety approach throughout the project, covering all relevant aspects of ISO9001 with respect to the scope of the IKC delivery and all specified reliability, quality assurance, and safety requirements A quality assurance process was established to track the fabrication and nonconformities of components. ESS has developed a database to collate the engineering documentation to operate and repair the cryomodules once installed in the ESS tunnel.

CEA-IRFU partners have prepared a consistent and comprehensive project quality plan (based on the template) for its contribution, and submitted it for approval by the ESS WP leader. The aim was for the quality plan to comply in general with the recommendations of the ISO 10005:2005 Standard. The project quality plan supported the requirements of the in-kind deliveries and ESS equipment to be integrated in a safe manner in order for the ESS to obtain the relevant licenses from the authorities. The equipment design and assembly complied with harmonized standards, and were integrated in the final system assembly, in compliance with the Swedish legislation. Hence, a certificate from the relevant Swedish authorities is needed before the LINAC's operation can be permitted. Ultimately, the SSM (Swedish radiation authority) will issue the permit to operate.

The production of the equipment to be assembled at ESS was to follow the steps described in the Engineering Handbook. In fact, a series of reviews guaranteed the integrability of the equipment produced. Still, it is also worth noting that, due to project schedule constraints, the series manufacture of the components often had to be started before the prototyping phase was completed. Every six-month a risk analysis has been organized in collaboration with CEA-IRFU and ESS.

Even more important is the fact that as a green-field project, and as early as 2011, ESS had to base the implementation of the ESS requirements on the competencies and expertise of ESS, CEA-IRFU, and CNRS IJCLab engineers. These review deliverables were validated by an accredited body. Numerous lessons learned based on other international accelerator development have enable the identification of the optimal project management, e.g., a safety review have been conducted since 2013 in order to prefer a technical cryomodule design, complying with the EU Pressure Equipment Directive Sound Engineering Practice (SEP). Indeed, the design of the spoke and elliptical cryomodules evolved from those high-level requirements. This early decision has strongly benefited the success and growth of the SRF Collaboration, which INFN and STFC joined in 2014.

In order to support the progress of the ESS, ESS Rules Implementation for Equipment Compliance has been created, which applies to all ESS equipment supplied by in-kind partners or procured by ESS. This document aims to clarify the ESS rules for equipment compliance/CE marking (conformité européenne, i.e., European conformity) in alignment with the European Legislation[1] (regulations and directives) for products, by complying to the Essential Safety Requirements within them (i.e., the given equipment[2]). This means that the Swedish authorities grant ESS a permit to operate the facility, as pictured in figure 8.4.

The rules in this document apply to the following product Directives: Machinery Directive (MD), Low Voltage Directive, Electro-Magnetic Compatibility Directive, Pressure Equipment Directive (PED), ATmosphères EXplosibles Directive (ATEX), and the Restriction of Hazardous Substances. The legal aspect, as stated in these Directives, only concerns the safety of the product, which means (for example) that it does not cover financing aspects. Application of the Construction Product Regulation (CPR) is not covered in this document, even if the CPR applies together with one or more of the above Directives (e.g., a motor-driven industrial gate: the CPR and MD). These rules show the required steps for different types of equipment routes, as applied at ESS. The required technical documentation, mentioned in the text, mainly refers to document(s) regarding safety and, in the MD, refers to the technical file.

All operator national safety laws and legislation applicable to the design, development, manufacture, installation, testing, and operation of the item supplied shall be followed and fulfilled, as defined in the requirement document for the facility element by ESS.

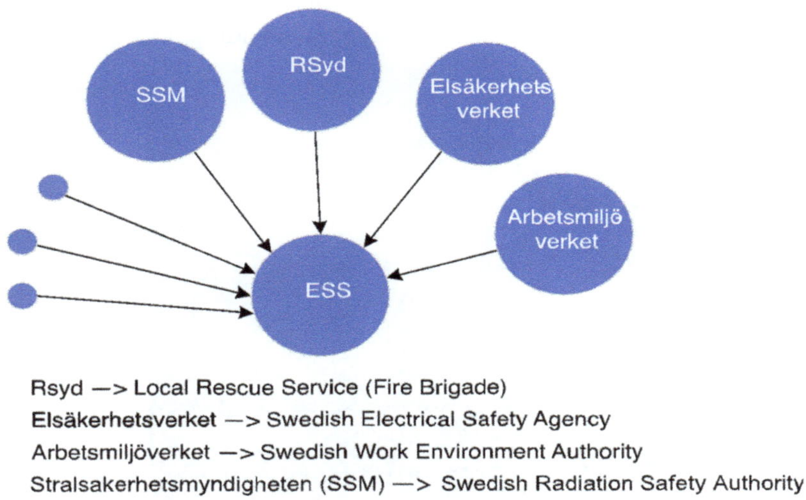

Figure 8.4. Applicable authorities at ESS. (Courtesy of ESS)

[1] The official EU guideline to interpret the approach to the above Directives, with the exception of CPR, can be found in the Blue Guide ['Blue Guide' on the implementation of EU product rules].

[2] Equipment: a component, an assembly of component, system or any subsystem delivered by the in-kind partners or procured by ESS.

8.3 Integration of the elliptical cryomodule components

8.3.1 Elliptical cryomodule description

The design of the cryomodule is based on the SNS/CEBAF concept, with an aluminum space frame and titanium alloy (TA6V) tie rods holding the cavity string and the thermal shield inside the vacuum tank. Figure 8.2 shows the content of an elliptical cryomodule.

Each cryomodule houses four cavities operating at 2 K, and there is no focusing magnet coil inside the vacuum tank. Figure 8.5 shows 3D views of ESS medium-beta elliptical cryomodules.

The main characteristics of the elliptical cryomodules are:

- Overall length 6.584 m.
- Beam axis height 1500 mm.
- Overall height 2826 mm from the ground.
- Weight 5.8 tons.
- Distance between couplers 1500 mm.

Two prototype cryomodules were developed in order to qualify the technology before initiating the production of the series of elliptical cryomodules [40]. These two prototypes are named M-ECCTD (for the medium-beta section at β=0.67) and H-ECCTD (for the high-beta section at β=0.86).

To illustrate the main components contained in the elliptical cryomodule, figure 8.5 shows the design of the high-beta cavity package, composed of the five-cell high-beta cavity, its helium vessel, its magnetic shielding, its PC, and its CTS, as designed by CEA-IRFU.

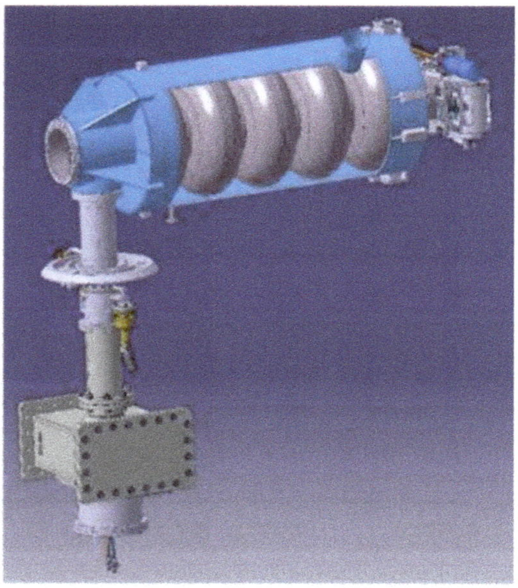

Figure 8.5. High-beta cavity package. (Courtesy of ESS)

Figure 8.6 shows a view of an elliptical-cavity string before it was assembled inside the cryomodule space frame and vacuum vessel. The cavity string contains four cavity packages.

The PCs are fixed longitudinally and the thermal shrinkage of the external conductors of about 0.6 mm is kept free, allowing the cavity axis to be kept fixed during cooling cycles.

The magnetic shielding thickness was enlarged from 1.5 to 2 mm in order to increase the margins on the material permeability at cold temperature, while achieving the same shielding efficiency of 35 (B_{ext} = 1.4 μT max.). For the CTS, the piezo frame design was simplified and its stiffness was optimized. A piezo frame stiffness of 14.7 kN mm^{-1} was adopted, which is compatible with both cavity types.

The thermal aluminum shield is cooled by helium gas at 50 K and 19 bar. Figure 8.7 shows the thermal shield and the space frame of the elliptical cryomodules.

In a change from the spoke cryomodule, the cryogenic heat exchanger of the elliptical cryomodule is located inside the cryomodule vacuum (figure 8.8), and the two cryogenic valves are placed on the cryomodule: one for cooling and one for helium level regulation. The heat exchanger is a HAMPSON type that is able to deliver up to 5 g s^{-1} and adapted in-house at CEA to match the ESS requirements. Table 8.4 shows the heat load calculated and estimated for all three types of ESS cryomodules.

The other cryogenic valves are located on the cryogenic valve box. The 100 mm diameter bi-phase cryogenic pipe above the cavities is made of titanium. It is welded in place, and three bellows in titanium along this pipe compensate for cavity misalignments and allow for thermal shrinkage. This bi-phase tube is fitted with two burst disks at its two extremities, ensuring protection against accidental events. The geometry of the bi-phase tube was optimized to limit the pressure increase in the

Figure 8.6. A cavity string ready for installation in the space frame and vacuum vessel. (Courtesy of ESS)

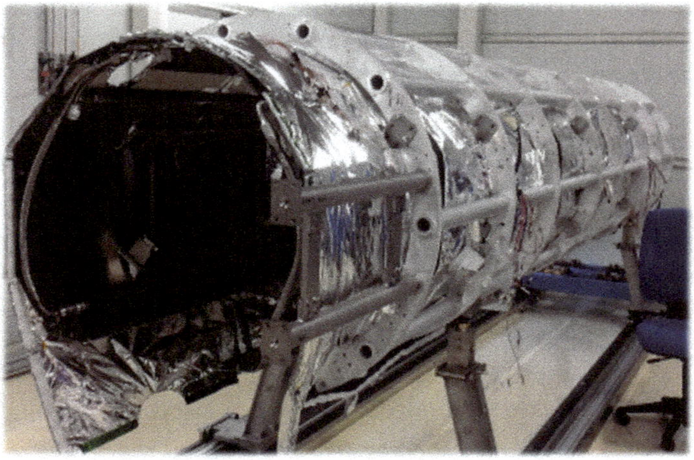

Figure 8.7. Elliptical cryomodule space frame and thermal shield. (Courtesy of ESS)

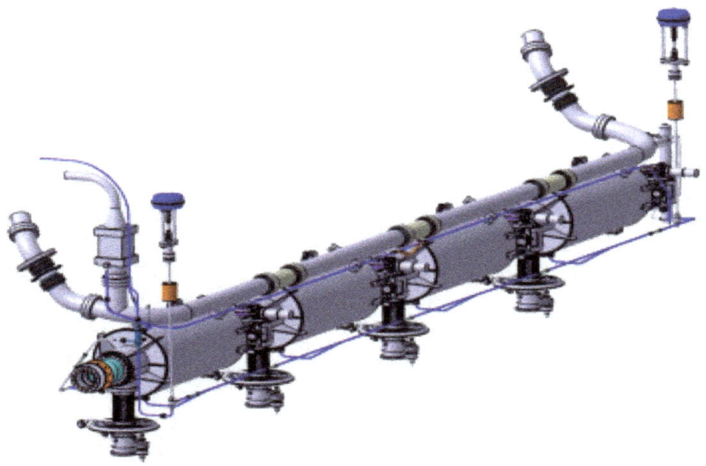

Figure 8.8. Cavity string equipped with the cryogenic pipes. (Courtesy of ESS)

worst event of a beam vacuum rupture accident. The maximum pressure was then determined by the burst disc's value, which was selected at 0.99 ± 0.05 bar.

The service pressure of the cryomodule vacuum vessel is P_s=1.04 bar, which ensures that these cryomodules are compliant with article 3.3 of the European PED (97/23/EC).

8.3.2 Assembly steps of elliptical cryomodule

The assembly steps of the elliptical cryomodule have been prepared by CEA-IRFU team using the M-ECCTD to benchmark the assembly of the 30 series medium and high-beta cryomodules. A purpose of the technology demonstrators is to validate the assembly procedures and the performance of each of the SRF's components.

Table 8.4. Heat load calculated and estimated on the elliptical cryomodules. (Courtesy of ESS)

	Medium-beta				High-beta			
	50 K		2–5 K		50 K		2–5 K	
	Stat.	Dyn.	Stat.		Stat.	Dyn.	Stat.	
Cavity string								
Beam losses (1 W m^{-1})				3.25				3.25
Rff losses				20				20
Radiations (14 m^2)			0.7				0.7	
Cold warm transition (×2)	3		2		3		2	
Supporting system	6		0.25		6		0.25	
Helium piping								
Supporting system	0.2		0.4		0.2		0.4	
Bursting disks (×2)	3		0.15		3		0.15	
Helium valves (×2)	2		0.2		2		0.2	
Safety relif valves (×2)	0.3		0.03		0.3		0.03	
Thermal shield raditions (21 m^2)	31.5				31.5			
Couplers (×4)								
Sleeve cooling(4×23 mg She at 5 k)			4	−4			4	−4
Radiation antenna to cavity			2.8				2.8	
Instrumentation, heaters, and actuators	1.5		2.7		1.5		2.7	
TOTAL static load	46.23		13.23		46.23		13.23	
TOTAL dynamic load				23.25				27.65
TOTAL	46.5		37		46.5		41	

The elliptical cryomodule equipment arrives at the CEA Assembly Hall after a careful validation of their requirement compliance. The nine workstations installed at CEA will permit the completion of the delivery of the cryomodules to ESS in Lund.

Each component is carefully instrumented and calibrated thanks to designated test-stand areas, prior to being assembled on the cavity test string.

The main components of the cavity package are the SRF cavity (EMR), the PC, and the CTS.

Then, the PC is assembled on the cavity in first workstation.

A dedicated clean room with workstations has been constructed for this purpose, as shown in figure 8.9.

Numerous verifications and strict toll gates are applied at all stages to ensure the successful delivery of the cryomodules.

Following the cavity package installation in the clean room, the steps of the assembly are illustrated in figure 8.10, showing the four main steps of the cryomodule assembly.

Figure 8.11 shows the assembly of the cavity string before insertion inside the thermal shield of the M-ECCTD. In this case, the three CEA cavities and the LASA prototype cavity were assembled to PCs and the cold-tuning system before being inserted into the M-ECCTD. Then, the M-ECCTD was tested at the end of 2017 in Saclay and at the beginning of 2018 at Lund Test Stand 2 (TS2).

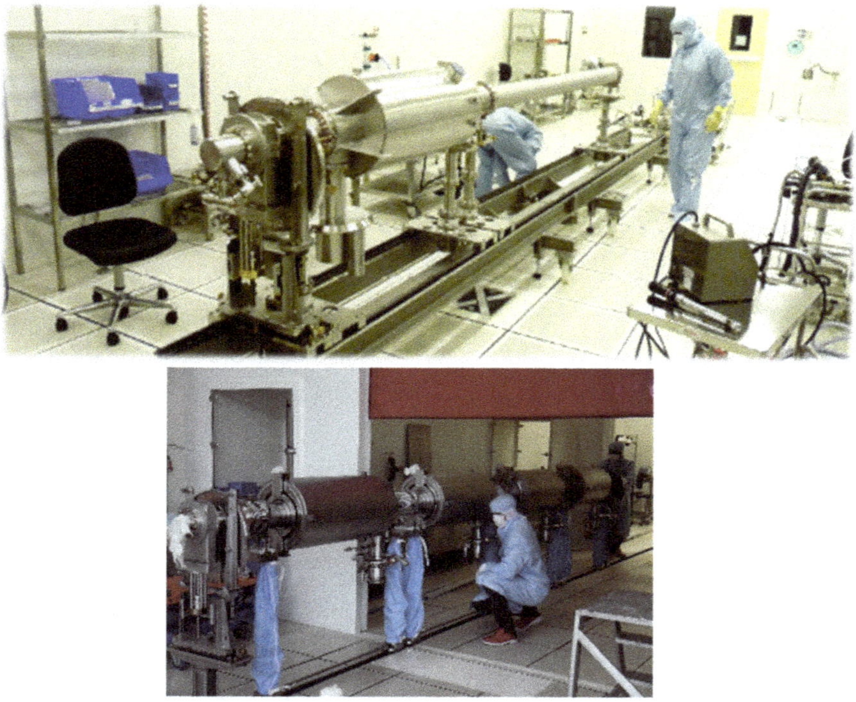

Figure 8.9. (a) Top: CEA-IRFU clean room with one elliptical cavity. (b) Bottom: Transfer of completed cavity string from the clean room to the cryomodule assembly area. (Courtesy of ESS)

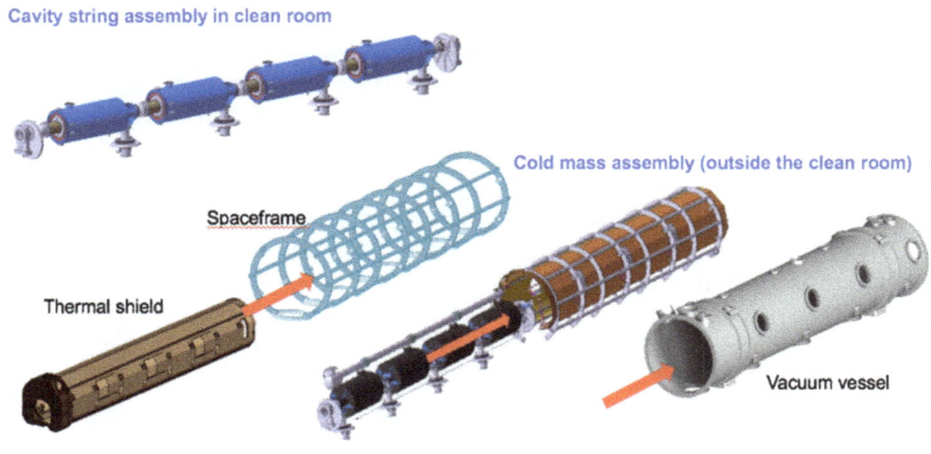

Figure 8.10. Steps of the elliptical cryomodules installation. (Courtesy of ESS)

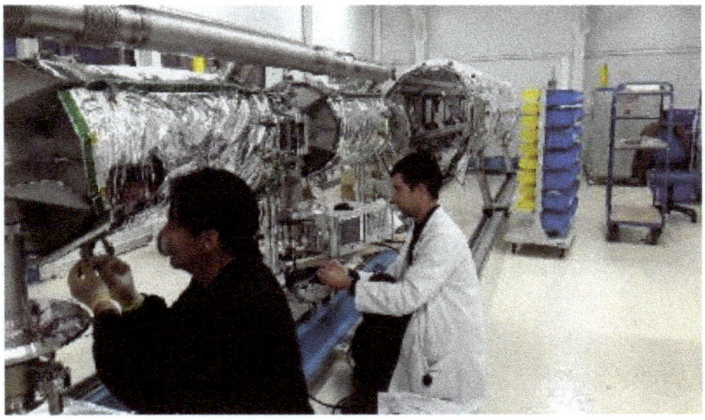

Figure 8.11. Elliptical cavity cryomodule string assembly. (Courtesy of ESS)

It is worth noting, that the medium-beta series cavities were tested at DESY using the European-XFEL vertical test infrastructures and STFC Daresbury has contributed to this SRF momentum with the commissioning of new infrastructures to test their high-beta cavities.

In a similar manner, during the course of the ESS SRF cold LINAC preparation, the new SRF collaboration broad competencies have enabled the testing of essential SRF components in different locations to limit delaying the challenging ESS schedule.

ESS led the effort to coordinate the interfaces between each partner institution, and the technical requirements to integrate those components in the cold LINAC.

8.3.3 Functional analysis and operational modes

CEA-IRFU validated the diverse control and operation parameters during the commissioning of the demonstrator and the first series cryomodules. A functional analysis was prepared in the earlier stage of the SRF design. The resulting operational logic has been implemented and finally tested as the elliptical cryomodules were tested at ESS test stand (TS2), as described in chapter 10. The functional analysis explains the relation between the different operational modes. Each operational mode of the ESS LINAC is a combination of one machine mode and one beam mode. The operation of the 704 MHz SRF LINAC is based on the experimental physics and industrial control system. High-level operating modes were identified, describing the required sequences to operate the CM. More information is given in [16].

8.3.4 Instrumentation and control

The process and instrumentation diagram (P&ID) in figure 8.12 shows the equipment and instrumentation required to control, monitor, and safely operate the cryomodule, which also serve as sensors and actuators for the PLC IO. A list summarizing the necessary instrumentation was produced, where each instrument is classified by location and type, such as: temperature sensors (64), heaters (16), pressure sensors (9), vacuum gauges (8), level gauges (2), flow meters (8), control valves (18), gate valves (2), step motors (2), piezo actuators (8), arc detectors (8), electron charge detector (4), RF pickup coils (4), pumps, etc The instrumentation list allows not only the assessment of the type and quantity of each instrument but also the signal and cable types, space allocation, and interfaces with other systems.

Process abstract and function tables were created to allow for a more detailed description of each operating mode. Here, the status of each controlled object was

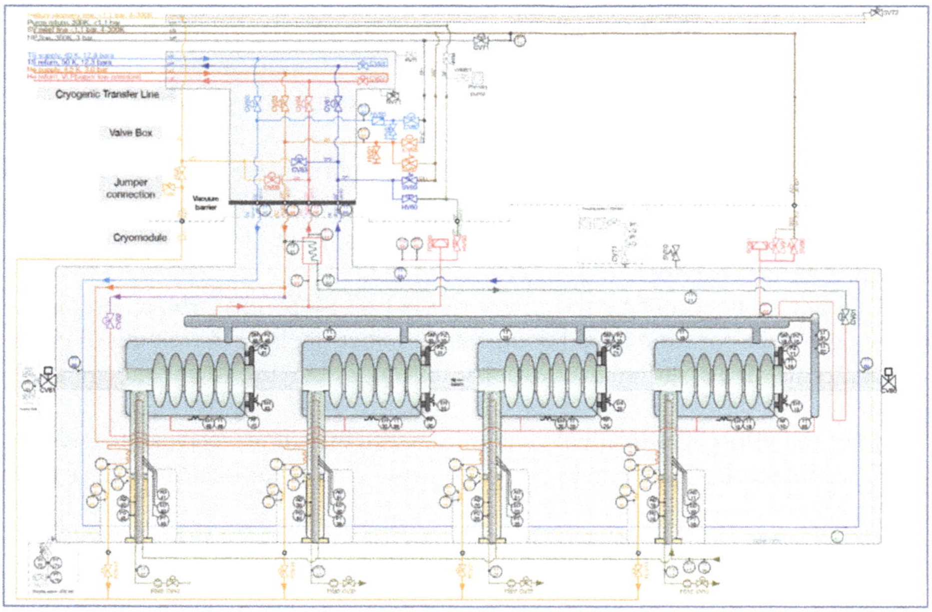

Figure 8.12. Series Cryomodule and Valve Box PFD. Courtesy of ESS. Courtesy N. Elias

defined for each moment during the operating sequence, along with the necessary interlocks and thresholds for process variables to allow for a safe operation.

This diagram depicts the different cryomodule subsystems (beam vacuum, cryostat vacuum, cavity cooling, PC cooling, CTS, etc)

The operation of the cryomodules depends on the numerous variables and equipment. Of particular importance to a high-power facility such as ESS is the operation of the machine protection system, which is handled very well through the requirements process and allows the necessary functionality of the system to be clearly specified without making assumptions regarding the architecture used to deploy this system. This functionality is documented at a high level, with requirements then placed on certain aspects of accelerator equipment (i.e., detectors and actuators) at a lower level.

8.3.5 Interfaces for the elliptical cryomodules

8.3.5.1 Interface with cryogenic system

The cryogenic requirements define the cooling capacity needed to maintain the proper operation of the superconducting cavities. The SRF cryomodules receive a distributed cooling capacity from the Accelerator Cryoplant (ACCP—chapter 3) through the CDS (chapter 4). For each family of cryomodules, the mass-flow, temperatures, and pressures are required to optimize the LINAC operation. The static and dynamic heat loads are controlled and used to define the overall cooling capacity of the cryogenic plant. The cryogenic requirements for the SRF LINAC also capture the restricted needs for the operation of the cryogenic plant.

The cryogenic interfaces of the cryomodule consist of four helium circuits (inlet and outlet for cavities and thermal shielding cooling) integrated in a 90° jumper connection. The base of the jumper hosts the 2 K heat exchanger, while most of the cryogenic valves are placed in the cryogenic distribution line (CDL) valve box.

The function of the interface between the cryomodule and the CDS is to provide the required cooling power at given temperature levels, as well as to regulate the cryogenic cooling of the niobium SRF cavity helium tank and the cooling of the PC double wall [9].

The helium provided by the ACCP is delivered to the cryomodules through the CDS. It consists of the CDL running alongside the LINAC for transfer, which supplies/returns from/to the valve boxes [38], interfacing with each individual cryomodule. Each valve box is equipped with a set of valves, allowing the warm up of the linked cryomodule while keeping the rest at cryogenic conditions. The CTL provides the cryomodules with supercritical helium at 4.5 K and 0.3 MPa. During cooldown and filling the helium is expanded to 0.14 MPa, while during nominal operation the required 2 K temperature level is created by precooling the 4.5 K helium in a heat exchanger and throttling in a Joule–Thompson (JT) valve, consisting of the vapors pumped back by cold compressors in the ACCP at 31 mbar. The heat exchangers and JT valves are located in the valve boxes for the spoke cryomodules and in the cryomodules for the elliptical cryomodules. The regulation of the thermal operating conditions of the SRF cavities is performed by a control loop acting upon cryogenic controlled valves and dedicated process variables.

8.3.5.2 Interface with RF and LLRF systems (incl. klystron and modulator)
We recall that the ESS elliptical superconducting LINAC consists of two families of elliptical cavities. The medium-beta cavities (β=0.67) accelerate the beam from the spoke LINAC at 216 MeV up to the 571 MeV at the entrance of the high-beta elliptical cryomodule section, which is composed of 84 elliptical cavities (β=0.86) [5]. The high-beta cavity was designed by CEA, and the medium-beta cavity was finalized by INFN Milano based on an original design by CEA in close collaboration with Lund University.

The RFS requirements define the operating conditions of the high-power RF sources and the low-level RF (LLRF) for the three families of SRF cavities. The RF sources provide the amplitude and the phase to the forward voltage wave and to the beam through the PCs and the SRF cavities. The RFS system is capable of handling 100% reflected power for the entire pulse length. The characteristics of the RF pulse are monitored and open loops permit the synchronization of the cavity frequency tuning with the beam pulse.

Each cavity and PC are supplied by one dedicated RF power source (Pmax = 1.5 MW) that is defined elsewhere [7–9]. The physical interface between the RF system and the cryomodule is located at the doorknob, which is installed on each PC at the lower side of the cryomodule. Figure 8.13 shows an elliptical PC.

The amplitude and phase of the electromagnetic field inside each cavity is controlled by a low-level RF (LLRF) system that uses the RF signal of the cavity through its pickup antenna probe. In-kind contributions to this effort included work by Atomki (Hungary) and Lund University

The PC transferring the high power (1.1 MW max) to the cavity and to the proton beam is one of the most critical components of the cryomodule. It needs an efficient local protection system that uses three diagnostics signals: two arc detectors,

Figure 8.13. Elliptical PC. (Courtesy of ESS)

installed on each side of the PC ceramic window, and one vacuum gauge plus one electron detector used to interlock the RF system and the machine beam operation.

The RFS system will be capable of handling 100% reflected power for the entire pulse length. The characteristics of the RF pulse are monitored and open loops allow synchronization of the cavity frequency tuning with the beam pulse.

8.3.5.3 Interface with the vacuum system

Similar to the cryogenic requirements, the vacuum requirements ensure safe operation of the cryomodules [9].

The insulating vacuum inside the cryomodules are pumped down to 1×10^{-6} mbar at the time of the site acceptance test, i.e., the testing of each specific cryomodule in TS2.

The beam vacuum is controlled to limit any possible contamination of the beam vacuum. The cryomodule is fitted with two warm gate valves at each extremity, isolating the beam vacuum and connected to the warm LINAC units. The physical interface between the cryomodule and the beam vacuum relates to the operation of the vacuum isolating gate valves, which are installed at the extremities of the cryomodules [15]. Vacuum failure scenarios are mitigated and minimized with the use of fast-acting valves located at the inlet and outlet of the SRF LINAC sections. During maintenance modes, the vacuum isolating gate valves are closed in order to protect the cavities from possible pressure increases.

Finally, it is worth noting that the PCs are cooled by pressurized water and compressed air is also used for the operation of the controlled valves.

8.4 Elliptical cryomodules testing

The aim of this section is to present the different components of the elliptical cryomodules and to give examples of performance measurement. Hence, the performance of the elliptical cryomodules depends on the validation and calibration of each cryomodules component.

8.4.1 Cryomodule testing

By the end of August 2018, a new test stand has been prepared at CEA-IRFU (figure 8.14) and the first operation was the RF commissioning of the PCs up to the maximum power of 1.1 MW at room temperature. The goal of this test-stand cryomodule was to validate the elliptical cryomodule assembly and that the RF losses were as required.

The successful result of the M-ECCTD tested in the CEA test stand has validated the technologies used for the ESS cryomodules of the series.

Field emission was measured in this test area and compared with the cavity vertical test [13].

Heat load measurements have been completed in the CEA test-stand area, but with a different environment than the one of the ESS tunnel. Indeed, the thermal shields were cooled by N_2 around 80 K and no supercritical helium was available, but rather helium was boiled-off at atmospheric pressure. A summary of the results

Figure 8.14. CEA test stand. (Courtesy of ESS)

obtained during the RF power tests performed with the M-ECCTD in CEA Saclay between September and December 2018 is available elsewhere [21].

The final measurement of the cryomodule's performance at ESS Test Stand 2 (TS2) is detailed in chapter 10

8.4.2 Cryomodule component testing

The demonstrator phase allowed the validation of the cryomodule components testing. For instance, two prototypes of high-beta elliptical cavities and six medium-beta cavities were manufactured and tested in a vertical cryostat at Saclay [21]. Note that all higher order modes (HOMs) are at least 5 MHz away from integer multiples of the beam-bunching frequency (352.21 MHz) for any HOMs whose resonant frequencies are below the cut-off frequency of the beam pipe. In this way, a stringent follow-up is applied to the medium and high-beta cavity manufacture sequences.

For the series cavities, the manufacture and testing were carried out by INFN/LASA for the medium-beta cavities, and by STFC Daresbury for the high-beta cavities. It should be noted that for risk reduction purposes, the niobium sheet material for the manufacture of the cells underwent eddy-current scanning and a subsequent analysis of inclusions/defects at DESY.

8.4.2.1 SRF cavities performance

As an example of elliptical cavity performance, figure 8.15 shows the achieved medium-beta cavities accelerating gradients, the Q-curve after chemical and heat treatment with $1.9\bullet10^{10}>Q_0>3.2\bullet10^{10}$, and surface resistances ranging from 3.14 to 7.09 [17].

The two prototype high-beta cavities exceeded the ESS expected nominal gradient of 19.9 MV m^{-1} and Q_0 of 5.10^9, as shown in figure 8.16.

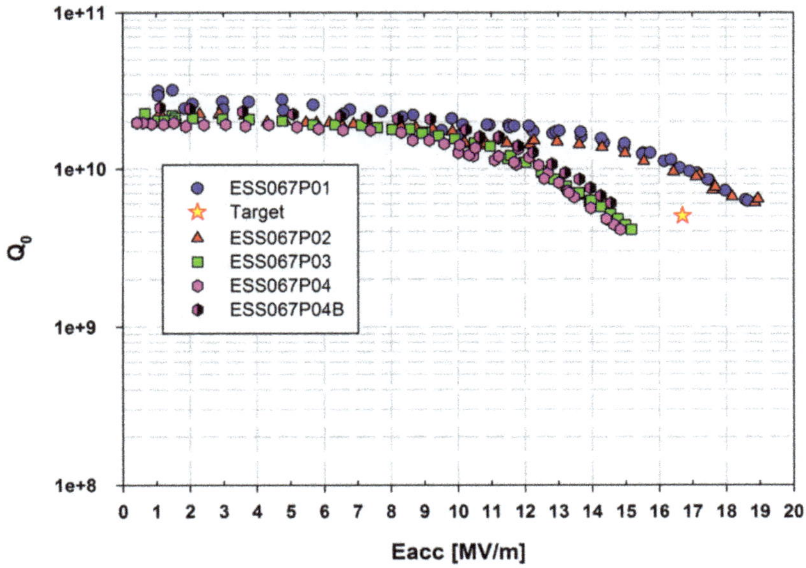

Figure 8.15. Vertical test ESS medium-beta cavities (prototypes) with tank (CW mode @ 2 K). Reprinted with permission from [20]. Copyright © 2017 by JACoW

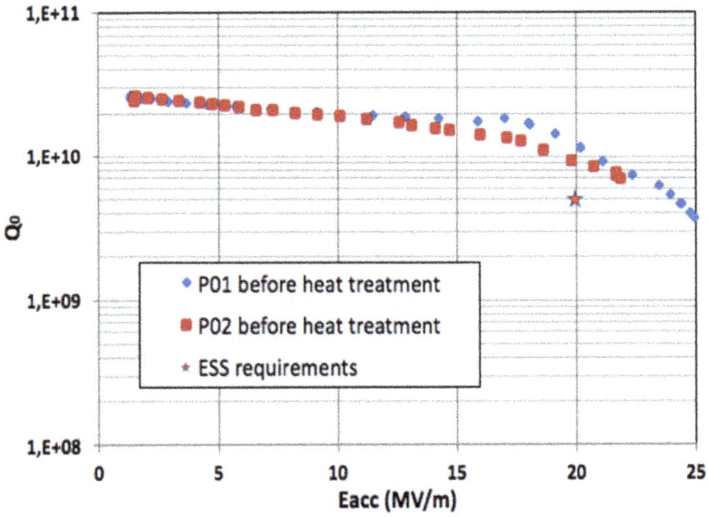

Figure 8.16. Vertical test ESS high-beta cavities (prototypes). Reprinted with permission from [20]. Copyright © 2017 by JACoW

8.4.2.2 PC performance

The ESS PC that was designed for the elliptical cavity is an adaptation of the 704 MHz, 1.2 MW peak power FPC developed in the framework of the European program CARE/HIPPI [21].

The ESS prototype PC is composed of three main parts, as shown in figure 8.17:

- A window with an antenna to allow RF power coupling to the cavity and to isolate the cavity vacuum from the atmosphere, thanks to an alumina disk.

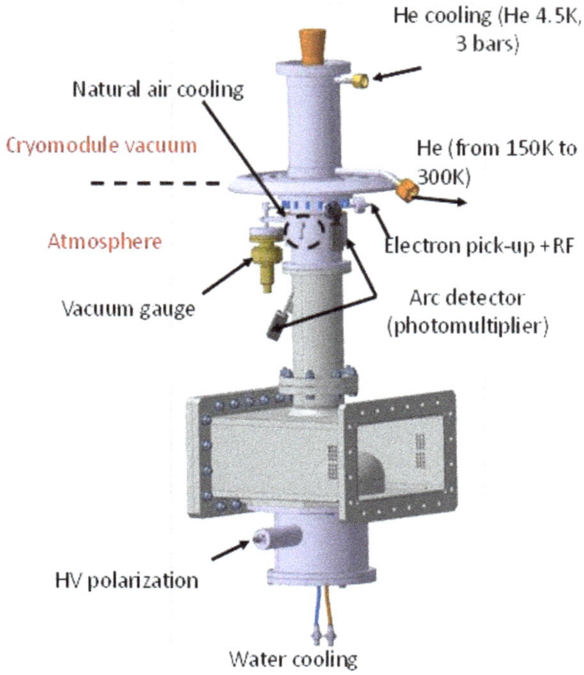

Figure 8.17. ESS PC interfaces and cooling circuits. Reproduced from [1].

- A cooled, double-walled tube to keep a coaxial configuration between the window and cavity, and allow thermal transition between ambient and cold cavity temperatures,
- A doorknob transition to allow RF matching between the coupler and the RF power source. The doorknob transition is equipped with a bias system.

The antenna is cooled by a water flow, and the external conductor is a double wall with an inner chicane, cooled by supercritical helium gas entering the tube at the cavity side flange at 3 bar and 4.5 K.

The ceramic windows and antennas are the same for both types of cavities. Their Q_{ext} is adjusted by using external conductors with two different lengths, one for the medium and one for the high-beta cavities.

Three diagnostics were used for the protection of the ceramic window: a pressure gauge, an electron pickup antenna, and an arc detector. Figure 8.18 presents the PC system and its cooling circuits.

The coupler conditioning was performed at CEA Saclay SupraTech-Cryo-HF facility, first in TW with low power, short pulses, and low repetition rate. Depending on the couplers' vacuum multipactor effect and arc breakdown diagnosis, the average power was increased up to 1.2 MW using 3500 μs pulses with a repetition rate of 14 Hz (nominal duty cycle). The power was then maintained at nominal value for at least 15 min before being decreased to check for multipacting zones. With regard SW RF conditioning, the measurement was rather similar to that of the TW.

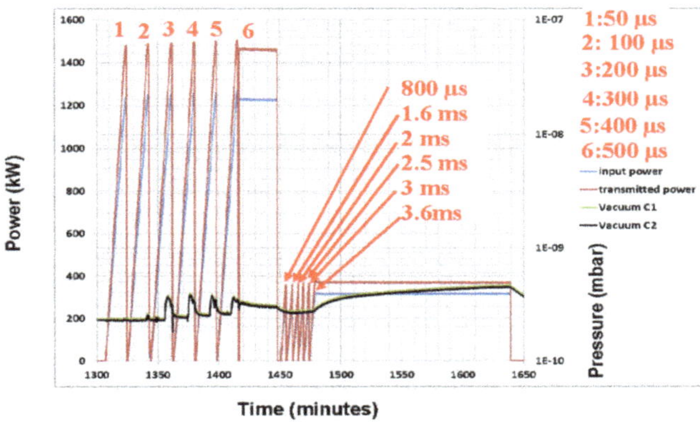

Figure 8.18. Power and vacuum (SW, 14 Hz). Reprinted with permission from [20]. Copyright © 2017 by JACoW

However, the RF is totally reflected by a movable short circuit, allowing the maximum field position sweeping along the coupler, particularly near the ceramic. This process is handled by an automated control system with human–machine interface supervision.

As an example, six PCs were conditioned in forward and reflected power up to 1.1 MW at 1 Hz and at 14 Hz [33]. Figure 8.18 shows the conditioning at 14 Hz standing wave up to 3.6 ms, which corresponds to the maximum achievable electrical field on the ceramic windows. No outgassing occurred during these measurements.

8.4.2.3 CTS performance

Since these cavities are intended to work in pulsed mode, it is important to minimize their sensitivity to Lorentz detuning. As such, stiffening rings are placed between the cavity cells. The static Lorentz coefficient, $K_L = \Delta f / E^2_{acc}$, measures the resonance frequency shift produced by the mechanical deformation induced by the electromagnetic field in a continuous wave. The positions of the stiffening rings and the cavity thickness were optimized in order to minimize $|K_L|$ when the cavity has fixed extremities. This resulted in the use of niobium cell sheets, 3.6 mm thick, and an 84 mm stiffening ring radius for the high-beta cavity. With these values, $|K_L|$ is equal to 0.36 Hz $(\text{MV m}^{-1})^{-2}$ for the cavity with fixed ends. An exhaustive list of the corresponding RF and mechanical parameters is given in table 8.5.

The CTS is based on the proven mechanical design principles of Saclay's tuners (figure 8.19). The Saclay V type tuner operates at cryogenic temperatures inside the insulating vacuum tank. All the surfaces subject to friction were treated with a dry lubricant coating efficient at cryogenic temperatures.

The CTS combines a slow and a fast tuner. The slow tuner is a double-lever system with eccentric shafts, actuated by a motor gearbox and screw. This mechanism operates under vacuum at cryogenic temperatures. The fast tuner is made up of two piezo actuators that can act simultaneously, or not, in order to compensate for the Lorentz force detuning. The tuning range of this slow tuner is ±600 kHz.

Table 8.5. Mechanical characteristics of the high-beta cavity. (Courtesy of ESS)

K_L fixed ends	Hz (MV m^{-1})$^{-2}$)	−0.36
K_L free ends	Hz (MV m^{-1})$^{-2}$	−8.9
Stiffness	kN mm^{-1}	2.59
Df/Dz	kHz mm^{-1}	197
Maximum VM stress/1 mm elongation	MPa	25
K_P fixed ends	Hz/mbar	4,85
K_P free ends	Hz/mbar	−150
Maximum VM stress/1 bar fixed	MPa	12
Maximum VM stress/1 bar free	MPa	15

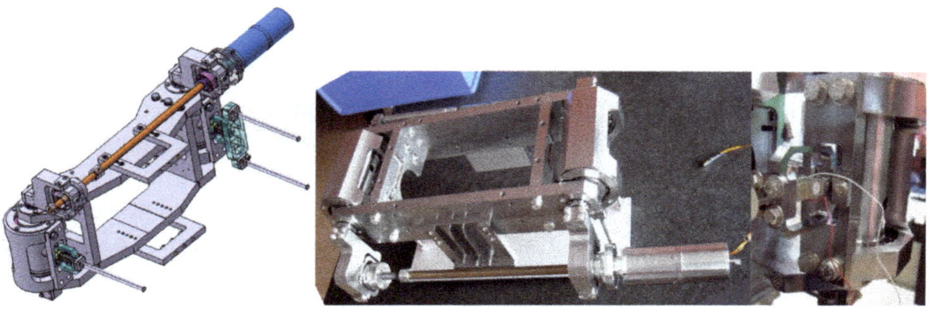

Figure 8.19. Elliptical CTS. Reproduced from [1]. © IOP Publishing Ltd. CC BY 3.0.

In order to improve reliability by simplifying maintenance, access ports planned for the maintenance of the CTS are available on the elliptical cryomodule vacuum vessel. This will permit some repairs of the CTS while the cryomodule is still in the tunnel.

8.5 Lessons learned

The following section lists the numerous transfers of knowledge and technology, and lessons learned during the duration of the SRF project and elliptical cryomodule design, production, and testing. Since the ESS is an example of in-kind collaboration, many of these lessons relate to communication and collaboration among partners.

8.5.1 Transfer of knowledge and technology for SRF collaboration

- For the sake of ESS performance and for sustainable operation, the ESS staff must be able to operate reliably and repair each component. Hence, the knowledge and technology transferred with equipment are stored in the ESS CHESS system. Technical document files, operating manuals, and possible risk analysis agreed between parties are part of the extensive materials transferred from in-kind to ESS.

- Transfer of knowledge, lessons learned were primarily obtained from the staff in charge of the design, fabrication, installation, and operation of the SRF LINAC. In-kind partner institutes and the ESS staff originating from physics laboratories brought their unique technical skills and experience to the ESS SRF LINAC design.
- Transfer of knowledge and lessons learned are crucial in the domain of particle accelerator in order to build state-of-the-art equipment. Team spirit together with exchange of knowledge and technology have been the core of the organization across the world. For instance, transfer of technology has been permitted thanks to the open-minded ESS SRF collaboration. The high-beta cavity designs prototyped by CEA-IRFU have been transferred as a blue-print to STFC Daresbury in order to conduct the large series production. This transfer of knowledge between national institutes could not have been possible without a respectful and reciprocal attitude of the SRF collaborators.
- Transfer of knowledge and training of collaborating staff was also enhanced by testing the elliptical cavity in a multifunction cryostat available in Uppsala University—FREIA laboratory. This extra testing sequence contributed to fostering a global approach between the commissioning of the spoke and elliptical cavities.
- From the contractual point of view, it has been important to refer to bilateral agreement even if the series of elliptical cryomodules has been produced by at least five institutes. Meaning that the two parties involved for the production of the series medium-beta cavities were INFN-LASA and ESS, even if the medium-beta cavities were shipped to CEA-IRFU for their installation inside the cryomodules. The technical files related to the specific medium-beta cavity are passed to CEA, and then to ESS.
- The way of working using clear Technical Requirements stored in a program such as DOORS is essential to prevent mistakes and miscommunications.

8.5.2 Lessons learned from earlier projects for elliptical cavities and cryomodules, and contributions to future projects

At the time of the ESS establishment, very few laboratories could build everything in-house, but this was not an option for a green-field site such as ESS. In-kind provides access to intellectual property, competence, and qualified manpower (including procurement staff) at partner laboratories [4]. The following aspects can be considered as lessons learned:

- The main lesson-learned is the importance of a fluid communication addressing technical and administrative competency. For WP04, WP05, and WP19, the in-kind partner and ESS have successfully channeled the flow of information thanks to rigorous weekly meetings and action items follow-up. Dedicated calibration databases, web pages, and agile communication tools have been used to exchange and lead the SRF project. When possible, open-source tools and databases should be used to facilitate the communication and to prevent miss-leading information.

- The ESS SRF cryomodule technical design has gained expertise from lessons learned from earlier collaborations. In turn, it has generated more lessons learned to enable the final technical design and to support similar future accelerator designs. In addition, the commissioning and operation phases benefited from personnel expertise on similar SRF equipment.
- ESS has shared its lessons learned and best practices with other large scale SRF collaborations and research infrastructures. For instance, conferences, such as the International Particle Accelerator Conference, the Particle Accelerator Conference, SRF Conference, and the TESLA Technology Collaboration meetings, or more specific meetings such as the SRF the Superconducting LINACs for High Power Proton Beams (SLHiPP meetings), encourage a collaborative spirit to enhance the SRF lessons learned and best practices, e.g., SNS downtime was reported.
- The use of SRF cavities has been instrumental for the recent operation of particle accelerators in order to reach larger energy for collider physics or for basic science physics.
- There are currently three to four accelerator-driven spallation sources worldwide. The J-PARC (1 GeV) in Japan is entirely normal conducting, whereas the SNS (upgrade 1.4 GeV) in the US and C-SNS in China also have SRF portions.
- Using the SNS expertise and lessons learned, ESS opted for a segmented layout for the cryomodules section instead of the continuous European-XFEL LINAC. This segmented layout allows the use of normal conducting quadrupole magnets between each cryomodule to focus and defocus the beam in accordance with the Focusing–Defocusing (FODO) concept. For the SNS, the cryomodules are connected to the CDS by bayonets. However, welded connections were chosen for these ESS interfaces in order to limit the global heat load and improve reliability while keeping maintenance and CM isolation possible.
- Prior to the ESS component installation in Saclay, the CEA also gained expertise from the FNAL clean-room infrastructure, which has been in service since 2001. The CEA clean room and expertise gained from their collaboration with FNAL have permitted the French contribution to the fabrication of the European-XFEL in DESY. Prior to this, FNAL expertise for SRF cavities was established with the Tesla Technology Collaboration. Indeed, FNAL inherited eight SRF cavities at 3.9 GHz from DESY, which were tested up to 20 MV m^{-1} at FNAL in 2005, and then assembled and tested in a cryomodule in FNAL new test infrastructures in 2011. Such knowledge and technology transfers are key to large collaborations and push the frontiers of physics.
- In a similar fashion, the experience of FNAL and European-XFEL, with licensing authorities, such as PED classification, has guided the ESS project in the engineering design choices. Indeed, according to DESY and other experiences in the US, in anticipation of the constraining category 4 of the PED imposed on the European-XFEL design, ESS has designed its spoke and

elliptical cryomodules to comply with the less demanding category of SEP. It is worth noting that the fact that European-XFEL fell under the category 4 requirement resulted in an extended project delay and extra cost, e.g., CEA had to perform extensive x-ray weld inspections due to having to comply with the legislation. The cautious ESS SRF cryomodule design initiated in 2012 meant that these requirements could be relaxed and allowed the participation of the INFN-LASA.

- CEA Saclay produced the 101 European-XFEL cryomodules. As such, they gained experience with series production, as well as of quality management tools and rigor. The quality assurance choice and the quality management system were set up in order to produce the large series of ESS cryomodules. The medium and high-beta cavities are different shapes; however, the industries able to manufacture the cavities were similar to those for the-European-XFEL cavities. The manufacturing processes, BCP and EP, followed similar specifications. The personnel, leaders, and quality engineers in these industries were also the same, hence limiting the industrial risk from a new series production. The Nb sheets were systematically scanned with the high-beta cavity as a result of the experience from DESY personnel on the European-XFEL. Less than 3% of the Nb sheets were rejected.
- Based on the unfortunate experience from European-XFEL, CEA insisted that the transportation of the cryomodules be the responsibility of ESS.
- Transfer of knowledge and technology, e.g., SRF cavities and front-end injectors, has been very valuable experience gained from CEA personnel in order to secure French support of further national and international projects, e.g., GAN, SONATE, PIP-II.
- As the ESS Environment Safety Health and Quality (ESH&Q) guidelines progress, benchmarked with other projects, we are also witnessing the lessons learned from ESS at CERN and in the Fermilab PIP-II project. Indeed, the in-kind management has been shaped using more rigorous guidance from FNAL, supported by the experience gained from the in-kind partners, ICJ Lab, CEA, and STFC for the PIP-II SRF accelerator.
- The collected lessons learned should be communicated to the relevant stakeholders for their use and be taken into account in the preparation of the TAs. References to experts from the given projects/experiments and research institutes may also be provided in order to facilitate their use, i.e. our Confluence pages enabled open access for the SRF communities for years. All documents have been available to the public, to the in-kind partners, or limited in access to relevant parties.
- ESS is collecting generic and specific lessons learned in an internal database. This tool helps each stakeholder to address issues and to improve their way-of-workings.
- In the same spirit, the European Commission has granted the ESS with additional financial support to strengthen the in-kind collaborative aspects. The BrightnESS initiatives enabled to organize workshops and gather experts during the preparation and construction phases of the ESS project.

- The core of the lessons learned are also applied on a daily basis as the ESS team has been leading the collaborative work. Web-related platforms are gathering technical information and requirements, which have been discussed on a daily basis with the different in-kind collaborations weekly and fortnightly meetings have generated synergies with the different in-kind partners, resulting in a global collective team building approach. This friendly spirit, completed by official agreement, was capable of enabling transfer of technology and transfer of knowledge.
- From the design to the operation of each component, the integration shall be carefully considered. Each activity should be considered during the life cycle of each product and should be integrated on the basis of similar experiment lessons learned. Meanwhile, 3D integration tools and installation sequences should be reviewed by relevant experts, based on their own lessons learned.
- Equipment interfaces and interconnections should be welded when appropriate in order to limit the overall system leak rate, but (dis)connection must be carefully studied to guarantee the equipment's reliability.
- Neutron activation due to field emission is a limiting factor. Lessons learned and communication channels with laboratories such as SNS and FNAL are essential to guarantee an effective integration and performing cavities operation.
- Plasma processing has been studied and implemented by different institutions to limit carbon perturbations, e.g., SNS experts have developed a stringent process.

8.5.3 Training and outreach activities

The dissemination of knowledge thanks to training and outreach is an effective tool to support and communicate on design choices. The following aspects can be considered:

- Training of staff for the installation and test of key equipment is included in the TA agreement, which enables us to build up and strengthen further collaborations between competent laboratories across Europe for the benefit of ESS.
- Training and outreach activities enhance the transfer of knowledge and the successful completion of the project. In the case of the SRF collaboration at ESS, training of the management system (ESS MS) became available online for the ESS WP leaders and the in-kind partners. Hence, ESS's ways-of-working became more comprehensible.
- Technical knowledge is also important to successfully design and operate the SRF LINAC. A series of summer schools followed by massive open-online courses explain the function of particle accelerators, light source, and neutron source. A grant by the European Commission have been used to develop such expertise, e.g., the EU Horizon 2020 Innovation—Fostering Accelerator and Science and Technology (I.FAST), Challenged based innovation (CBI).
- The dissemination of knowledge thanks to training and outreach is an effective tool to support and communicate on design choices.

- In the framework of the Nordic Particle Accelerator Project (NPAP) and the ARIES, massive open-online courses were developed to support sustainable training [49].

8.6 Conclusion

The ESS SRF collaboration was established between CEA, CNRS, INFN, STFC, DESY, and Uppsala and Lund universities, and provided its first deliverables with the manufacture, assembly, and testing of the spoke and elliptical prototype cryomodules. Extensive SRF activities demonstrated the proof of concept of the series SRF LINAC.

Deliverables of the cooperation agreements established the design criteria for the spoke and elliptical cavities and PCs. The resulting prototypes allowed the validation of the technical feasibility of key components and permitted the signature of in-kind contracts. The ESS elliptical-cavity cryomodules have been shown to meet the requirements of the ESS project. To date (2023), 10 series elliptical-cavity cryomodules have been built and sent to ESS for testing (chapter 10). This is sufficient for the first phase of ESS science operations.

Acknowledgments

It should be clear from this chapter that the success of the elliptical-cavity cryomodules stems from the work of many people at a number of institutions. Please find a list of other contributors from CEA, IJCLab, INFN, STFC, and FREIA, in the Contributors section of the book.

References

[1] Garoby R *et al* The European Spallation Source design *Phys. Scr.* **93**

[2] Peggs S *et al* 2013 *ESS Technical Design Report* Released 2.63

[3] Darve C *et al* 2017 ESS superconducting RF collaboration *Presented at IPAC'17, Paper THOBB3* (Copenhagen, Denmark)

[4] Pisent A *et al* 2023 uropean collaboration for the realization of ESS (Presented at IPAC)

[5] Darve C, Eshraqi M, Lindroos M, McGinnis D, Molloy S, Bosland P and Bousson S 2013 The ESS superconducting linear accelerator *SRF13 (Paris, France)*

[6] Garoby R *et al* Progress on the ESS project construction *Proc. IPAC17, Paper MOXBA1*

[7] McGinnis D *et al* 2014 New design approaches for high intensity superconducting linacs: the new ESS linac design *Proc. IPAC'14 (Dresden, June 2014)* p 23

[8] Eshraqi M Beam physics design of the optimus+ SC linac *ESS AD Technical Note ESS/AD/0050*

[9] Darve C *et al* 2014 Requirements for ESS superconducting radio frequency linac *IPAC'14* (Germany: Dresden) p 3311

[10] Devanz G *et al* 2013 Cryomodules with elliptical cavities for ESS *SRF'13* (Paris, France)

[11] Olivier G *et al* 2013 ESS cryomodules for elliptical cavities *MOP084, SRF'13* (Paris, France)

[12] Darve C, Bosland P, Devanz G, Olivier G, Renard B and Thermeau J P 2013 The ESS elliptical cavity cryomodules *CEC2013* (Anchorage, Alaska)

[13] Olivier G *et al* 2014 Technical Note, Cryomodule for Elliptical Cavity—Global Heat Balance—3

[14] Piquet O *et al* Final Design of the LB650 Cryomodule for the PIP-II Linear Accelerator

[15] Darve C *et al* The superconducting radio-frequency linear accelerator components for the european spallation source: first test results *Proc. of LINAC15*

[16] Elias N *et al* 2014 Preliminary functional analysis and operating modes of the ESS 704 MHz superconducting radio-frequency linac *THPME042, Proc. of IPAC'14 (Germany)* (Dresden)

[17] Peauger F *et al* 2014 Status and first test results of two high-beta prototype elliptical cavities for ESS *Proc. IPAC'14* (Dresden)

[18] Peauger F *et al* 2015 Progress in the elliptical cavities and cryomodule demonstrators for ESS *TUPB007, SRF'15* (Canada: Whistler)

[19] Darve C *et al* 2016 The SRF linac components for the ESS: first test results *LINAC2016* (USA: East Lansing)

[20] Darve C, *et al*SRF E S S Linear accelerator components preliminary results and integration *Proc. IPAC17*

[21] Peauger F *et al* Preliminary test results of the first ESS elliptical cryomodules demonstrator *Proc. IPAC19*

[22] Bosland P *et al* 2017 Status of the ESS elliptical cryomodules at CEA saclay *Presented at IPAC'17, Paper MOPVA040* (Denmark: Copenhagen)

[23] Darve C *et al* 2014 The ESS superconducting RF cavity and cryomodule cryogenic processes *ICEC25* (Netherland: Enschede)

[24] Asensi Conejero E *et al* 2017 The cryomodule test stands for the european spallation source *Presented at IPAC'17, Paper MOPVA089* (Denmark: Copenhagen)

[25] Cenni E *et al* 2017 Vertical test results on ESS medium and high-beta elliptical cavity prototypes equipped with helium tank *Presented at IPAC'17, Paper MOPVA041* (Denmark: Copenhagen)

[26] Arcambal C *et al* 2017 Conditioning of the RF power couplers for the ESS elliptical cavity prototypes *Presented at IPAC'17, Paper MOPVA044* (Denmark: Copenhagen)

[27] Michelato P *et al* 2017 Vertical tests of ESS medium-beta prototype cavities at LASA *Presented at IPAC'17, Paper MOPVA063* (Denmark: Copenhagen)

[28] Monaco L *et al* 2017 Fabrication and treatment of the ESS medium-beta prototype cavities *Presented at IPAC'17, Paper MOPVA06* (Denmark: Copenhagen)

[29] Bertucci M *et al* 2017 Quench and field emission diagnostics for the ESS medium-betaprototypes vertical tests at LASA *Presented at IPAC'17, Paper MOPVA061* (Denmark: Copenhagen)

[30] Chen J *et al* 2017 Multipacting studies in ESS medium-betacavity *Presented at IPAC'17, Paper MOPVA064* (Denmark: Copenhagen)

[31] Pirani S *et al* 2017 Investigation of HOM frequency shifts induced by mechanical tolerances *Presented at IPAC'17, Paper MOPVA091* (Denmark: Copenhagen)

[32] Sertore D *et al* 2017 Experience on design, fabrication and testing of a large grain ess medium-beta prototype cavity *Presented at IPAC'17, Paper MOPVA068* (Denmark: Copenhagen)

[33] Arcambal C *et al* 2015 Status of the power couplers for the ess elliptical cavity prototypes *THPB078, SRF'15* (Canada: Whistler)

[34] Cenni E *et al* 2015 ESS medium-beta cavity prototypes manufacturing *THPB028, SRF'15* (Canada: Whistler)

[35] Bousson S, Darve C, Duthil P, Elias N, Molloy S, Reynet D and Thermeau J-P 2014 The ESS spoke cavity cryomodules *Adv. Cryog. Eng* **59A**
[36] Li H, Santiago Kern R, Jobs M, Bhattacharyya A, Goryashko V, Hermansson L, Gajewski K, Lofnes T, Fransson K and Ruber R 2017 First High Power Test of the ESS Double Spoke Cavity *FREIA* Report 2017/10
[37] Reynet D *et al* Design of the ESS spoke *Proc. of SRF2013*
[38] Reynet D *et al* 2015 ESS spoke cryomodule and test valve box *THPB109, SRF'15* (Canada: Whistler)
[39] Duchesne P *et al* 2013 Design of the 352 MHz, beta 0.50, double-spoke cavity for ESS *SRF'13* (Paris, France) p. 1212
[40] Olry G *et al* 2015 Recent progress of ESS spoke and elliptical cryomodules *TUAA06, SRF'15* (Canada: Whistler)
[41] Li H *et al* 2017 ESS spoke cavity conditioning *Presented at IPAC'17, Paper MOPVA094* (Denmark: Copenhagen)
[42] Li H *et al* 2015 Test characterization of SC Spoke cavities at Uppsala University *TUPB083, SRF'15* (Canada: Whistler)
[43] Santiago Kern R *et al* 2015 Cryogenic performance of the HNOSS test facility at Uppsala University *TUPB026, SRF'15* (Canada: Whistler)
[44] Fransson K *et al* 2017 The EPICS-based control system at the FREIA laboratory *Presented at IPAC'17, Paper TUPIK08* (Denmark: Copenhagen)
[45] Lanfranco G *et al* 2014 Constructing the ESS linear accelerator: pragmatic approaches to design and system integration at the european spallation source *WEPRO076, Proceeding of IPAC*
[46] Rampnoux E *et al* 2013 Design of 352.21 MHz RF power input coupler and window for the European Spallation Source (ESS) *SRF'13* (Paris)*(September 2013)* p 1069
[47] Gandolfo N *et al* 2014 Fast tuner performance for a double spoke cavity *LINAC'14* (Switzerland: Geneva) p 1034
[48] Fydrych J *et al* 2014 Cryogenic distribution system for the ESS superconducting proton linac *ICEC25* (Netherland: Enschede)
[49] (IJCLab) N D 2022 New online course to initiate undergraduate students to particle accelerators *CERN Accelerating News*

Chapter 9

Spoke cavity cryomodules

C Darve, S Bousson, G Olry, P Duthil and P Pierini

This chapter describes the design and construction of the spoke cavity cryomodules. It also provides results from initial cavity and cryomodule testing. Details and history of the in-kind collaboration that produced these cryomodules are also given. Lessons learned from this spoke cavity cryomodules effort and the related elliptical cavity cryomodules are described.

9.1 ESS SRF collaboration related to spoke components

The spoke cryomodules contain niobium superconducting radio-frequency (SRF) cavities operating at 2 K. The cavities and cryomodules were produced by an extensive in-kind collaboration.

9.1.1 In-kind management for spoke cavities and cryomodules

The European Spallation Source (ESS) SRF collaboration has been developed to achieve the challenging SRF requirements of the 2013 SRF linear accelerator (LINAC) layout [1–9]. The process described in this chapter is very similar to the one of WP05 [10–34]. Letters of Intent (LoI) have been collected during the initial phase of the ESS establishment.

As with the elliptical cavity cryomodules (chapter 8), and in unison with the ESS local workforce, the spoke SRF LINAC has been prototyped and constructed based on a collaboration with European research institutes including Centre National de la Recherche Scientifique and Laboratoire de Physique des 2 Infinis Irène Joliot (CNRS-IJCLab) and Uppsala University [35–48].

Since the number of required spoke cryomodules is smaller than the number of required elliptical cryomodules, and since the competencies of IJCLab could fulfil the broad assignment, the whole contribution of the spoke cryomodule package fabrication has been assigned to that unique institute, i.e. the design and fabrication of the double spoke cavities and components, up to the assembly of each spoke cryomodule. In addition, IJCLab committed to the design and fabrication of the spoke

doi:10.1088/978-0-7503-3223-1ch9

cryomodule associated valve boxes and cryogenic transfer line in the ESS tunnel, the cryogenic distribution system (CDS; see chapter 4). Therefore, the functional interface between the spoke cryomodule and the CDS has been completed by the same institute design office, hence limiting the requirement interpretation error.

In addition, the spoke cryomodule demonstrator and series spoke cryomodule high power tests have been conducted by Uppsala University (UU) at the FREIA lab. A complete and comprehensible description of this latest testing phase is presented in chapter 11. The prototype spoke SRF components designed by IJCLab were measured and validated in order to successfully manufacture the series cryomodule components, after numerous improvements at several steps of the component's lifecycle.

9.1.2 Towards an ESS collaborative model for the spoke cryomodules

Since IJCLab has driven the cryostat design of both elliptical and spoke cryomodules, that resulting know-how expedited the integration and standardisation of cryomodule components interfaces. IJCLab personnel working on spoke design have exchanged valuable expertise with CEA-Saclay on numerous topics of common interest.

On one hand, as for the elliptical cryomodule, and in a top-down approach, collaboration agreements were initiated by ESS in 2011 in the 'Cooperation Agreement in the field of neutron and accelerator sciences to the ESS Design Phase', which was based on utilising the experience of Alternative and Atomic Energies Agency (CEA) and Institute Recherche Sur Les Lois Fondamentales De L'univers (IRFU; CEA-IRFU) and IJCLab in the design and construction of accelerating structure. The main purpose of this agreement was to enable an early start of the design, prototyping, and testing of key parts of the ESS accelerator before in-kind contracts and other contributions could be secured from the future member states. In addition, a collaborative agreement was established with Uppsala University to test prototype SRF cavities and develop the radio-frequency (RF) amplifier stations at FREIA Laboratory.

On the other hand, in a bottom-up approach, two work-packages (WP) have been created in the accelerator project (ACCSYS) to manage the elliptical (WP05) and the spoke (WP04) cryomodules fabrication, respectively. WP04 covers the construction and testing of the spoke cryomodule prototype and the design, construction and assembly of the series spoke cryomodules with all their components at the ESS in-kind partner, as well as transport of the cryomodules to the FREIA test facility in Uppsala. It is worth noting that the ESS WP leader for WP04 was also leading WP05. Hence, creating even more synergies and common ground for the whole SRF collaboration and ESS communities.

When the cryomodules were fabricated, a new WP19 was created. This work package covers activities relating to superconducting cavities and cryomodules once they have been delivered from in-kind partners to ESS in Lund. This includes the SRF part of acceptance tests at FREIA for spoke cryomodules and at TS2 in Lund for the elliptical cryomodules, as well as installation in the tunnel, commissioning and future operation.

In complement and in a bottom-up approach, the work-package for spoke cavities and cryomodules fabrication, WP04, has been supported by ESS in parallel with the WP05 for elliptical cavities and cryomodules fabrication. In addition, IJCLab committed to the fabrication of the CDS interface (WP11). Chapter 4 gives a description of the CDS and fabrication of those components.

At the earlier stage of the ESS project, technology demonstrators were designed, manufactured, and tested for the spoke cavities and cryomodules. These demonstrators validated the technologies to be implemented in the ESS SRF LINAC [17–20]. This prototyping phase has included the design, manufacture, and testing of three spoke cavities [40], and power couplers [46], which were assembled and tested in the cryomodules.

9.1.3 The SRF technology and spoke cryomodule

As explained in chapters 1 and 8, the choice of SRF technology was a key element in the development of the ESS accelerator. The choice of 90 MeV as the transition energy between the warm and the cold LINAC, is a unique specificity of the ESS accelerator layout. For the sake of size and efficiency, ESS has raised the LINAC performance by proposing to use more SRF spoke cryomodules rather than extending the use of a drift tube LINAC (DTL), while using a more conventional and lower transition energy. The output energy of the 13 spoke cryomodules is 216 MeV at the inlet of the nine medium-beta elliptical cryomodules. Table 9.1 shows the specification of the cryomodule and the electromagnetic resonator (EMR) requirements to obtain the challenging ESS 5 MW capacity.

Table 9.1 shows the specification of the spoke cryomodule and figure 9.1 shows a 3D view of the spoke cryomodule. Each spoke cryomodule contains two double-spoke bulk niobium cavities contained in titanium grade 2 helium tanks.

The ESS LINAC is fully segmented and cavities are grouped into cryomodules. The spoke cavities and elliptical cavities are gathered two-by-two and four-by-four in their cryomodules, respectively. Accordingly, the superconducting LINAC is

Table 9.1. ESS cavity and cryomodule requirements (recall from chapter 8 table 8.3). (Courtesy of ESS)

	Spoke	Medium-β	High-β
Proton energy range, MeV	90–216	216–571	571–2000
Cryomodules/sector	13	9	21
Cavities/cryomodule	2	4	4
Cavities/scetor	26	36	84
Sector length, m	55.9	76.7	178.9
Operating frequency, MHz	352.21	704.42	704.42
Operating temperature, K	2	2	2
Cavities optimum β	0.5		
Cavities geomatric β		0.67	0.86
E_{ace}, MV m^{-1}	9	16.7	19.9

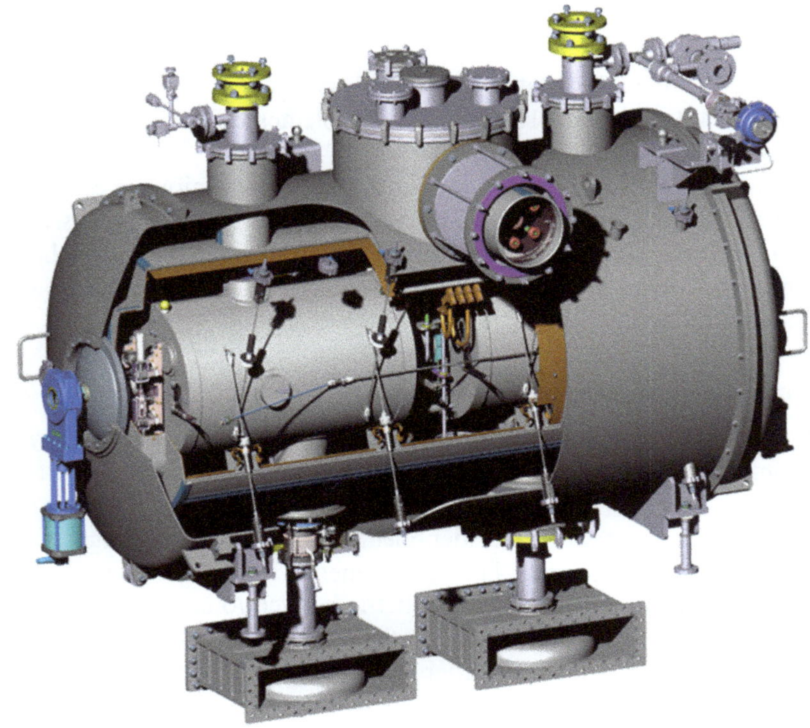

Figure 9.1. View of a spoke cavity cryomodule. (Courtesy of ESS)

composed of 26 double-spoke cavities, 36 medium-beta elliptical cavities, and 84 high-beta elliptical cavities. Each double-spoke cavity has three gaps, with an accelerating gradient of 9 MV m^{-1}.

9.1.4 Manufacture cost estimate and schedule for spoke cavities and cryomodules

The cost estimates for the spoke (WP04) design and construction were developed in collaboration with our main in-kind partners, IJCLab/CNRS and UU. The detailed cost estimate and the different breakdown structures were included in the 2013 cost book. Any deviations from the cost book from 2013 followed a strict process entailing change control requests (CCRs), and were tracked accordingly.

In the same way as the cost estimate, the schedule estimate was captured in the Technical Annexes (TAs), and monitored by relevant WP leaders over the years. ESS planners supported the WP with the overall estimate. Any deviations and changes were tracked and captured in the rebaselines of 2016 and 2021.

9.2 Spoke cryomodule technical management

9.2.1 Spoke cryomodule project management

The choice of spoke cryomodule to be implemented as a SRF technology has been a key element in the development of the ESS LINAC. We will refer here to the description in chapter 1 combining the three families of cavity strings, operating at 2 K.

A specific work breakdown structure (WBS) is tracking the product phases and progress of WP04. It has been elaborated and implemented in an ESS planning tool, so-called Primavera (P6).

In synergies with WP05, the continuous and harmonious weekly meetings permitted ESS to track the details of the progress and address all possible issues, including at the interfaces. Lessons learned and status reports of each partner's technical activities and integration were followed-up on a weekly basis, to ensure an effective collaboration.

9.2.2 Technical requirements applied to spoke cryomodules

This SRF LINAC accelerates the proton beam from 90 MeV, as input energy of the 13 spoke cryomodules, to 216 MeV at the inlet of the nine medium-beta elliptical cryomodules.

Hundreds of requirements were identified in order to best integrate the SRF components in the ESS tunnel. The requirements were also used to define the geometrical and functional interfaces with the conventional facilities, the control system, and to define the operating modes for the ESS SRF LINAC [9]. The high-level requirements were defined in order to optimise the layout of the ESS super-conducting LINAC [7, 8].

The technical requirements of components and subcomponents formed the bonding agreement between ESS and each in-kind partner institution and were reported in the TAs. Institutions were chosen based on their capacity to deliver the operating components.

9.2.3 Quality management and equipment compliance applied to spoke cryomodules

The equipment quality process to be operated at ESS has been described in the Project Quality Plan (PQP) prepared by IJCLab. The performance of each spoke cryomodule is validated with a Factory Acceptance Test (FAT), while testing the cryomodule in Uppsala.

Finally, the equipment will be fully validated by the individual test, with the site acceptance test (SAT), when the spoke cryomodules will be fully installed in the tunnel, as planned in the Engineering Handbook.

The equipment design and assembly complied with harmonised standards, and were integrated in the final system assembly, in compliance with the Swedish legislation. Hence, a certificate from the relevant Swedish authorities is needed before the LINAC's operation can be permitted. For instance, the design, construction and operation of the spoke section have to comply with the European Pressure Equipment Directive (PED 97/23/EC) and with Swedish national requirements (ordinance AFS 2005:2). The cryo-distribution of the spoke cryomodule of the cold mass is made of stainless-steel pipes and material transition with the titanium helium tanks of the cavities is done by the use of niobium-titanium and stainless-steel flanges. To simplify the commissioning of the cryomodules of the

LINAC, a decision was taken to classify helium tanks in the article 3.3 of the PED. The volume of each cavity tank is hence reduced to 48 l, yielding a maximum allowable pressure, p_s (cf EN13445), of 1.04 barg. Ultimately, the SSM (Swedish radiation authority) will issue the permit to operate.

The risk matrix is based on the risk levels that were judged acceptable for the ESS facility by the project team. A methodology that allows for an assessment in three stages has been used, as described in IEC 61508.

A safety review was conducted validated with the support of an accredited body. Numerous lessons learned based on other international accelerator development have permitted to prefer a technical cryomodule design, complying with the PED Sound-Engineering Practice (SEP).

Safety consideration from IJCLab can be summarised as follows [35]: *'From a standpoint focusing on in the spoke cavities cryomodules, additional guidelines are being completed to provide guidance through design, manufacture, conformity assessment and operation. This documentation is compiled taking into account the best possible engineering practices and incorporating expertise from similar projects occurring at CERN and FNAL* [12]. *For instance, several spoke cavity prototypes have been tested to support the R&D program from which the transfer of knowledge can be directly re-applied to mitigate fabrication and operating potential issues of the ESS spoke cavity cryomodules. Risk analysis, audit, reliability studies and technical safety reviews are part of the engineering design of the ESS cryomodules. The cryomodule design accounts for safety equipment, like active components for overpressure and interlock systems. The Machine Protection System (MPS) is developed to guarantee a safe operation of the cryomodules. The cryomodule control and monitoring systems ensure a high degree of reliability and fail-safe emergency scenarios. The design follows the 'As Low As Reasonably Achievable' (ALARA) principle. In addition, the Radio- Protection and Rad-hard equipment are used by the ESS to provide passive and active safety measures (safety barrier), Personnel Protection System, Machine Protection System (MPS).'*

9.3 Spoke cryomodule component integration

9.3.1 Spoke cryomodule description

The ESS spoke superconducting LINAC consists of 26 double-spoke cavities (β=0.5) to accelerate the beam from the DTL at 90 MeV up to the 216 MeV at the entrance of the elliptical cryomodule section.

As with the elliptical cavities, the spoke cavities provide: (1) the capacity to transfer energy from RF system to the beam, (2) the capacity to confine the protons longitudinally, and (3) the capacity to steer the protons longitudinally [8].

Figure 9.2 shows a 3D drawing of the prototype spoke cryomodule, which contains: a cavity string, consisting of two SRF spoke cavities inside their helium tanks, two RF power-couplers, two cold tuners, two UHV gate valves, and optionally including two small diameter dished ends for the cryostat closure and two cold-warm transitions; one cold magnetic shield per cavity; a thermal shield; multilayer insulation blankets covering the cold mass and the thermal shield; several

Figure 9.2. View of a spoke cavity cryomodule. Reproduced from [1]. © IOP Publishing Ltd. CC BY 3.0.

rods for the mechanical anchoring and positioning of the cavity string inside the vacuum vessel; cryogenic piping for the helium distribution inside the cryomodule; vacuum, cryogenic, and RF instrumentation for the cryomodule operations; and a vacuum vessel that is equipped with dished ends for the insulation vacuum closure, and alignment devices and several ports for interfacing with the vacuum pumping system, the control and measurement instrumentation, as well as safety equipment (relief valves, burst disk, etc). Figure 9.3 shows the spoke cavity string.

In order to facilitate its maintenance, all the cryogenic valves are placed in the valve box. Hence, superfluid helium is produced within the valve box from the saturated helium provided by the test facilities. The subcooler is a plate heat exchanger of the same type as those used in the LHC. A phase separator is also placed inside the prototype valve box to channel the saturated helium supply.

IJCLab executed the design and prototyping work required for the development of the spoke section of the ESS accelerator. The challenge was to achieve a robust and reliable design of this new type of accelerating structure that was compatible with the ambitious operating parameters of the ESS accelerator. The cavity design was achieved in 2012, with an optimised electromagnetic and mechanical configuration, allowing efficient acceleration of the proton beam at 9 MV m^{-1}. Three prototype spoke cavities were tested at IJCLab and in Uppsala University in dedicated cryostats. In parallel, four power couplers and cold-tuning systems (CTSs)

Figure 9.3. String of two spoke cavities. (Courtesy of ESS)

were tested. A string of two equipped cavities with two power couplers were finally assembled at IJCLab before being delivered and tested in Uppsala at the end of 2017.

Differing from the elliptical cryomodule, where the heat-exchanger is located in the elliptical cryomodule, the cryogenic process of the spoke cryomodule is managed by the valve box. The spoke cryomodule valve box contains all of the process controllable equipment necessary to control and regulate the fluids flowing through the cryomodule and taking into account process variables located inside the cryomodule (helium level, pressure). It also contains additional components to change fluids temperature (heat exchanger), pressure (throttle valve), and phase state (combination of both). The cryogenic connection between the valve box and the cryomodule is done with a cryogenic jumper. For ESS, a vacuum barrier is located within this connection, on the valve box side, and it is thus possible to disconnect the cryomodule without affecting the vacuum within the valve box and the cryo-distribution.

9.3.2 Assembly steps of spoke cryomodules

The assembly of the cryomodule was performed in two stages: (1) assembly of the cavity string in an ISO4 cleanroom, and then (2) assembly of the remaining cryomodule elements inside the vacuum vessel as described in [18]. In the framework of the ESS-IJCLab collaboration, a dedicated clean-room has been established to assemble the spoke cavities string. Figure 9.4 shows the clean room with a specific cavity string.

Figure 9.5 shows the components, which are assembled to complete the spoke cryomodule. The thermal shield is cooled by a 50 K helium flow from the accelerator cryoplant (ACCP; chapter 3).

The thermal shield of the spoke cryomodule is made of 2 mm thick aluminium alloy 6082 sheets and covered with 30 layers of MLI. It consists of two half cylinders and four half dish ends [5]. Four extruded aluminium profiles are welded along the half cylinders and connected together to form a continuous circuit in which

Figure 9.4. IJCLab clean room. (Courtesy of ESS)

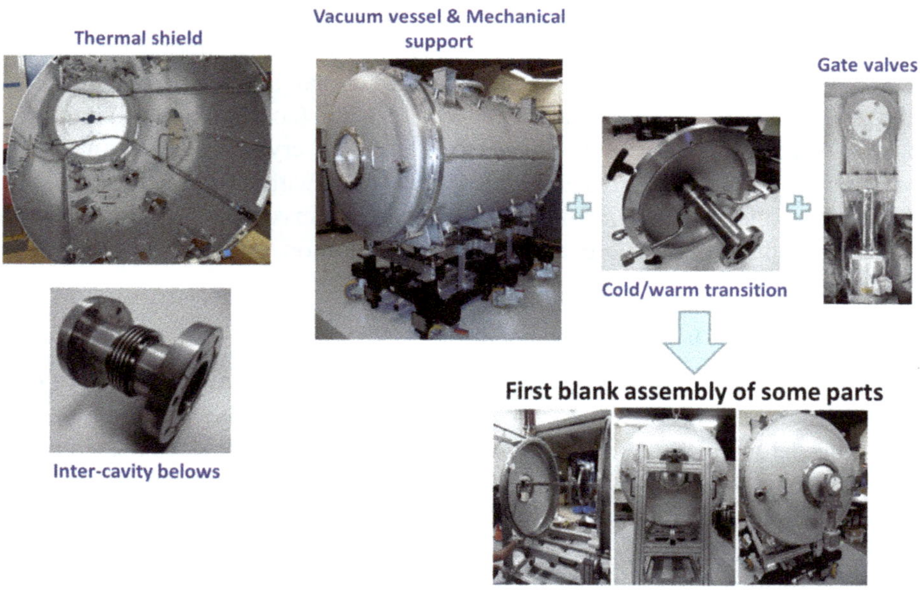

Figure 9.5. Components used for the spoke cryomodule assembly. (Courtesy of ESS)

cryofluids flows. At each side of the cryomodule, two cold-to-warm transitions link the string of cavities to the room temperature ultra-high vacuum gate valves forming the beam pipe. To limit conductive heat loads to the cavities' bath, heat interceptions are performed on these two heat transitions by making the thermal shield cryofluid circulate into a small exchanger box welded around each transition. Each cold-to-warm transition is thus part of the thermal shield.

These components are then assembled outside the clean-room, as shown in figure 9.6.

Figure 9.7 shows the final assembly of the spoke cavity string inserted in its vacuum vessel. Figure 9.8 shows the first spoke cryomodule ready to be shipped from IJCLab to Uppsala.

The spoke cryomodule functional analysis and operational modes, as well as the instrumentation and control, are similar to the elliptical ones and have been implemented to operate the spoke cryomodule in the ESS tunnel. A broad variety and quantity of data are stored to calibrate and identify each specific spoke cryomodule.

Figure 9.6. Sequences for the installation of the spoke cryomodules. (Courtesy of ESS)

Figure 9.7. IJCLab spoke cryomodule assembly. (Courtesy of ESS)

Figure 9.8. IJCLab spoke cryomodule assembly and ready for transportation. (Courtesy of ESS)

The spoke cryomodule interfaces are identical to the elliptical ones. They are also composed of the CDS, the RF system, and the vacuum system.

9.4 Spoke cryomodules testing

The aim of this section is to present the different components of the spoke cryomodules and to give some examples of performance measurement. Hence, the performance of the spoke cryomodules depends on the validation and calibration of each cryomodules component.

9.4.1 Spoke cryomodule testing

A spoke cryomodule contains: a cavity string, consisting of two SRF double-spoke cavities inside their helium tanks, two RF power couplers, two cold tuners, two UHV gate valves, and optionally including two small diameter dished ends for the cryostat closure and two cold-warm transitions; one cold magnetic shield per cavity; a thermal shield; multilayer insulation blankets covering the cold mass and the thermal shield; several rods for the mechanical anchoring and positioning of the cavity string inside the vacuum vessel; cryogenic piping for the helium distribution inside the cryomodule; vacuum, cryogenic, and RF instrumentation for the cryomodule operations; and a vacuum vessel, equipped with dished ends for the insulation vacuum closure, and alignment devices and several ports for interfacing with the vacuum pumping system, the control and measurement instrumentation, and the safety equipment (relief valves, burst disk, etc). All cavities operate at 2 K.

The critical SRF LINAC technologies were prototyped in order to validate the component design performance and their assembly phase before transferring knowledge to industry for the series production. The functional analysis of the spoke and elliptical cryomodules were validated and the operation modes were tested and verified. Figure 9.9 shows the test area at IJCLab, where the spoke cavity cryomodules were cooled down and their assembly validated.

Figure 9.9. IJCLab spoke cavity prototype cryomodule test area. (Courtesy of ESS)

Then, the spoke cavity cryomodules have been tested at high power in Uppsala University since 2014,as reported in chapter 11. Indeed, in the framework of the early collaboration agreements, the spoke cavities and cryomodule were assembled and tested at different stages in IJCLab and Uppsala. The spoke SRF configuration and cavity package (cavity, power-coupler, and CTS) were validated in IJCLab and in the Uppsala horizontal cryostat (HNOSS) [41–44].

9.4.2 Spoke cryomodule components and interfaces

9.4.2.1 Double-spoke cavity design, fabrication and test results

The spoke cavity electromagnetic (EM) design was driven by the frequency, the optimum beta, and the optimisation of the peak fields [40]. Figure 9.10 shows the resulting spoke cavity design.

SRF resonators must be sheltered from any magnetic field to obtain a large quality factor ($>10^9$): from the Earth's magnetic field and from possible local sources. Hence, the cryomodule integrates magnetic shields.

Whereas the most important parameter for the beam is the accelerating field or the voltage seen by the particle, the most important optimisation sounds criteria is the ratio of surface fields to peak fields, i.e., to minimise the peak fields keeping the accelerating field constant. Another important factor to optimise is the cavity's

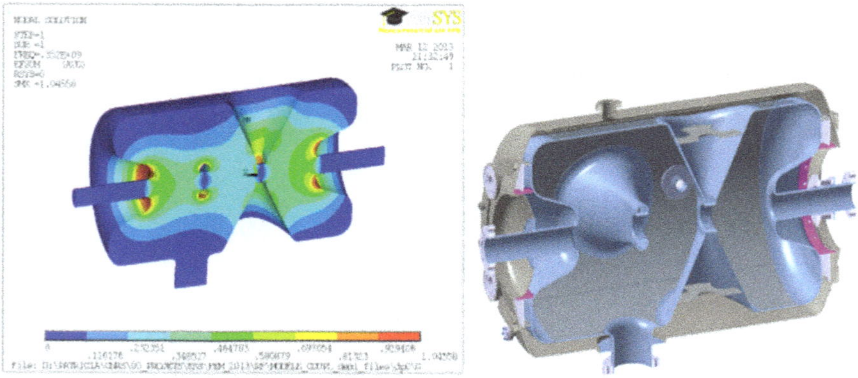

Figure 9.10. Spoke cavity design. Reprinted with permission from [20]. Copyright © 2017 by JACoW

Table 9.2. Double-spoke main parameters and requirements. (Courtesy of ESS)

Beam mode	Pulsed (4% duty cycle)	
Frequency	352.2	MHz
Optimal beta	0.50	
Temperature	2	K
Bpk	62.9 (max)	mT
Epk	39.6 (max)	MV m^{-1}
Accelerating gradient, Eacc	9	MV m^{-1}
Lacc	0.639	m
Beam tube diameter	56 (min)	mm
Pmax	335 (max)	kW

Note: Eacc = Electrical accelerating gradient

overall length: a spoke cavity has a re-entrant shape, and the size of the re-entrant part can be increased to give more volume to store the energy, thereby resulting in a peak field decrease. The RF and mechanical designs are presented in [35]. Following the beam dynamics simulations, a set of parameters was established to fulfil the requirements, which are summarised in table 9.2.

The first three double-spoke prototypes were fabricated and delivered at the end of 2014 by European suppliers and delivered fully jacketed with their 4 mm thick titanium helium vessel. The integration of the helium vessel around the bare cavity was one of the most critical steps in the manufacture. All three cavities were etched by a buffered chemical polishing (BCP) process. Figure 9.11 shows the two types of BCP tested.

Then, the cavities were high-pressure rinsed four times at 100 bar, through all ports in a vertical position. Each high-pressure rinsing (HPR) pass lasted three hours. A total of 6000 l of ultra-pure water was necessary for the entire process. The cavities were then dried for a minimum of 48 h.

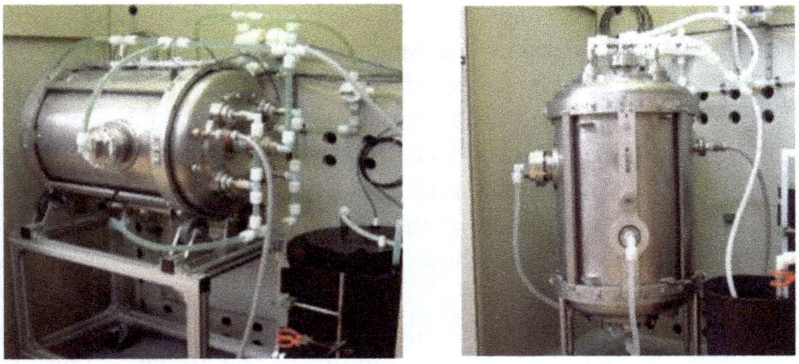

Figure 9.11. BCP set-up: horizontal (left-hand panel) and vertical (right-hand panel). (Courtesy of ESS)

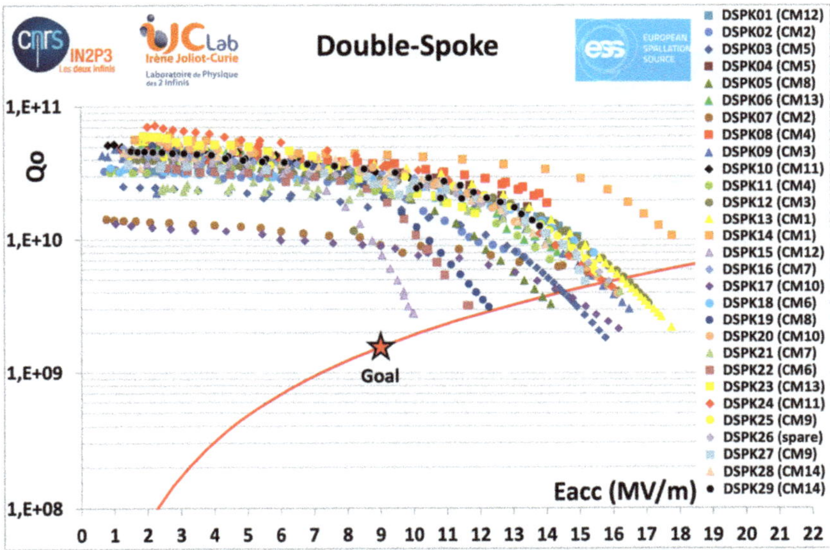

Figure 9.12. Double-spoke cavity results. (Courtesy of ESS)

The cavities were tested at 2 K in vertical position inside the vertical cryostat due to the lack of space (radially). As shown in figure 9.12, all cavities exceeded the ESS requirements (i.e., $Q_o > 1.5 \times 10^9$ at 9 MV m^{-1}). Multipacting barriers were observed at 1 MV m^{-1} (narrow and soft) and between 6 and 8.5 MV m^{-1} (wide and harder). The processing time for the hardest barriers lasted about one hour. Similar results were reproduced to validate the Uppsala high power test stand test capacity [15, 16].

9.4.2.2 Power coupler for spoke cavities

The spoke cryomodule power coupler has the double role of supplying RF power to the cavity while insulating the cavity vacuum from the atmospheric pressure inside the RF line [46]. The design is based on the coupler developed for the superconducting spoke cavities in the framework of the EURISOL Design Study. To adapt that design to the ESS power coupler requirements, a water-cooling system is integrated in the inner antenna and the water-cooling system of the 100 mm diameter ceramic window has been modified to direct the water flow more effectively. The outer conductor is cooled with supercritical helium. The ESS spoke power coupler has been designed for a peak power of 400 kW (335 kW nominal) and its doorknob transition from coaxial to half-height WR2300 waveguide. The design of the RF power coupler for the ESS spoke cavities is presented in [35]. The conditioning of the power-coupler for the prototypes and for the series have been completed in IJCLab. Figure 9.13 shows the high power RF test area used for the power-coupler conditioning.

The first four prototype power couplers were manufactured by two French companies (a pair for each) and delivered in 2015. The antenna is electropolished and the RF window is coated with TiN. Both the antenna and window will be cooled with water.

Figure 9.13. Spoke power coupler and its high-power RF test stand. Reproduced from [1]. © IOP Publishing Ltd. CC BY 3.0.

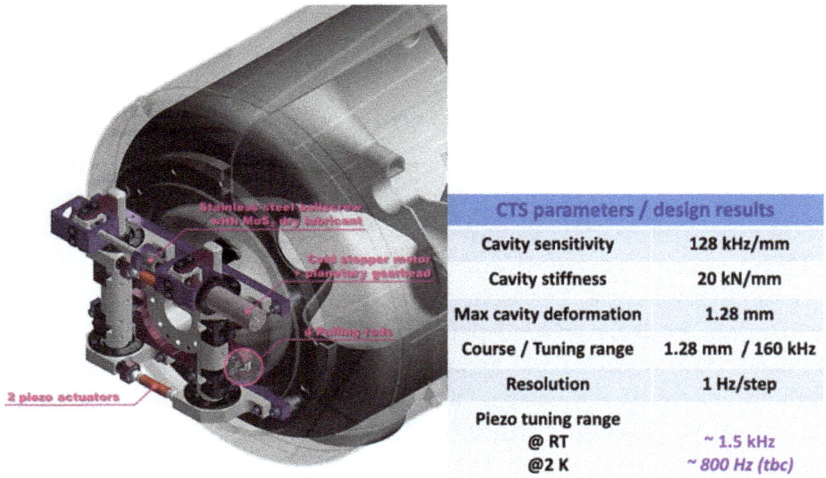

Figure 9.14. 3D model of the CTS for spoke cavities. Reproduced from [1]. © IOP Publishing Ltd. CC BY 3.0.

The first pair of couplers were assembled on the RF conditioning cavity in an ISO4 cleanroom at IPN. Then, the RF conditioning test bench was backed up to 155 °C for 24 h after a ramp-up period of 24 h.

9.4.2.3 CTSs for spoke cavities

Each ESS double-spoke cavity is tuned with a 'double-level arm' type tuner with an eccentric shaft actuated by a cold motor and equipped with two piezo stacks (figure 9.14). The coarse tuning range is about +170 kHz with a maximum stroke of 1.3 mm, and the fast-tuning range will be +675 Hz minimum to compensate for the dynamic Lorentz force detuning. The detailed design and the first measurements performed at room temperature can be found in [47]. Two versions of the CTS were designed in order to test three different lengths of piezos, and two sets of each version were manufactured. CTS version 1 (nominal configuration) can host 36 mm and 50 mm long piezo, whereas version 2 (optional design) can host 50 mm and 90 mm long ones. Several configurations of CTS versions 1 and 2 were investigated during vertical tests of the cavities.

IJCLab tested and validated the design of the spoke cavity, power-coupler, and CTS. The prototype spoke cryomodule and its valve box were cooled to validate the cryogenic design.

9.5 Lessons learned

This section lists the numerous transfers of knowledge and technology and lessons learned during the duration of the SRF project and spoke cryomodule design, production, and testing. Since the ESS is an example of in-kind collaboration, many of these lessons relate to communication and collaboration among partners. Note that because the spoke cavity cryomodule and elliptical cavity cryomodule projects are, by design, very tightly linked, the lessons learned below are very similar to those in chapter 8.

9.5.1 Transfer of knowledge and technology for SRF collaboration

- For the sake of the ESS's performance and for sustainable operation, the ESS staff must be able to operate reliably and repair each component. Hence, the knowledge and technology transferred with equipment are stored in the ESS CHESS system. Technical document files, operating manuals, and possible risk analysis agreed between parties are part of the extensive materials transferred from the in-kind partners to ESS.
- Transfer of knowledge and lessons learned were primarily obtained from the staff in charge of the design, fabrication, installation, and operation of the SRF LINAC. In-kind partner institutes and the ESS staff originating from physics laboratories,brought their unique technical skill and experience to the ESS SRF LINAC design.
- Transfer of knowledge and lessons learned are crucial in the domain of particle accelerators in order to build state-of-the-art equipment. Team spirit together with exchange of knowledge and technology have been the core of the organisations across the world. For instance, transfer of technology has been permitted thanks to the open-minded ESS SRF collaboration. The high-beta cavity design prototyped by CEA-IRFU has been transferred as a blue-print to STFC-Daresbury in order to conduct the large series production. This transfer of knowledge between national institutes could not have been possible without a respectful and reciprocal attitude of the SRF collaborators.
- Transfer of knowledge and training of collaborating staff were also enhanced by testing the spoke cavity cryomodules at the FREIA Laboratory in Uppsala University. This arrangement not only made available the excellent staff and facilities at FREIA but also enhanced the capabilities of FREIA for future projects, and has supported the testing of elliptical cavities too.
- The way of working using clear technical requirements stored in a program, such as DOORS is essential to prevent mistakes and miscommunications.

9.5.2 Lessons learned from earlier projects for spoke cavities and cryomodules and contributions to future projects

At the time of the ESS's establishment, very few laboratories could build everything in-house, but for a green field site such as ESS this was not an option. In-kind procurement provides access to intellectual property, competence, and qualified manpower (including procurement staff) at partner laboratories [4]. The following aspects can be considered as lessons learned:

- The main lesson-learned is the importance of a fluid communication addressing technical and administrative competency. For the WP04, WP05, and WP19, the in-kind partner and ESS have successfully channelled the flow of information thanks to rigorous weekly meetings and action items follow-up. Dedicated calibration databases, web pages, and agile communication tools have been used to exchange and lead the SRF project. When possible, open-source tools and databases should be used to facilitate the communication and to prevent miss-leading information.

- The ESS SRF cryomodule technical design has gained expertise from lessons learned from earlier collaborations. In turn, it has generated more lessons-learned to enable the final technical design and to support similar future accelerator designs. In addition, the commissioning and operation phases benefited from personnel expertise on similar SRF equipment.
- ESS has shared its lessons learned and best practices with other large scale SRF collaborations and research infrastructures. For instance, conferences, such as IPAC, PAC, SRF, and TTC, or more specific meetings such asthe SRF the Superconducting LINACs for High Power Proton Beams (SLHiPP meetings) encourage a collaborative spirit to enhance the SRF lessons learned and best practices, e.g., Spallation Neutron Source (SNS) downtime was reported.
- The use of SRF cavities has been instrumental for the recent operation of particle accelerators in order to reach larger energy for collider physics or for basic science physics.
- There are currently three or four accelerator-driven spallation sources worldwide. The J-PARC (1 GeV) in Japan is entirely normal conducting, whereas the SNS (upgrade 1.4 GeV) in the US and C-SNS in China also have SRF portions. Using the SNS expertise and lessons learned, ESS opted for a segmented layout for the cryomodules section, instead of the continuous European XFEL LINAC. This segmented layout allows the use of normal conducting quadrupole magnets between each cryomodule to focus and defocus the beam, in accordance with the focusing–defocusing (FODO) concept. For the SNS, the cryomodules are connected to the CDS by bayonets. However, welded connections were chosen for these ESS interfaces in order to limit the global heat load and improve reliability, while keeping maintenance and CM isolation possible.
- The experience of Fermi National Accelerator Lab (FNAL) and the European-XFEL,with licensing authorities, such as Pressure Equipment Directives (PED) classification, has guided the ESS project in the engineering design choices. Indeed, according to DESY and other experience in the US, in anticipation of the constraining category 4 of the PED imposed on the European-XFEL design, ESS has designed its spoke and elliptical cryomodules to comply with the less demanding category of Sound Engineering Practice (SEP). It is worth noting that the fact European-XFEL fell under the category 4 requirement resulted in an extended project delay and extra cost, e.g., CEA had to perform extensive x-ray weld inspections due to having to comply with the legislation. The cautious ESS SRF cryomodule design initiated in 2012 meant these requirements could be relaxed, and allowed the participation of the INFN-LASA.
- Transfer of knowledge and technology, e.g., SRF cavities and front-end injectors, has been very valuable experience gained from CEA personnel in

order to secure French support of further national and international projects, e.g., GAN, SONATE, PIP-II.

- As the ESS ESH&Q guidelines progress, benchmarked with other projects, we are also witnessing the lessons learned from ESS at CERN and in the Fermilab PIP-II project. Indeed, the in-kind management has been shaped using more rigorous guidance from FNAL, supported by the experience gained from the in-kind partners, IJCLab, CEA and STFC for the PIP-II SRF accelerator.
- The collected lessons learned should be communicated to the relevant stakeholders for their use and be taken into account in the preparation of the TAs. References to experts from the given projects/experiments and research institutes may also be provided in order to facilitate their use, i.e., our Confluence pages have enabled open access for the SRF communities for years. All documents have been available to the public, to the in-kind partners or limited in access to relevant parties.
- ESS is collecting generic and specific lessons-learned in an internal database. This tool helps each stakeholder to address issues and to improve their way-of-workings.
- In the same spirit, the European Commission has granted the ESS with additional financial support to strengthen the in-kind collaborative aspects. The BrightnESS initiatives enabled to organise workshops and to gather experts during the preparation and construction phases of the ESS project.
- The core of the lessons-learned are also applied on a daily basis as the ESS team has been leading the collaborative work. Web-related platforms are gathering technical information and requirements, which have been discussed on a daily basis with the different in-kind collaborations Weekly and bi-weekly meetings have permitted to generate synergies with the different in-kind partners, resulting in a global collective team building approach. This friendly spirit, completed by official agreement, was capable of enabling transfer of technology and transfer of knowledge.
- From the design to the operation of each component, the integration shall be carefully considered. Each activity should be considered during the life-cycle of each product and should be integrated on the basis of similar experiment lessons-learned. Meanwhile, 3D integration tools and installation sequences should be reviewed by relevant experts, based on their own lessons-learned.
- Equipment interfaces and interconnections should be welded when appropriate in order to limit the overall system leak rate but (dis)connection must be carefully studied to guarantee the equipment's reliability.
- Neutron activation due to field emission is a limiting factor. Lessons-learned and communication channels with laboratories such as SNS, FNAL are essentials to guarantee an effective integration and performing cavities operation.
- Plasma processing has been studied and implemented by different institutions to limit carbon perturbations, e.g., SNS experts have developed a stringent process.

9.5.3 Training and outreach activities

The dissemination of knowledge thanks to training and outreach is an effective tool to support and communicate on design choices. The following aspects can be considered:

- Training of staff for the installation and test of key equipment is included in the TA agreement, which enables us to build up and strengthen further collaborations between competent laboratories across Europe, for the benefit of ESS.
- Training and outreach activities enhance the transfer of knowledge and the successful completion of the project. In the case of the SRF collaboration at ESS, training of the ESS Management System (ESS MS) became available online for the ESS WP leaders and the in-kind partners. Hence, ESS's ways of working became more comprehensible.
- Technical knowledge is also important to design and operate successfully SRF LINAC. A series of summer schools followed by massive open-online courses explains the function of particle accelerators, light source, and neutron source. More European grants have been used to develop such expertise, e.g., the EU Horizon 2020 Innovation—Fostering Accelerator and Science and Technology (I-FAST), Challenged-based Innovation (CBI).
- In addition, numerous schools could be mentioned, such as the CERN Accelerator School (CAS), Joint University Accelerator School (JUAS), and US-PAS, where knowledge is collected to train personnel.
- In the framework of the Nordic Particle Accelerator Project (NPAP) and the ARIES, massive open-online courses were developed to support sustainable training [49].

9.6 Conclusion

The ESS SRF collaboration was established between CEA, CNRS, INFN, STFC, DESY, Uppsala and Lund Universities, and conducted the manufacture, assembly and testing of the spoke and elliptical prototype cryomodules. Extensive SRF activities demonstrated the proof of concept of the series SRF LINAC.

Deliverables of the cooperation agreements established the design criteria for the spoke cavities and power couplers. The resulting prototypes allowed the validation of the technical feasibility of key components, and permitted the signature of in-kind contracts. The ESS spoke cavity cryomodules have been shown to meet the requirements of the ESS project. To date (2023), all 13 of the required spoke cavity cryomodules have been produced, with 11 of them successfully tested at the FREIA faculity at Uppsala University (chapter 11). A spare spoke cavity cryomodule is currently under construction at IJC Lab. Its worth noting that once the ESS LINAC is commissioned, it will represent the first significant use of double spoke SRF cavities in an accelerator.

Figure 9.15. Spoke cavity package collaboration (2014). (Courtesy of ESS)

Acknowledgments

The authors thank the commitment of the CNRS IJCLab, CEA-IRFU, LASA, STFC, ESS, DESY, and the Uppsala and Lund Universities teams who designed and tested the SRF cavity and cryomodule prototypes, and were the actors of the SRF series cryomodule production. ESS SRF accelerator achievements were driven by the collaborative effort of talented engineers and scientists. A complete list of all contributors to the ESS cryogenic system is provided in the front matter of this book.

Figure 9.15 shows the spoke collaboration team in front of the HNOSS. A complete description of the test of spoke cryomodules in Uppsala University is given the Contributors section of the book.

References

[1] Garoby R *et al* The European spallation source design *Phys. Scr.* **93** 1–121

[2] Peggs S *et al* 2013 ESS Technical Design Report *Released 2.63*

[3] Darve C *et al* 2017 ESS superconducting RF Collaboration *Presented at IPAC'17, Paper THOBB3* (Denmark: Copenhagen)

[4] Pisent A *et al* 2023 European Collaboration for the realization of ESS *Presented at IPAC*

[5] Darve C, Eshraqi M, Lindroos M, McGinnis D, Molloy S, Bosland P and Bousson S 2013 The ESS superconducting linear accelerator *SRF13* (Paris, France)

[6] Garoby R *et al* Progress on the ESS project construction *Proc. IPAC17, Paper MOXBA1*

[7] McGinnis D *et al* 2014 New design approaches for high intensity superconducting linacs—the new ESS linac design *Proc. IPAC'14 (Dresden, June 2014)* p 23

[8] Eshraqi M Beam physics design of the Optimus+ SC linac *ESS AD Technical Note ESS/AD/0050*

[9] Darve C *et al* 2014 Requirements for ESS superconducting radio frequency linac *IPAC'14* (Germany: Dresden) p 3311
[10] Devanz G *et al* 2013 Cryomodules with elliptical cavities for ESS *SRF'13* (Paris, France)
[11] Olivier G *et al* 2013 ESS cryomodules for elliptical cavities *MOP084, SRF'13* (Paris, France)
[12] Darve C, Bosland P, Devanz G, Olivier G, Renard B and Thermeau J P 2013 The ESS elliptical cavity cryomodules *CEC2013* (Alaska: Anchorage)
[13] Olivier G *et al* 2014 Technical Note, cryomodule for elliptical cavity—global heat balance—3
[14] Piquet O *et al* Final design of the LB650 cryomodule for the PIP-II linear accelerator
[15] Darve C *et al* The superconducting radio-frequency linear accelerator components for the European spallation source: first test results *Proc. LINAC15*
[16] Elias N *et al* 2014 Preliminary functional analysis and operating modes of the ESS 704 MHz superconducting radio-frequency linac', THPME042 *Proc. IPAC'14* (Germany: Dresden)
[17] Peauger F *et al* 2014 Status and first test results of two high-beta prototype elliptical cavities for ESS *Proc. IPAC'14* (Dresden)
[18] Peauger F *et al* 2015 Progress in the elliptical cavities and cryomodule demonstrators for ESS *TUPB007, SRF'15* (Canada: Whistler)
[19] Darve C *et al* 2016 The SRF linac components for the ESS: first test results *LINAC2016* (USA: East Lansing)
[20] Darve C *et al* ESS SRF Linear accelerator components preliminary results and integration *Proc IPAC17*
[21] Peauger F *et al* Preliminary test results of the first ESS elliptical cryomodules demonstrator *Proc IPAC19*
[22] Bosland P *et al* 2017 Status of the ESS elliptical cryomodules at CEA Saclay *Presented at IPAC'17, Paper MOPVA040* (Denmark: Copenhagen)
[23] Darve C *et al* 2014 The ESS superconducting RF cavity and cryomodule cryogenic processes *ICEC25* (Netherland: Enschede)
[24] Asensi Conejero E *et al* 2017 The cryomodule test stands for the European spallation source *Presented at IPAC'17, Paper MOPVA089* (Denmark: Copenhagen)
[25] Cenni E *et al* 2017 Vertical test results on ESS medium and high-beta elliptical cavity prototypes equipped with helium tank *Presented at IPAC'17, Paper MOPVA041* (Denmark: Copenhagen)
[26] Arcambal C *et al* 2017 Conditioning of the RF power couplers for the ESS elliptical cavity prototypes *Presented at IPAC'17, Paper MOPVA044* (Denmark: Copenhagen)
[27] Michelato P *et al* 2017 Vertical tests of ESS medium-beta prototype cavities at LASA *Presented at IPAC'17, Paper MOPVA063* (Denmark: Copenhagen)
[28] Monaco L *et al* 2017 Fabrication and treatment of the ESS medium-beta prototype cavities *Presented at IPAC'17, Paper MOPVA06* (Denmark: Copenhagen)
[29] Bertucci M *et al* 2017 Quench and field emission diagnostics for the ESS medium-betaPrototypes vertical tests at LASA *Presented at IPAC'17, Paper MOPVA061* (Denmark: Copenhagen)
[30] Chen J *et al* 2017 Multipacting studies in ESS medium-betacavity *Presented at IPAC'17, Paper MOPVA064* (Denmark: Copenhagen)
[31] Pirani S *et al* 2017 Investigation of HOM frequency shifts induced by mechanical tolerances *Presented at IPAC'17, Paper MOPVA091* (Denmark: Copenhagen)
[32] Sertore D *et al* 2017 Experience on design, fabrication and testing of a large grain ESS medium-beta prototype cavity *Presented at IPAC'17, Paper MOPVA068* (Denmark: Copenhagen)

[33] Arcambal C *et al* 2015 Status of the power couplers for the ESS elliptical cavity prototypes *THPB078, SRF'15* (Canada: Whistler)
[34] Cenni E *et al* 2015 ESS medium-beta cavity prototypes manufacturing *THPB028, SRF'15* (Canada: Whistler)
[35] Bousson S, Darve C, Duthil P, Elias N, Molloy S, Reynet D and Thermeau J-P 2014 The ESS spoke cavity cryomodules *Adv. Cryog. Eng* **59A** 665–72
[36] Li H, Santiago Kern R, Jobs M, Bhattacharyya A, Goryashko V, Hermansson L, Gajewski K, Lofnes T, Fransson K & Ruber R 2017 First High Power Test of the ESS Double Spoke Cavity *FREIA* Report 2017/10
[37] Reynet D *et al* Design of the ESS spoke cryomodule *MOP089*
[38] Reynet D *et al* 2015 ESS spoke cryomodule and test valve box *THPB109, SRF'15* (Canada: Whistler)
[39] Duchesne P *et al* 2013 Design of the 352 MHz, beta 0.50, double-spoke cavity for ESS *SRF'13* (Paris, France) p 1212
[40] Olry G *et al* 2015 Recent progress of ESS spoke and elliptical cryomodules *TUAA06, SRF'15* (Canada: Whistler)
[41] Li H *et al* 2017 ESS spoke cavity conditioning *Presented at IPAC'17, Paper MOPVA094* (Denmark: Copenhagen)
[42] Li H *et al* 2015 Test characterization of SC spoke cavities at Uppsala University *TUPB083, SRF'15* (Canada: Whistler)
[43] Santiago Kern R *et al* 2015 Cryogenic performance of the HNOSS test facility at Uppsala University *TUPB026, SRF'15* (Canada: Whistler)
[44] Fransson K *et al* 2017 The epics based control system at the FREIA laborator *Presented at IPAC'17, Paper TUPIK08* (Denmark: Copenhagen)
[45] Lanfranco G *et al* 2014 Constructing the ESS linear accelerator: pragmatic approaches to design and system integration at the European spallation source *WEPRO076, Proc IPAC*
[46] Rampnoux E *et al* 2013 Design of 352.21 MHz RF power input coupler and window for the European Spallation Source (ESS) *SRF'13* (Paris) September 2013 p 1069
[47] Gandolfo N *et al* 2014 Fast tuner performance for a double spoke cavity *LINAC'14* (Switzerland: Geneva) p 1034
[48] Fydrych J *et al* 2014 Cryogenic distribution system for the ESS superconducting proton linac *ICEC25* (Netherland: Enschede)
[49] Delerue N(IJCLab) 2022 New online course to initiate undergraduate students to particle accelerators *CERN Accelerating News*

Chapter 10

Cryogenics testing of elliptical cryomodules at ESS

Nuno Elias

The cryomodule qualification test stand (TS2) in Lund was fully commissioned in 2021 and is performing site acceptance testing of all the elliptical cryomodules (CMs) before installation in the linear accelerator (LINAC) tunnel.

This chapter describes the infrastructure available at TS2 and the workflow leading to the acceptance of cavities and cryomodules received from the different partners. We will then focus on the cryogenic operation aspects of TS2 and the cryogenic acceptance testing of elliptical cryomodules. Finally, we will draw some lessons learned about the construction and operation of TS2 (figure 10.1).

10.1 Cavities and cryomodules life cycle

The superconducting part of the LINAC is composed of different families of cryomodules organized in three sectors containing: 13 spoke cryomodules, 9 elliptical medium-β cryomodules, and 21 elliptical high-β cryomodules. Each of the aforementioned cryomodules contains superconducting cavities with a similar name (figure 10.2).

As described in chapters 8 and 9, both the medium-β and high-β cryomodules each house four superconducting cavities; however, the cavity geometry has been optimized to maximize the proton beam acceleration, resulting in medium beta and high-beta cavities of different geometries (medium beta cavities with beta equal to 0.67 and six cells, while high-beta cavities with beta equal to 0.86 and five cells) (figures 10.3 and 10.4).

Despite the cavities having different geometry resulting in different length, the internal geometry of the medium-beta and high-beta cryomodule and cryogenics circuits has been optimized such that most internal components are similar and/or have the same interfaces.

This unified design solution for internal and external interfaces carries multiple advantages. Worth highlighting here is the sharing of assembly tooling, common

doi:10.1088/978-0-7503-3223-1ch10

Figure 10.1. Wide angle picture of the TS2 premises with 1 cryomodule inside the test bunker, and preparation area for receiving, incoming verification tests and preparation for dispatch. (Courtesy of ESS)

	Spoke	Medium- β	High-β
Proton energy range, MeV	90 to 216	216 to 571	571 to 2000
Cryomodules/Sector	13	9	21
Cavities/Cryomodule	2	4	4
Cavities/Sector	26	36	84
Sector length, m	55.9	76.7	178.9
Operating freq., MHz	352.21	704.42	704.42
Operating temp., K	2	2	2
Cavities optimum β	0.5		
Cavities geometric β		0.67	0.86
E_{acc}, MV·m^{-1}	9	16.7	19.9

Figure 10.2. ESS superconducting LINAC specifications for each family of cavities and cryomodules. (Courtesy of ESS)

Design Requirements	Spokes	Medium-β	High-β
E_{acc}, MV·m^{-1}	9.0	16.7	19.9
Q_0 at E_{acc}	>1.5x10^9	>5x10^9	>5x10^9
Iris Diameter, mm	56	94	120
E_{pk}, MV·m^{-1}	39	40	44
B_{pk}/E_{acc}, mT·MV^{-1}·m	6.80	4.79	4.3
E_{pk}/E_{acc}	4.28	2.36	2.2
G (Ω)	130	196.6	241
R/Q max, Ω	425	367	435
RF peak power, kW	335	1100	1100

Figure 10.3. Summary of cavity design requirements. (Courtesy of ESS)

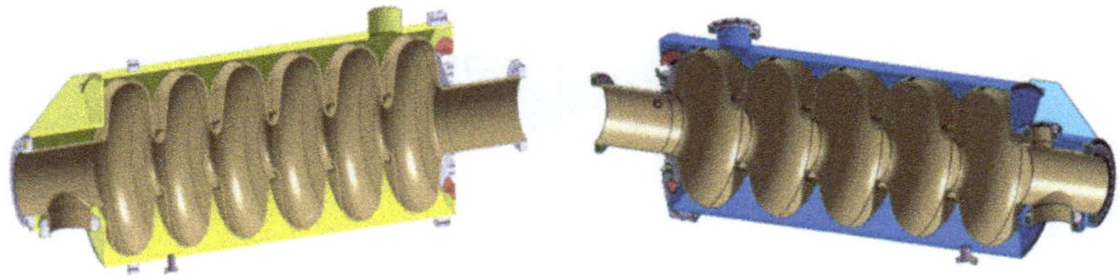

Figure 10.4. Illustration of the ESS six-cell medium-beta (left-hand) and five-cell high-beta (right-hand) superconducting cavities and their helium tanks. (Courtesy of ESS)

assembly workplan, and installation procedures, and also common testing and commissioning procedures.

10.1.1 SRF collaboration

The complexity of the European Spallation Source (ESS) project requires an array of collaborating institutes to bring it to completion. The accelerator components, and in particular superconducting cavities and cryomodules, follows the same line, where design, construction and testing are performed by various institutes and at various locations under ESS coordination and oversight.

Due to the vast distribution of responsibilities and the multitude stakeholders, each step of the project life cycle follows well-defined acceptance criteria, where collaboration and efficient information exchange are paramount. A synoptic table summarizing some of the selected steps of the project life cycle for each family of cryomodules and the collaborating institute responsible for each deliverable follows.

Acronyms used in figure 10.5 of in-kind partners, collaborating institutes, and participating industrial partners:

- IPNO: Institut de Physique Nucléaire Orsay, FRANCE (recently renamed IJC-LAB: Laboratoire De Physique Des 2 Infinite Irène Joliot-Curie, FRANCE).
- INFN: Istituto Nazionale di Fisica Nucleare, ITALY.
- STFC: Science and Technology Facilities Council, UK.
- CEA: Commissariat à l'énergie atomique et aux énergies alternatives, FRANCE.
- UU: Upsala University, SWEDEN.
- DESY: Deutsches Elektronen-Synchrotron, GERMANY.
- ZANON: Zanon Research and Innovation SRL, ITALY.
- RI: Research Instruments GmbH, GERMANY.
- ESS: European Spallation Source, SWEDEN.

10.1.2 Life cycle of cavities and cryomodules

During the niobium production, samples are taken to assure the correct characteristics and documented prior to shipment (e.g., chemical composition, RRR, mechanical properties, and absence of surface defects).

	SPOKES	MEDIUM-BETA	HIGH-BETA
Nb Procurement*	IPNO	INFN	STFC
Nb Eddy scanning* (location)	-	ESS (DESY)	ESS (DESY)
Cavity manufacturer*	IPNO	INFN (ZANON)	STFC (RI)
Cavity Testing* (Location)	IPNO (IPNO)	INFN (LASA and DESY)	STFC (STFC and DESY)
Cryomodule Assembly	IPNO	CEA	CEA
CM Testing (prototype)	IPNO	CEA	CEA
CM Testing (series)	UU (FREIA)	ESS (TS2)	ESS (TS2)

Figure 10.5. Simplified life cycle of cavities and cryomodules showing the responsibilities of each collaborating institute. (Courtesy of ESS)

The selection of the correct radio frequency (RF) side for the niobium sheets is done by a combination of visual inspection, dimensional controls, and eddy current scanning (elliptical cavities only).

The different cavity parts (e.g., hall cells, dumbbells, and end-groups) are also subject to quality controls in order to achieve the correct final frequency and overall dimensions. Once the cavities have been produced, surface treatments take place (chemical and heat treatments). At each step of the way, strict quality controls (e.g., acid temperature, flow, oven temperature, gas analysis) and reports are created. The cavities are then tuned to achieve the correct field flatness and fundamental mode frequency, and are integrated with the helium tank, followed by pressure tests and leak tests.

Cavity tests are performed at 2 K in vertical cryostats, nevertheless some important checks are done at incoming inspection. Here fundamental mode frequency and bandwidth is remeasured and compared, and checks are made for completeness and damage. At cold, the fundamental frequency is remeasured, cold leak checks are done, and the power rise curves are derived (Q vs. E), as well as the maximum achievable gradient (quench limit).

The accepted cavities are then sent for string assembly where again they undergo incoming inspections which should match the previous outgoing inspections. After that, readiness for cryomodule installation is granted and sorting is done if possible. String assembly is done in a clean room where fundamental power couplers are mounted. In addition, the full cryomodule is assembled, encompassing the cold tuning system, magnetic and thermal shields, cryogenic circuitry, instrumentation, cabling, and cryostating in the vacuum vessel. Warm checks are done (electrical, pressure, and leak checks) and quality documentation is gathered. Further information on cavity and cryomodule assembly may be found in chapters 8 and 9.

The first three elliptical cryomodules of each series (pre-series) undergo cold testing at CEA-Saclay in slightly different conditions than that of ESS (both in terms of cryogenic operation as well as RF conditions) to validate the assembly procedure.

The final site acceptance tests at ESS allow for the determination of the readiness for installation of all cryomodules. A thorough inspection protocol is followed to access that all requirements are met. This allows for the correct validation of previously measured parameters, as well as to validate the correct functioning of other systems (e.g., dimensional control and final integration of components, absence of damage during transport, cold tuning system operation associated with tuning sensitivity measurements, power coupler conditioning, cryogenic heat loads, and absence of leaks).

Once the final site acceptance is granted, the cryomodule's status becomes ready for installation and will follow ESS installation, testing, and commissioning plan.

As highlighted, ESS is responsible for CM acceptance testing at TS2 of all elliptical cryomodules (medium beta and high beta), while the Freia Facility (chapter 11) is responsible for acceptance testing of the spoke cryomodules.

10.1.3 The ESS TS2

TS2 [1] is located at the ESS site in Lund, located in the contingency area of the klystron gallery building, also referred to as the gallery technical area. TS2 consists of a radio-protection bunker with the same cross section as the accelerator tunnel, a test stand cryoplant and a modulator, and two klystrons as RF power sources. The test stand provides facilities for testing the elliptical cavity cryomodules of both varieties: medium beta and high beta.

10.1.3.1 General layout

The overall layout of TS2 is shown in figure 10.6. Two rows of together 24 electrical racks are placed in the test stand to include the control and instrumentation equipment for RF, cryogenics, vacuum, cooling water, timing system, motion control, personnel protection system, and oxygen deficiency hazard detectors. A number of spare racks will allow the test stand to be flexible and expandable.

10.1.3.2 TS2 radio-protection bunker

The main function of the bunker is to protect the surrounding area from *x*-rays produced during the cryomodule tests. It consists of 1200 t of heavy, magnetite loaded concrete, arranged to form 1 m thick walls all around the cryomodule, while allowing services such as cryogenics, RF, and cabling to enter through chicanes. A

Figure 10.6. Simplified layout of TS2 RF systems and RF distribution. Reprinted with permission from [1]. Copyright © 2017 by JACoW

personnel access chicane and a full width access door for the cryomodules are also part of the bunker. The radio-protection bunker has the same cross section as the accelerator tunnel. In addition, the layout of the various systems inside the bunker is similar to that of the tunnel because it allows us to practice installation routines and identify shortcomings that can be timely mitigated and overcome.

10.1.3.3 TS2 RF system

The RF system at TS2 consists of RF sources and conditioners (klystrons and modulator), an RF distribution system, and the low-level RF (LLRF) system, as shown in figure 10.7. At TS2, the modulator powers two klystrons and one klystron per two cavities. The modulator converts electrical power to the klystrons' cathodes from a standard low voltage distribution grid to an electrical pulse of 1 MW at the accelerator's pulse frequency of 14 Hz. The klystrons convert the pulsed power coming from the modulator to RF waves at 704.42 MHz. The phase and amplitude of the electrical field in the cavity is controlled by the LLRF system controls using a PI-controller.

10.1.3.4 TS2 cryoplant (TICP) and cryogenic distribution

The test stand cryoplant (TICP) is composed of a liquefier cold box, a recovery system, and an external purifier. It serves the following tasks:

- Provide cooling for TS2 at 2 K and 35 K with a liquefaction rate of 6 l h^{-1}.
- Provide liquid helium for ESS's neutron instruments and sample environments with a liquefaction rate of 7500 l/month.
- Recover, purify, and manage helium at the facility.

Although the TICP is only to be used for continuous cryomodule testing for the first few years, this operation defines the plant's design [2]. The cryogenic distribution system (CDS) is dedicated to transferring cooling power from the TICP to the ESS elliptical cryomodules under their site acceptance tests in the test stand bunker.

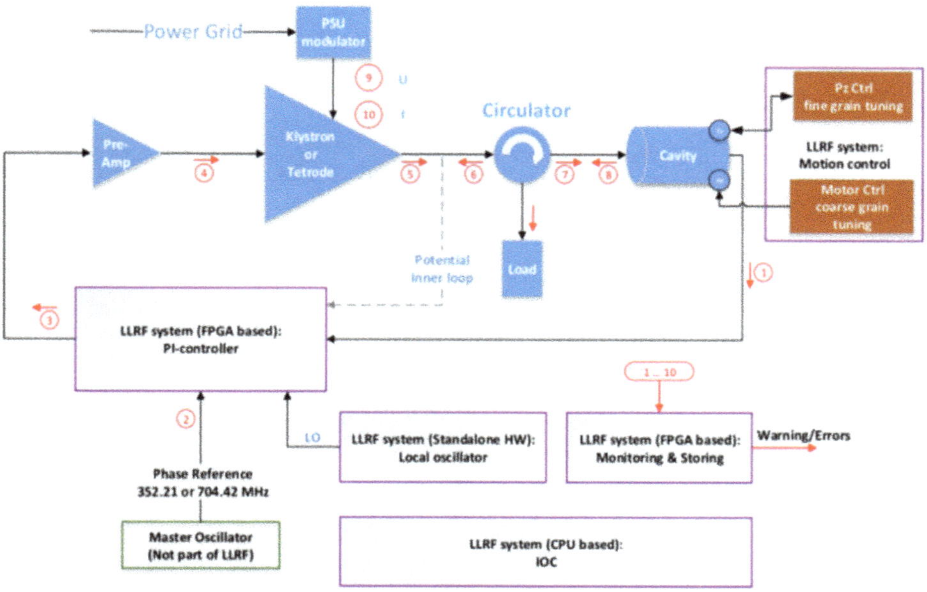

Figure 10.7. Block diagram of the TS2 LLRF system. (Courtesy of ESS)

The system includes a cryogenic transfer line (CTL), one valve box, and four auxiliary process lines. Its layout is shown in figure 10.8.

The CTL runs from the TICP cold box in the cold box building to the test stand bunker placed in the klystron gallery. The line is a vacuum insulated multichannel line and its vacuum jacket houses four cold process lines, thermal shield, supports, and thermal compensation system. The cryoline ends in the test stand valve box, in which four branch process lines connect to the cryomodule cold circuits.

The test stand cryoplant (TICP; explained in great detail in chapter 7), provides supercritical helium at 5 K and 3 bara for cooling of the cavities cold mass and helium at 35 K at 13.5 bara for cooling of the cryomodule thermal shield.

Each of the cryomodules under test interface with the CDS, leading to the TICP, through a dedicated valve box (VBox) that is also placed inside the bunker. The CDS used in TS2 is nearly identical to the system used in the accelerator tunnel (chapter 4). Figure 10.9 shows the layout of TS2.

Additional systems such as water-cooling skids, and data acquisition and equipment control racks are placed outside the radio-protection bunker.

10.1.4 The elliptical cryomodule

10.1.4.1 Main components

In chapter 8, a detailed description of the elliptical cryomodule was given. Here we will again illustrate a few important components.

In nominal conditions each elliptical cavity is immersed in superfluid liquid helium contained inside individual cavity tanks. The helium supply to the tanks can be done via the bottom of each tank through the CV02 valve directly from the

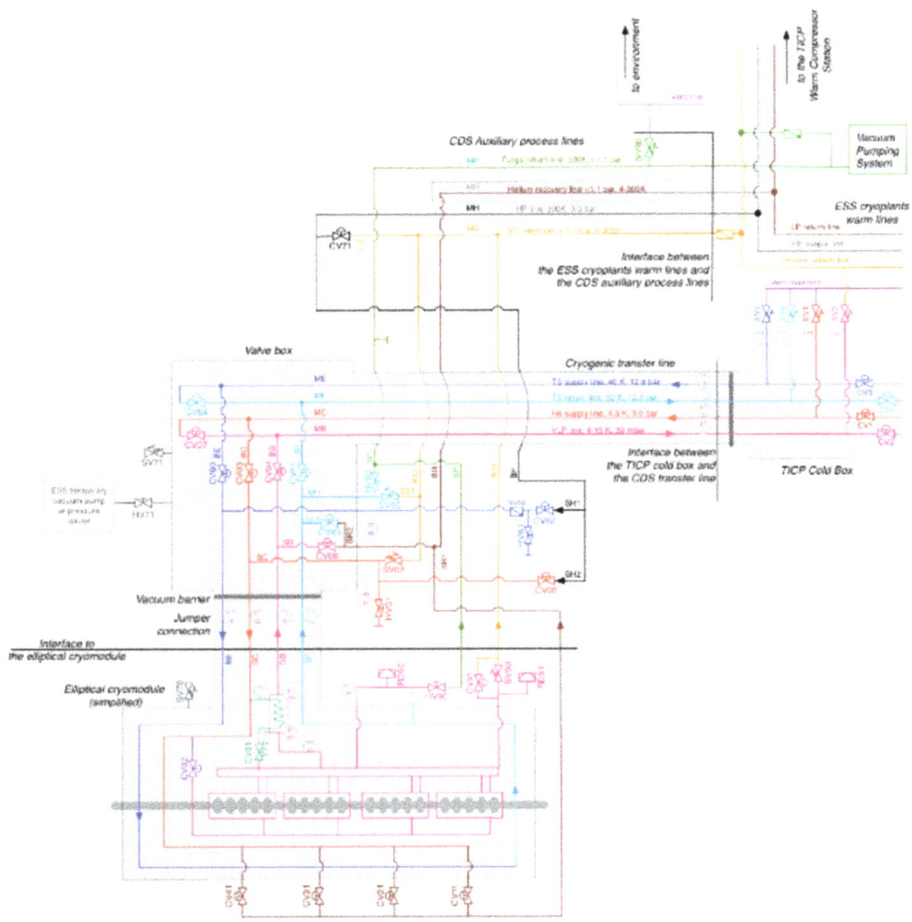

Figure 10.8. Simplified layout of the cryogenic distribution of TS2. Reprinted with permission from [1]. Copyright © 2017 by JACoW

helium supply line, or via the CV01 valve where supplied helium is first precooled at the heat exchanger and then passing through CV01 and supplied to the top of the tanks via the diphasic line interconnecting all the tanks.

Visible on figure 10.10 are the rupture disks (one placed at each side) that provide protection to the low-pressure circuit (helium tanks). The rupture disks are connected to a helium vent line (not shown), which directs helium vapors outside of the bunker in case of an accident scenario where over pressure occurs.

Visible on figure 10.10 are the four circuit and interfaces interconnecting to the valve box. These four circuits are the thermal shield supply and return, and the cavity tanks' supply and vapor return.

Also illustrated here are the power coupler component which provides RF power to the cavities. The RF distribution is interconnected at the doorknob transition where the rectangular waveguide (not shown) transitions to a coaxial line with the power coupler antenna at the center.

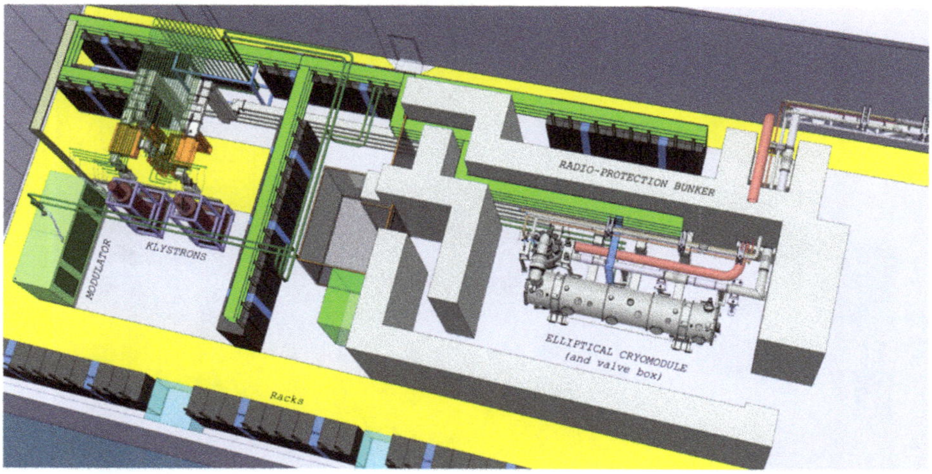

Figure 10.9. Layout of ESS Lund Test Stand 2 (TS2). The RF sources shown on left-hand side bunker with cryomodule under test on the center right-hand side and the cryogenic distribution leading to the TICP on the top right-hand side. (Courtesy of ESS)

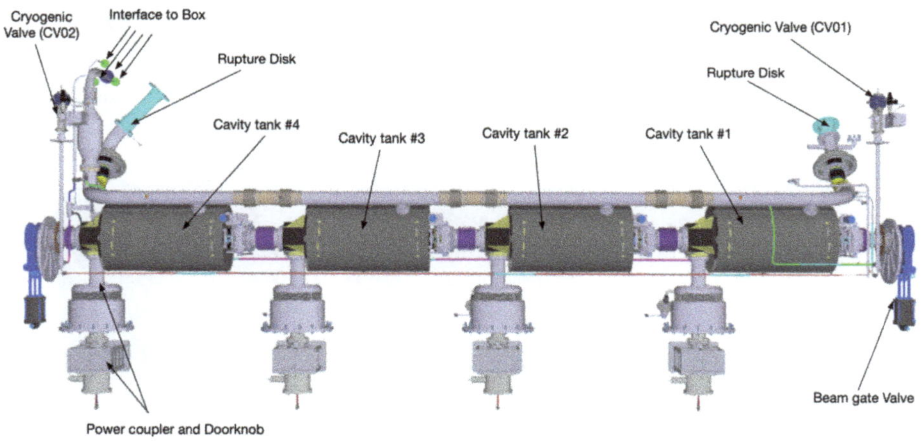

Figure 10.10. Front side view of the cavity string inside the vacuum vessel. (Courtesy of ESS)

Also shown in this picture are the two gate valves (one at each side) which provide isolation of the beam vacuum during transport.

In figure 10.11 we display the cavity string as seen from the back side—now the heat exchanger appears on the right-hand side. Also visible in this illustration are the cold tuning systems, which allow for the cavity frequency to be precisely adjusted.

This picture also shows the power coupler double wall cooling circuit outlet. Each of the power couplers is cooled with a fraction of the supply helium, which dramatically reduces the conduction heat from the ambient temperature to the flange interconnecting to the superconducting cavities.

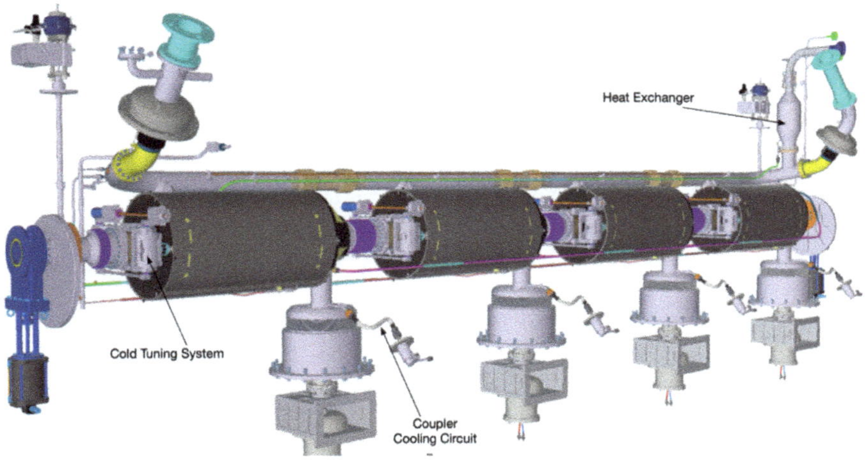

Figure 10.11. Back-side view of cavity string. (Courtesy of ESS)

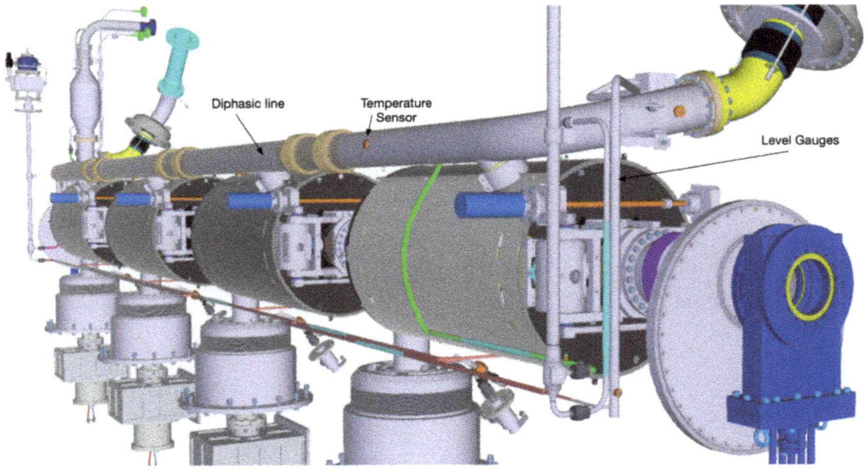

Figure 10.12. View near cavity #1 showing level gauges pipe and diphasic line on top. (Courtesy of ESS)

In figure 10.12 we show an up-close view from cavity #1 where CV01 is located and also the level gauge pipe. The diphasic line interconnecting the cavity tanks is visible at the top. A few temperature sensors are also visible in this view.

10.1.4.2 Cryogenics circuitry

Test Stand 2 was initially commissioned with the medium-beta elliptical cryomodule prototype. This cryomodule included additional instrumentation when compared to the final (Series) cryomodule. In figure 10.13 we present the process-flow diagram for the prototype cryomodule and in figure 10.13 we present the series cryomodule when connected to the test stand (figure 10.14).

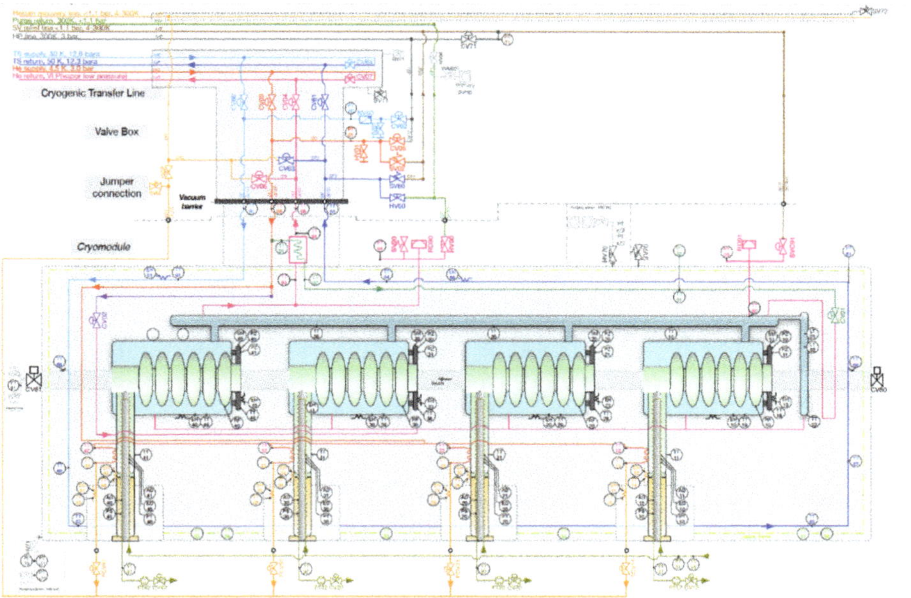

Figure 10.13. Prototype cryomodule and valve box PFD. (Courtesy of ESS)

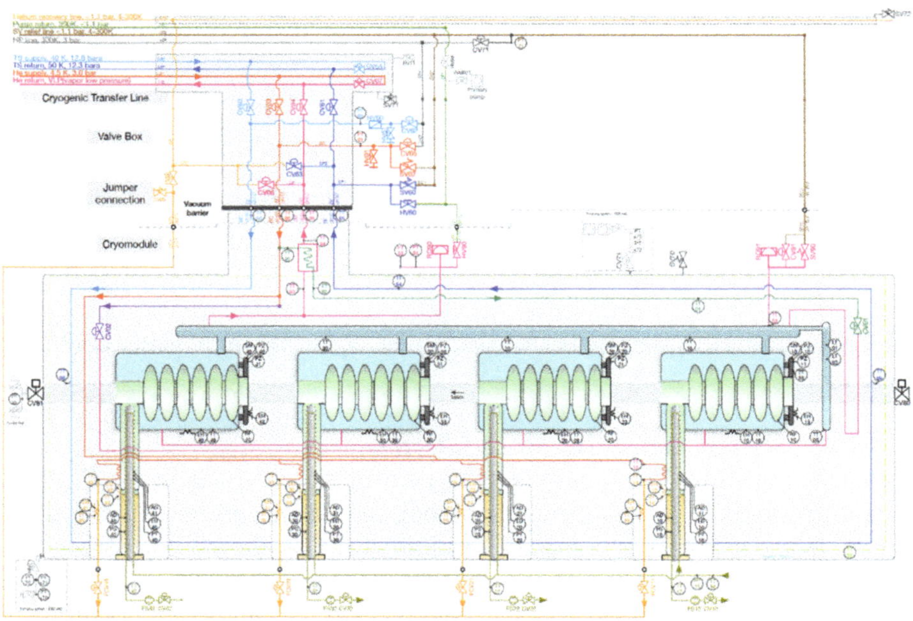

Figure 10.14. Series cryomodule and valve box PFD. (Courtesy of ESS)

10.1.4.3 Cryogenics instrumentation

In table 10.1 we list the instrumentation equipment in the cryomodules (prototype and series) and valve box at TS2. The equipment related to auxiliary systems are

Table 10.1. Cryogenic instrumentation in the prototype and series elliptical cavity cryomodules and valve box. (Courtesy of ESS)

Cryomodule	Prototype	Series
Cryogenic control valves (inside)	2	2
Cryogenic control valves (outside)	1	1
Cryogenics pressure relief valves	1	1
Rupture disks	2	2
Mass flow controllers	4	4
Level gauges	2	2
Temperature sensors	55	32
Pressure sensors (main)	4	2
Vacuum gauges	4	4
Heaters	18	16
Stepper motors	4	4
Piezo actuators	8	8
Arc detectors	8	8
Electron detector	4	4

Valve box	
Cryogenic control valves (inside)	8
Cryogenic control valves (outside)	3
Cryogenics safety valves	2
Other safety relief valves	2
Hand valves	7
Cryogenic check valves	1
Temperature sensors	4
Pressure sensors (main)	2

excluded from this list, such as beam vacuum, isolation vacuum, water cooling, and air instrument.

10.1.4.4 Cryogenics control system architecture

The control system in TS2 is developed to control the processes in the CDS, valve box, and cryomodule in TS2 in order to cool, maintain, and recover the elliptical cryomodule under test from cryogenic conditions and to allow the superconducting radio-frequency (SRF) cavities to test and operate the cavities in the cryomodule. A large portion of this section is extracted from the paper ‘Commissioning the Control System for Cryomodule Cryogenics Distribution System in Test Stand 2’ [3].

The architecture of the control system for TS2 includes sensors and actuators for cryogenics, cooling water, vacuum, motion control, and RF (figure 10.2). These last two disciplines are not programmable logic controller (PLC) controlled and are therefore out of this scope, all the previous disciplines are connected to signal conditioner elements or power supplies to be controlled by a PLC controller. The operator interface (OPI), archiver, and alarm services are developed and connected to the PLC through the ESS EPICS environment (EEE) [4].

As for any other automated PLC-based control system developed at ESS, the control system for the cryomodule cryogenic has been generated through PLC Factory [5] retrieving the information of the field devices from the Controls Configuration Database (CCDB) [6]. This method creates the base code to control field devices by passing the EPICS process variables (PVs) associated to each object type to both the PLC and IOC (input - output controller), allowing the communication of data through EPICS to visualize and manipulate the system using devices from the cryogenics library inside the OPIs developed in CS-Studio Display builder [7]. Figure 10.15 shows the architecture of the TS2 cryogenics control system.

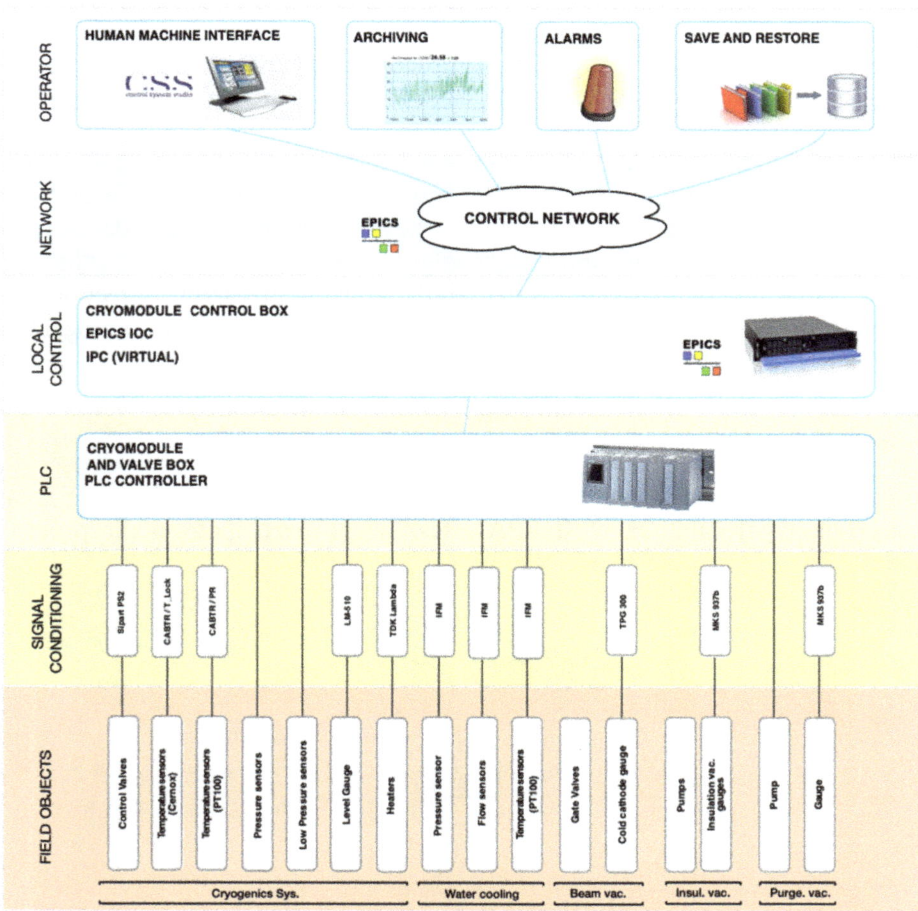

Figure 10.15. Architecture of the cryogenics control system present at TS2. Reprinted with permission from [1]. Copyright © 2017 by JACoW

10.1.4.5 Cryogenic operating modes

When operating all cryomodules in the tunnel together with the cryogenics distribution and accelerator cryoplant, the whole system will have to go through a series of steps before being able to be operated in nominal beam condition at 2 K.

Figure 10.16 shows the idealized operating modes and the steps required to operate the cryomodules, such as cool down from ambient temperature to 4.5 K, stand-by at 4.5 K, cool down to 2 K, stand-by at 2 K, and nominal operation at 2 K with RF and beam being turned on. Conversely, there is a series of operating modes to warm the SRF LINAC to ambient temperature (e.g., during scheduled maintenance periods), to warm up and cool down a single cryomodule, and also to bring a cryomodule into a ‘ready’ state after interventions.

The experience at TS2 is paramount in order to properly define the control process for each mode, together with the required conditions for transition between states, warnings, alarms, and interlocks, and to optimize each step.

At TS2 we follow a similar state diagram as for the full cryogenic distribution together with cryomodules, even though we only have one cryomodule and one valve box. This allows us to improve our processes with every cool-down/warm-up cycle and improve the definition of the programming logic for the LINAC operation.

The state diagram in figure 10.16 shows a sequence of ‘boxes’ corresponding to different modes of operation. Within each mode a sequence of actions is performed with the help of control logic in order to keep the system in the desired state. Entering or exiting a specific mode can only be done when specific conditions are satisfied (indicated by the green letter and following the arrow directions). An additional layer of complexity, not illustrated below, is needed during accidental scenarios in order to bring the system into a nominal state again.

10.1.5 Site acceptance testing of elliptical cryomodules

The elliptical cryomodules acceptance testing at TS2 is the responsibility of ESS with the important participation of IFJ-PAN team, who participate in the project as Poland’s in-kind collaborators to the project.

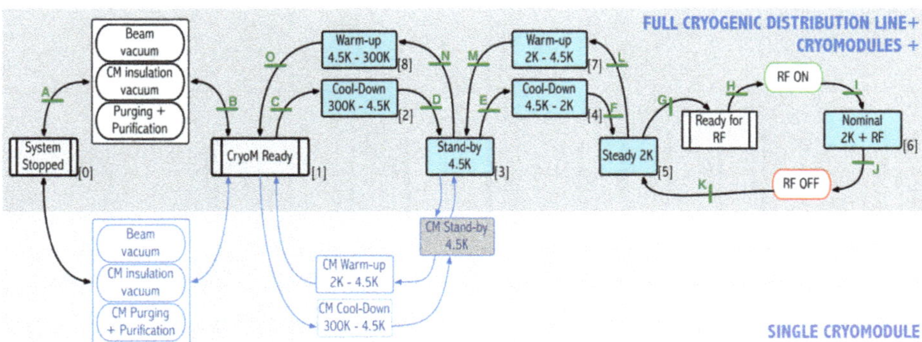

Figure 10.16. State diagram of cryomodule operating modes in the LINAC. (Courtesy of ESS)

#	Phase
1	Cryomodule reception
2	Cryomodule preparation
3	Cryomodule installation
4	Cryomodule Warm Validation
5	Cryomodule Cold Validation
6	Cryomodule Warm-up
7	Cryomodule Disconnection
8	Cryomodule Preparation for Dispatch
9	Cryomodule Dispatch

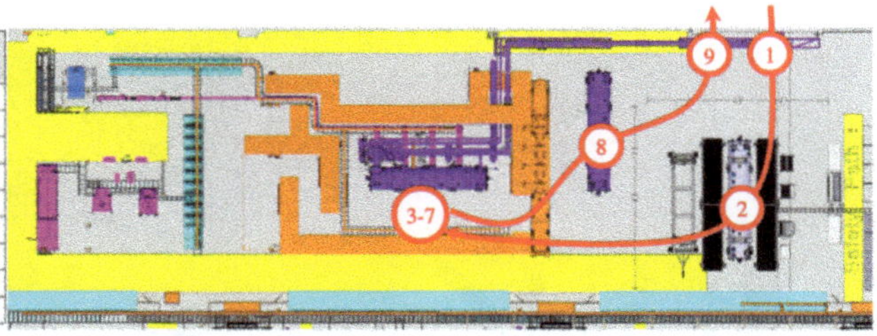

Figure 10.17. Phases and workflow of the cryomodule for SAT. (Courtesy of ESS)

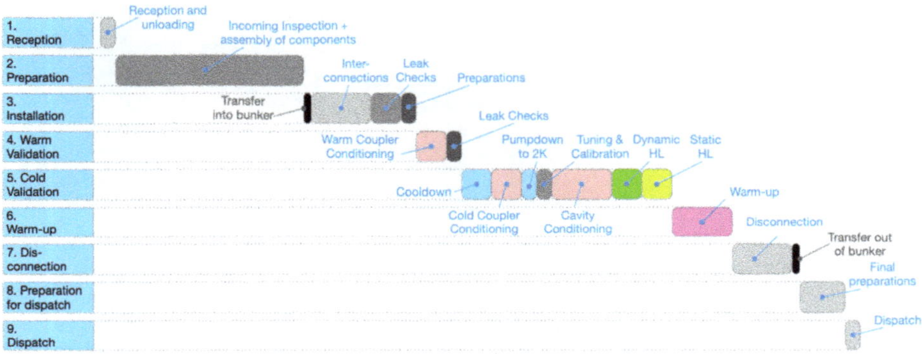

Figure 10.18. Net plan at each phase and main activities. (Courtesy of ESS)

The workflow for cryomodule site acceptance testing at ESS TS2 (shown in figure 10.17) is divided into nine distinct phases where the cryomodule is moved through various locations of the TS2 area [8]. Dedicated transportation and transfer equipment, as well as positioning stand allow for the cryomodule to be safely transferred and positioned at each of the various phases.

Figure 10.18 summarizes the main activities occurring at each of the phases during SAT. The bulk of the testing occurred when the cryomodule is inside the bunker (phases 3 through 7).

10.1.5.1 Cryomodule reception

The cryomodule arrives from CEA loaded on a dedicated transport container and mounted on a tailored shock absorbing frame. The cryomodule goes through initial inspection, before being unloaded with the help of a crane. During this phase the cryomodule and parts-in-circulation boxes are checked for completeness. Additionally, a rough visual inspection is done to verify that no equipment has been displaced due to transport vibration. A set of shock loggers (accelerometers), which were previously mounted at departure from the in-kind partners, is read to verify that the cryomodule was not subject to any excessive acceleration.

The cryomodule is then transferred onto the transportation cart for in-house displacements at TS2 (figure 10.19). This cart is equipped with a remotely operated

Figure 10.19. Picture of a cryomodule loaded on top of the transportation cart used front in-house displacements at TS2. (Courtesy of ESS)

skate (battery assisted), allowing for slow but precise and highly maneuverable rotations on one axis.

10.1.5.2 Cryomodule preparation

The cryomodule is transferred to the preparation stand (see figure 10.20). Here a detailed and meticulous inspection takes place. The cavity transport fixtures are removed and the doorknob transitions are installed. Additionally, a set of electrical verification tests is done to all instrumentation in order to check the electrical continuity.

Additional auxiliary circuits are attached to the module such as compressed air manifolds.

10.1.5.3 Cryomodule installation in thetest bunker

Once all the preparation work is done, the cryomodule is again transferred onto the transportation cart and transferred into the bunker, and finally onto the bunker support stand (see figure 10.21). Here all interconnections onto the bunker systems are done (similarly to what is expected to take place in the tunnel). A non-exhaustive series of interconnecting systems is listed below:

- RF distribution system.
- Isolation vacuum pumping system.
- Beam vacuum pumping system.
- Water cooling system.
- Cryogenics system (jumper-interconnection).

Figure 10.20. Picture of two cryomodules in the preparation area. One mounted on top of the transportation cart and one mounted on the preparation stands. (Courtesy of ESS)

- Cryogenics auxiliary lines.
- Instrument compressed air distribution.

At this stage a set of leak-checking campaigns takes place. Here each cryogenic circuit is checked individually.

10.1.5.4 Cryomodule warm validation

To achieve the desired power levels for the power couplers, coupler conditioning is necessary. This process takes place on two occasions: first with the cryomodule in 'warm' state and later when the cryomodule has been cooled-down to lower than 4.5 K.

This process has been automated through the use of an EPICS procedure. This procedure follows a predefined sequence of steps established at CEA, incorporating variations in forward power, pulse length, repetition rate, and amplitudes. The control variable employed is the vacuum, while the waveforms of the electron pickup and arc detector are monitored. During the CM test, when the system operates in standing wave mode without a beam, the total power transmitted to the coupler must not exceed 300 kW for pulse lengths longer than 0.5 ms, irrespective of the repetition rate. However, for shorter pulse lengths, the peak power can reach up to 1.2 MW, regardless of the repetition rate.

Figure 10.21. Picture of a cryomodule installed inside the test bunker on top of the stands. (Courtesy of ESS)

While the warm coupler conditioning is taking place, a set of control system verification steps takes place prior to the cool down. This includes loading of the temperature sensors' calibration points and verification of proper measurement at room temperature. Similarly, the pressure sensors are calibrated using a two-point method of different states of pressure, these steps are done while preparing the circuits for cool down, which include cleaning and purification using sequential pump and purge cycles. Additionally, all other control devices and diagnostics instrumentation are actuated and read via the control system and its correct behavior is noted.

10.1.5.5 Cryomodule cold validation

When all verification checks at room temperature have been completed and the warm coupler conditioning has been concluded, the cool down of the cryomodule is allowed to start. This is done with the coordination of two teams: the cryoplant operators, and the cryomodule and cryogenic distribution operators.

The cool down is typically done by cooling the CM thermal shield while simultaneously cooling the cold mass (cavity tanks). The cool down to 4 K typically takes one day and the initial goal during this process is to sustain liquid inside the helium tanks and check the proper behavior of the level gauges. The helium level is then raised to the nominal level and level control is kept active.

In parallel, the cold conditioning of the power couplers takes place.

Once the cold conditioning of the couplers is finished, the cryomodule is cooled-down to 2 K, this is done with the help of pumps in the TICP allowing the helium bath to be pumped to pressures lower than 25 mbar. The pressure in the helium bath is kept stable by regulating the valves to maintain a typical pressure of at 31 mbar with a stability better than 0.2 mbar during cavity operation.

Once stable operation at 2 K is achieved, cavity operations are started [8]. First the cavities are tuned to resonance. During the process of tuning the cavity to achieve resonance, several crucial tasks must be performed. These include determining the initial parking frequency of the cavity, calculating the expected shift in the pi-mode, and confirming that the tuner frequency sensitivity aligns with the anticipated nominal value of around 20 kHz per turn of the tuner shaft. A high-level tool has been developed to facilitate these tasks. The determined steps to achieve resonance and the parking frequency are stored for future LINAC commissioning activities.

For cavity calibration, TS2 possesses a notable advantage in the form of redundant RF power monitoring capabilities, which prove valuable for evaluating systematic calibration uncertainties. Multiple directional couplers positioned along the RF path, along with diverse sensors, enable the measurement of power readings (Pf, Pr, and Pt). The evaluation of the cavity gradient, Eacc, involves the utilization of several methods. These methods include computing the stored energy based on the reflection trace, utilizing the vertical test calibration coefficient, and employing over-coupled calculations derived from the forward power Pf.

Following the calibration process, the cavities undergo individual conditioning to attain the desired nominal gradient in an open-loop configuration. This conditioning is achieved by gradually increasing the pulse length and ramping up the power. To ensure optimal performance, the levels of cavity multipacting are closely monitored by observing arc and electron signals. For the majority of the cavities, the conditioning is extended to higher fields until the coupler reaches its administrative limit of 300 kW during long pulses. An example of cold conditioning is shown in figure 10.22.

The RF system at TS2 comprises two klystrons that supply power to two cavities each within the module through a variable power divider. Once each cavity has been conditioned to operate at the desired pulse level, all the cavities are operated simultaneously to measure the dynamic heat load (DHL). This measurement is achieved by manipulating the power distribution system.

The static heat loads (SHLs) to the cryogenic system are accurately measured when cavity operation is not taking place and after the cryomodule has fully thermalized. More details can be found in sections 10.1.6.1 and 10.1.6.3.

10.1.5.6 Cryomodule warm-up

Once all the tests on the superconducting cavities (e.g., driven them to nominal performance and above, characterizing the limiting mechanisms, tuning sensitivity, etc) have been completed, and all the cryogenics measurements and studies have taken place, the cryomodule can then be warmed to ambient temperature. First the helium level inside the tanks is evaporated and later a gradual temperature increase takes place by flowing a progressively warmer helium. This process is done in conjunction with the TICP operation.

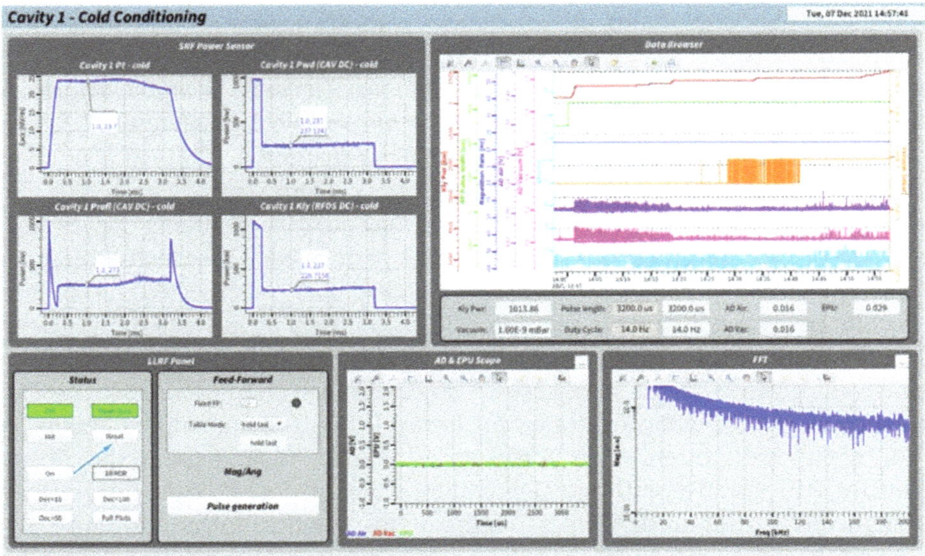

Figure 10.22. Cavity conditioning operator interface, showing LLRF power traces and other important monitoring trends. (Courtesy of ESS)

10.1.5.7 Cryomodule disconnection

In the phase of the acceptance, and once the cryomodule has been warmed, the disconnections from the bunker can start. Complementary leak tests are done after the warm-up in order to make sure the that the cool-down/ warm-up thermal cycle had no degrading effect on the initial measurement.

The systems to be disconnected are the same as mentioned in section 10.1.5.3 in the installation of the cryomodule in the test bunker.

A special procedural change-request is followed and recorded to allow pressure 'de-energization' of the cryogenic lines, which after successful completion allows for the access, disconnection, and exposure to the environment. This adds an extra layer of safety to the procedure where all the teams involved are required to exchange knowledge about the state of the system and approval from system leaders is needed, ensuring that the works have been properly completed. A similar procedure has been followed during cryomodule installation in the Test Bunker for allowing for 'energization' of the cryogenic circuits.

Once all the system disconnections are completed, the cryomodule is then transferred from the bunker stands to the transport cart and moved out of the bunker to the preparation area to initiate the final preparations for dispatch. The slot in the test bunker is then filled by a new cryomodule to be tested.

10.1.5.8 Cryomodule preparation for dispatch

At this stage the cryomodule is prepared for dispatch into storage before installation in the tunnel. Final preparations are done to prepare the cryomodule's interfaces for transport, storage, and later installation in the tunnel.

10.1.5.9 Cryomodule dispatch

Here the cryomodule is moved out of the TS2 area (see figure 10.23) and final reports about the cryomodule acceptance are completed.

In a few cases we expect the cryomodules not to fully reach the acceptance criteria. In these cases, various grades of nonconformities are issued, which in turn trigger different mitigation actions.

Some cryomodules will still be able to properly function with accepted minor nonconformities, while in the opposite case a small fraction of cryomodules are expected to have intrinsic faults that will not allow for proper function or have limited performance. For these cases, corrective actions will need to be discussed amongst the various stakeholders and the risks will need to be evaluated (e.g., scope, schedule, and budget) before bringing them to full scope.

10.1.6 TS2 cryogenics testing

An important portion of the acceptance tests consists of verifying the proper working behavior of the superconducting cavities and other cryomodule components at nominal working conditions. Another important part of the cryomodule acceptance tests is the measurement of cryogenic SHLs and RF-induced DHLs. A large portion of this section is extracted from the paper 'Heat Loads Measurements Methods for the Elliptical Cryomodules SAT at the Lund Test Stand' [9].

The cold test takes place inside the radio-protection bunker where the CM is connected to various auxiliary systems, such as radio-frequency distribution, beam

Figure 10.23. Picture of 2 cryomodules, one inside the test bunker (access door open) and another on the preparation area outside of the bunker getting ready for dispatch. (Courtesy of ESS)

vacuum, isolation vacuum, water cooling, and cryogenic distribution. The TS2 cryomodule cryogenics system is designed and constructed such that the nominal operating conditions for the accelerating cavities can be achieved. It also allows us to safely go through a series of operating modes [10] that are necessary (for example) for the cool down, warm up, or response to accidental states. At the TS2, a dedicated cryogenic plant, TICP [11] and distribution system, and CDS [12] supply refrigeration to the cryomodule at conditions similar to those present in the accelerator tunnel where each single cryomodule operates together with a dedicated valve box (VBox) and can be operated independently from other cryomodules.

At the interface between the CM and the VBox there are four cold process lines. Two of these lines are for the thermal shield cooling, TS (supply and return), while the other two lines are for the cold mass helium supply and vapor return. Table 10.2 shows the typical thermodynamic conditions at the Vbox-CM interface.

A set of diagnostics instrumentations (i.e., pressure, temperature, and flow, level) allows us to gather additional information about the conditions of each of the circuits relevant for operation. Figure 10.24 below shows a very simplified scheme of the cryogenic circuits at TS2 with selected relevant equipment.

The high-pressure (HP) supercritical helium is supplied at 3 bar and 5.3 K. When operating at 2 K, it first passes through the heat exchanger (HX), where it is precooled by heat exchange with the low-pressure (LP) evaporated vapors from the 2 K bath. The precooling of the HP line allows to recover the available frigories from the LP line and significantly enhance the efficiency of the Joule–Thomson (JT) expansion valve (CV01) feeding the 2 K bath.

Under nominal conditions, the superconducting cavities operate fully immersed in the superfluid liquid helium bath, with the helium level stabilized inside the diphasic line that interconnects the four helium tanks at the top.

The level of the helium bath is measured by two redundant level gauges, LT-01 and LT-02, and is controlled by regulating the JT valve (CV-01). The typical operating level is 87%±5% with liquid on the diphasic line.

The pressure of the saturated helium bath is measured by PT-04 sensor and maintained stable by controlling the very low pressure (VLP) return valve (CV-04) placed downstream of the HX.

The flow returned to the TICP is measured after the 2 K pumps by the flowmeters FT-20 and FT-21.

Table 10.2. Conditions at the interface between cryomodule and valve box during nominal operation. (Courtesy of ESS)

Circuit name	Pressure, temperature
Thermal shield supply, TSS	13.8 bar, 33 K
Thermal shield return, TSR	13.3 bar, 36 K
Helium supply line, HSL	3 bar, 5.3 K
Vapor return line, VLP	31 mbar, –

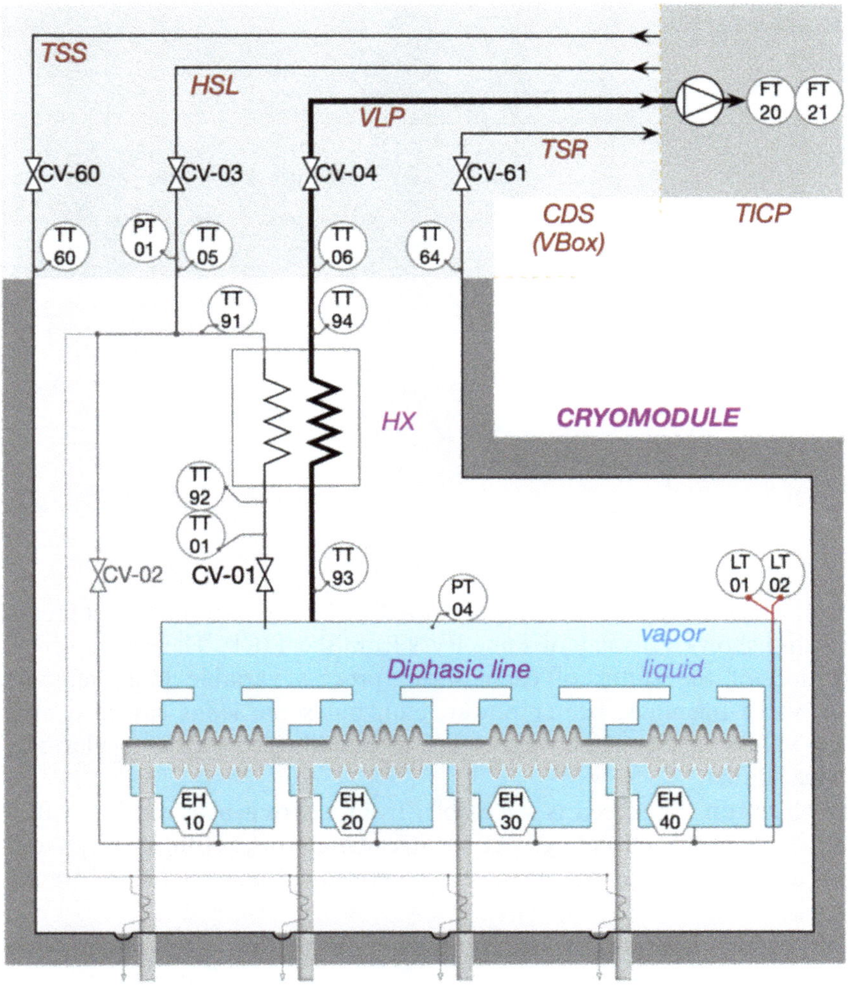

Figure 10.24. Simplified scheme of the TS2 cryogenic system including cryomodule and valve box. (Courtesy of ESS)

A fraction of the HP helium supply is used to cool the power coupler's double wall (23 mg s^{-1} per coupler), limiting heat loads by conduction from the external vessel parts through the power coupler structure into the cavities.

10.1.6.1 Typical cryogenic operation of elliptical cryomodules at TS2

The operation of the cryomodule cryogenics system and cryogenics distribution system at TS2 is done via a master OPI managed by the SRF team cryogenics experts, while the TICP operation is done by the cryoplant operators.

The master OPI (see figure 10.25), which was developed in-house in CS-Studio, is continuously going through revisions, making it more user friendly and free from implementation bugs. It depicts a visualization similar to the process and flow diagram and can be divided in various regions: at the bottom is the cryomodule

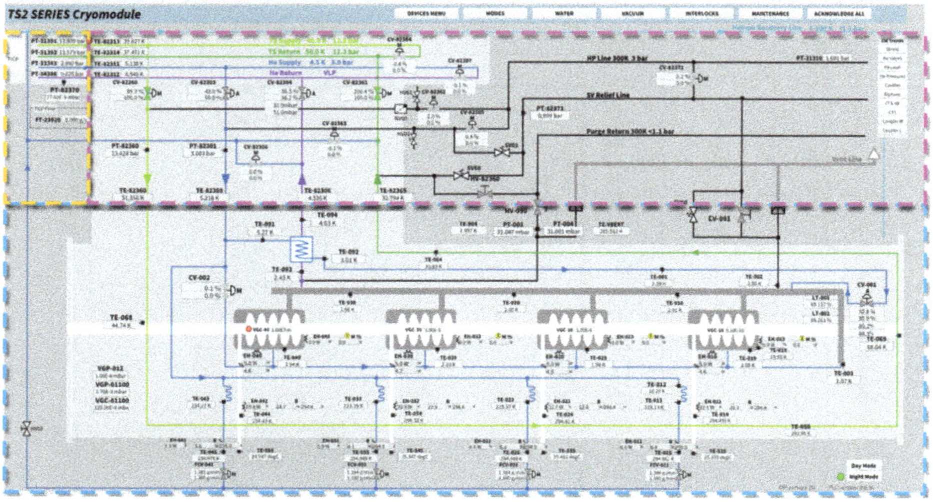

Figure 10.25. Master operator OPI. (Courtesy of ESS)

under test, while the top is for the cryogenics distribution part of the test facility. The top left-hand shows a display of some PV's from the TICP. The screen offers a fast visualization of the status of each object process variable (e.g., temperatures, pressure, valve openings, heater power, etc), but some icons can also appear to show that values have gone above specified thresholds (i.e., warnings, alarms, errors, and device status)

Each object on the screen is selectable, from where a new window will appear allowing further interaction (e.g., actions, trends, alarms, settings, etc.). Additional menu buttons allow for expert menus to be consulted and predefined trends to be visualized.

The cryogenics operation inside the bunker starts after all the interconnections have been finished and the cryogenics circuits have gone through the approval of pressure energization. The first actions after verification that all instruments are correctly read from the control system is to initialize the positioner of the control valves and start the purification of the circuits with clean helium gas through a pump and purge procedure. We also use this procedure to do two-point calibration of pressure sensors and cross-check gas flows through the system. We once again perform a thorough leak checking of the cryogenics circuits to ambient, isolation vacuum, and beam vacuum.

Once the conditioning at warm of the power couplers has finished, the cryogenics operators take over and the cool down to 4 K can start. Typically, it takes a day to cool down the cryomodule and cryogenics distribution to 4 K and to achieve helium liquid inside the tanks. Some equipment inside the cryostat takes a long time to fully thermalize, e.g., the cold tuning system takes about a week to reach a stable temperature of ∼20 K. This is due not only to poor thermal contact of the cold tuning system to the cavities cold mass but the tuners are also in disengage mode when the cool down takes place.

Once a stable helium level has been achieved through proportional, integral and differential (PID) control regulation with CV01 or CV02, the cool down to 2 K can start. This is managed by the TICP operators that trigger the start of the pumps, these will reduce the bath pressure in a gradual manner down to a minimum of ~25 mbar. The bath pressure inside of the tanks is then maintained stable at 31 mbar with PID control regulation of CV04 (see the simplified diagram in figure 10.24).

Cavity operations can then star. Most of the cryogenics operation is mainly focused on keeping the system as stable as possible so that cavity measurements can be done as accurately as possible.

While cavity operations are taking place, we also focus on gathering information about performance of the system and record behaviors that can then be used as cryogenics performance indicators when the cryomodules are installed in the tunnel. A list of a few of these studies follows:

- Heat exchanger (HX) performance: We analyze the temperature, flow, and pressure in the circuits surrounding the heat exchanger and perform a heat/mass exchange balance in order to evaluate the HX performance at various flow rates and also to evaluate Joule–Thomson effect efficiency. Along with this measurement, we perform an analysis of systematic temperature measurement errors.
- Power coupler double wall cooling efficiency: The design of the power coupler double wall cooling accounts for a nominal flow of 23 mg s^{-1} per circuit, with expected reduction to 1 W conduction heat per coupler. We perform studies at various flow rates and evaluate the effects on global heat loads. Along with these measurements, we preform studies of the temperature evolution of each circuit at various flow rates and the required heater output to avoid condensation onto external surfaces.
- Level meter referencing: An important diagnostic of the cryomodule status is the helium level gauges measurement. Each cryomodule comes with two gauges installed and we analyze discrepancy measurement between the gauges, and most importantly we reference the measured helium level with the real liquid height inside the tanks. For this we used two methods. The first method is to observe the change of level increase with the liquid rises from the top of the tanks to the diphasic line. The second method is to observe the behavior or the temperature of temperature gauges when superfluid liquid helium becomes in contact with pipe surfaces.
- Measurement noise: We continuously observe the behavior of each measured process variable and look for a sudden increase of noise or faults. This can usually indicate the development of unexpected phenomenon, such as spontaneous thermoacoustic oscillations, electrical noise pick-up on the acquisition chain, thermal sensors coming into contact with surfaces and causing deterioration, electrical wire breakage, or bad grounding, compressed air leak on valve actuators, etc. This study can in the long term help to identify failing components or faulty assembly. Moreover, it helps operators to get a deeper knowledge of the system.

Once all the studies have been performed (see sections 10.1.6.2, 10.1.6.3 and 10.1.6.4 on heat load measurements, which are major indicators of the cryomodule cryogenics performance), and the performance of the cavities and limiting mechanisms have been evaluated, the cryomodule can then be warmed. The warm up typically takes two to four days depending on the day of the week that the warm up can be started—we typically prefer to extend cavity studies and start the warm up before the weekends (using the time for a steady and slow warm-up). Figure 10.26 shows a full test cycle.

10.1.6.2 Measurement of cryomodule SHLs to the 2 K bath

The cryomodule is equipped with a variety of instrumentation for process control and diagnostics, these are complemented with additional instrumentation located in the Vbox and TICP. Taking advantage of this variety, we developed complementary methods to determine the SHLs.

The following methods can determine total heat loads (static plus dynamic) to the helium bath; however, they are typically used for static heat load evaluation.

The measurements can be performed with the cavity heaters (EH) turned off or on. In the scenario where the heaters are turned on, the final result of SHLs needs to be corrected by taking into consideration the power dissipated by the heaters.

10.1.6.2.1 Flowmeter method

The static heat load $\dot{Q}_s$ to the 2 K helium circuit is calculated by the direct measurement of the evaporated liquid mass flow rate with flowmeters FT-20 or FT-21. During this measurement the helium supply to the bath is stopped by closing

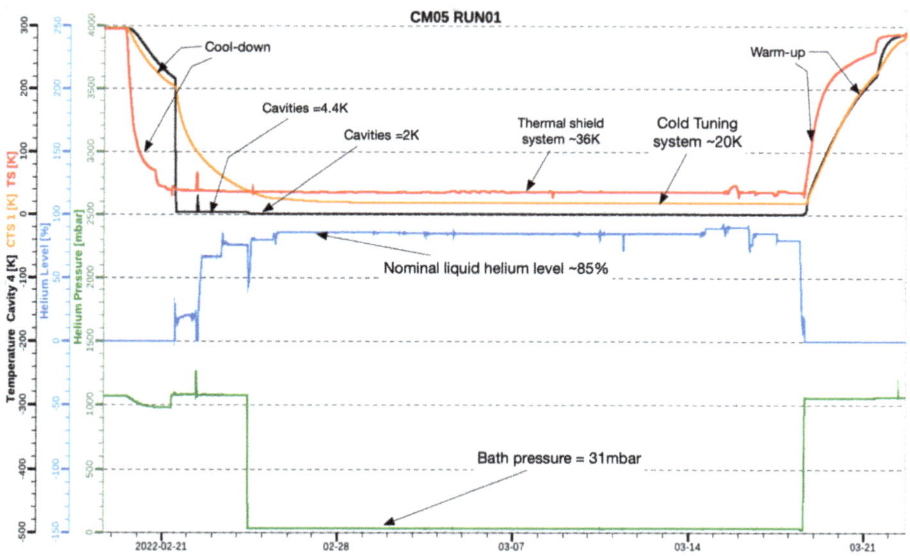

Figure 10.26. Selected PVs showing the full test cycle: cool down of the thermal shield and cold mass, stable operation at 4 K, level increase, cool down to 2 K, stable conditions at 31 mbar for cavity studies, and warm up. (Courtesy of ESS)

the valve CV-01. The helium level inside the diphasic starts at nominal height and decreases during the measurement as liquid evaporates. The bath pressure is kept stable at 31 mbar (regulated by CV-04), with a stability <0.2 mbar, for improved measurement accuracy (figure 10.27).

The SHLs are calculated by multiplying the measured mass flow rate $\dot{m}$, by the latent heat of evaporation L (23.06 J g^{-1}, for 31 mbar saturated bath pressure).

$$\dot{Q}_s = L \cdot \dot{m} \tag{10.1}$$

This method provides a direct measurement of the heat loads; however, its accuracy is directly affected by the flow measurement accuracy.

10.1.6.2.2 Level meter method

Similar to the flowmeter method, in this method the helium supply to the bath is stopped. The heat loads to the 2 K helium circuit are derived via the direct measurement of the helium level decrease using the level gauges, LT-01 and LT-02.

Thanks to computer-aided design (CAD) and the CM mechanical model, we can compute the volume of liquid for a given helium level measurement. The mass of liquid is calculated by applying the corresponding helium density ρ, and then we calculate the evaporated mass flow by simply subtracting the initial m_i, and final m_f, helium mass over the test period $dt = t_f - t_i$, and:

$$\dot{m} = (m_i - m_f)/dt$$

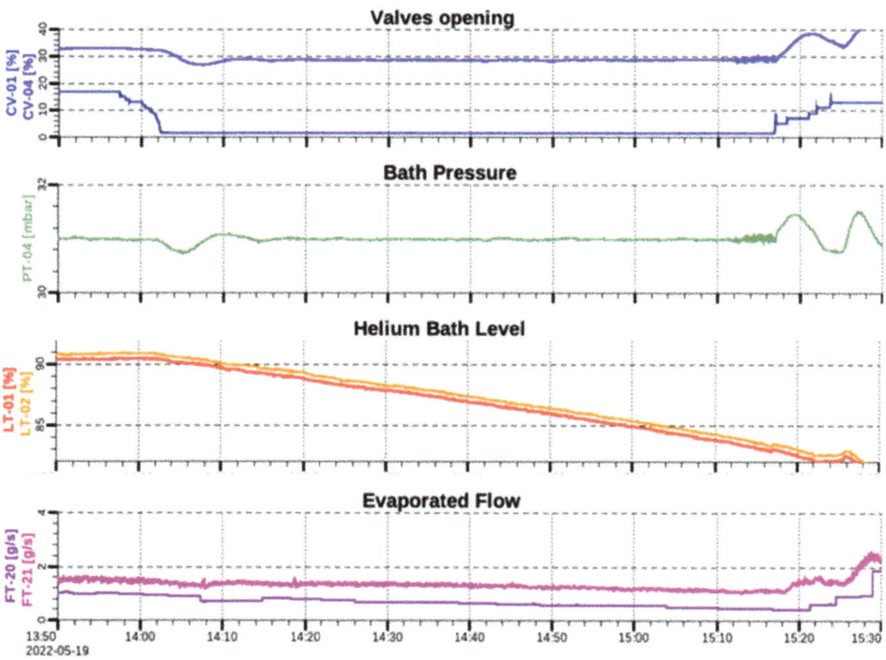

Figure 10.27. Selected PVs during static heat load measurement (flowmeter and level meter method). Reprinted with permission from [10]. Copyright © 2017 by JACoW

The SHLs are then calculated using equation (10.1).

Similar to the flowmeter method, because no helium is supplied to the bath, the level inside the diphasic line decreases for the duration the measurement as liquid evaporates. The accuracy of the measurement depends on a stable bath pressure and proper manufacturing according to design.

10.1.6.2.3 Pressure increase method

During this measurement both the helium supply to the bath and vapor return from the bath are stopped. This is done by closing CV-01 and CV-04 keeping the system sealed, thus no flow to TICP.

The heat loads to the 2 K helium circuit (or energy gain over time) are calculated by measuring, with PT-04, the resulting increase of pressure in the system (figure 10.28).

The SHLs (equation 10.2) are calculated from the change in energy, *dE* of the system 'liquid + vapor' over the test period, *dt*.

$$\dot{Q}_s = dE/dt \tag{10.2}$$

The energy of the system is computed by adding the liquid enthalpy (H_l) and vapor enthalpy (H_v), considering the measured thermodynamic properties for a given measured pressure of the saturated bath and vapor.

The mass of the liquid (m_l) and the mass of the vapor (m_v) are computed from the helium level measurement and considering the CM CAD model information. The SHLs equation takes the following form:

$$\dot{Q}_s = \frac{(m_l \cdot H_l + m_v \cdot H_v)_f - (m_l \cdot H_l + m_v \cdot H_v)_i}{dt} \tag{10.3}$$

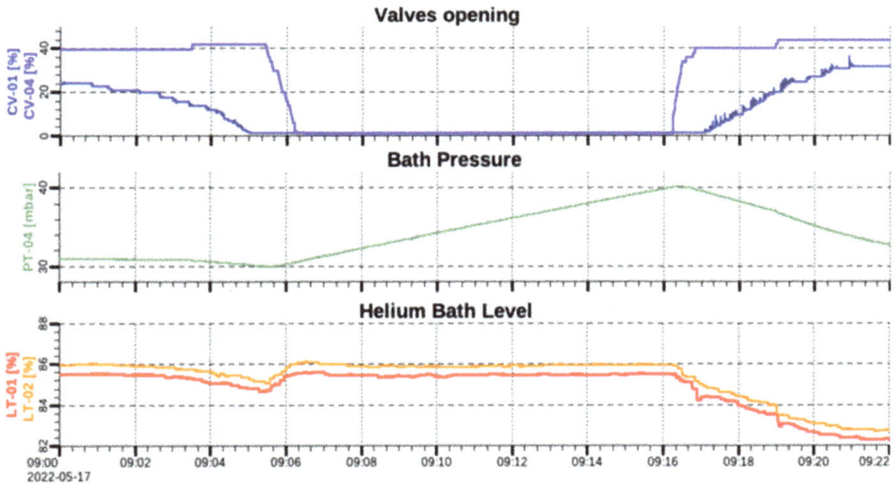

Figure 10.28. Selected PVs during static heat load measurement (pressure increase method). Reprinted with permission from [10]. Copyright © 2017 by JACoW

The contribution of the static heat load on the surrounding materials (i.e., niobium cavities, titanium helium tanks, and titanium bi-phasic line) can also be included by taking in consideration the mass of each component, the specific heat, and the sensible temperature increase.

10.1.6.3 Measurement of DHLs from cavity operation

The DHLs from SRF cavity operation are calculated by measuring how much power due to RF origin is dissipated into the helium bath.

10.1.6.3.1 Heater-compensation method

Before (or after) RF power is sent into the cavities the cryogenic system is stabilized and referenced. A steady helium flow is supplied to the bath by fixing the opening of CV-01. Heaters placed on the cavity tanks are powered and precisely adjusted to maintain a constant helium level, and the power supplied to the heaters is noted, (EH_{No_RF}). The bath pressure PT-04 is maintained constant at 31 mbar by regulation of CV-04.

The cavities are then supplied with RF power and the additional dissipated power from this source is compensated by means of decreasing the total power supplied to the cavity heaters (EH_{With_RF}), while precisely maintaining the same thermodynamic conditions on the helium bath as for the reference (i.e., level and pressure). The DHL is given by the expression:

$$\dot{Q}_{RF\quad dynamic} = EH_{No\_RF} - EH_{with\_RF} \tag{10.4}$$

10.1.6.3.2 Two-run variance method

This method consists in taking two measurement runs (one with RF and one without) using the methods described for SHLs to the 2 K bath, and calculating the variance of the results. We tend to exclude this approach because the heater compensation method is fast and more accurate.

10.1.6.4 Measurement of global heat loads to the CM cold mass

10.1.6.4.1 Flow-enthalpy method

The global heat loads to the CM cold mass are calculated by the direct measurement of the evaporated liquid mass flow with FT-20 or FT-21, and the enthalpy difference between helium supply (H_{supply}) and helium return (H_{return}), where the enthalpy is derived from the temperature and pressure measurements,

$$\dot{Q}_{\text{global}} = \dot{m} \cdot (H_{\text{return}} - H_{\text{supply}}) \tag{10.5}$$

During the measurement, the helium supply to the bath is kept constant through CV-01, while maintaining a steady helium bath level in the diphasic line.

The accepted level fluctuations are kept to a minimum throughout the measurement period. The helium bath pressure is maintained stable at 31 mbar ± 0.04 mbar.

A variation of equation (10.5) can be used to analyze the of heat loads on segments of the system, becoming a useful tool to identify the approximate location of heat sources or to evaluate systematic errors in the temperature measurement.

10.1.6.5 Results of the elliptical cryomodule qualification at TS2

At the time of writing more than 10 elliptical cryomodules have gone through full qualification at TS2. The paper submitted recently to the IPAC23 conference [13] summarizes some of our achievements. We will reproduce a few of the results here.

One indicator of the throughput of the cryomodule qualification is the time spent inside of the test bunker (figure 10.29). This time includes time required from transporting the cryomodule inside and outside of the bunker, interconnection and disconnection times, warm and cold conditioning of couplers, cavity operation, trips, and additional time for problem solving. The prototype CM, which is shown in the left-hand plot with the series, has been used to commission the cryogenic and test infrastructure and went through three cryogenic cycles and the RF tests, and therefore the test times increased. The right-hand plot shows the learning curve obtained after a few tests where an average of 79 days are required and a minimum of 46 days was achieved. The test of the second high-beta cryomodule (CM32) was extended due to three trips of the cryoplant, requiring to warm the infrastructure.

An important indicator of the performance of the cryomodule and cavities is the DHLs, which indicates the amount of RF power induced from cavity operation and dissipated to the cryogenic system. All the tested CMs show a low DHL when compared to the design specification (see figure 10.30).

For the last three tested CMs we were able to measure the DHL contribution from each cavity at various accelerating gradients using heater compensation method with heaters under regulation. We are introducing this measurement at for all subsequent modules as an additional gauge for long term cavity performance.

Finally, the SHLs to the cryogenic system at 2 K are an important indicator of the manufacturing quality of the cryomodule compared with initial design estimation. Some CMs record a SHL slightly above the design estimation but below the design value including margin (see figure 10.31). Moreover, analyzing the total heat load to

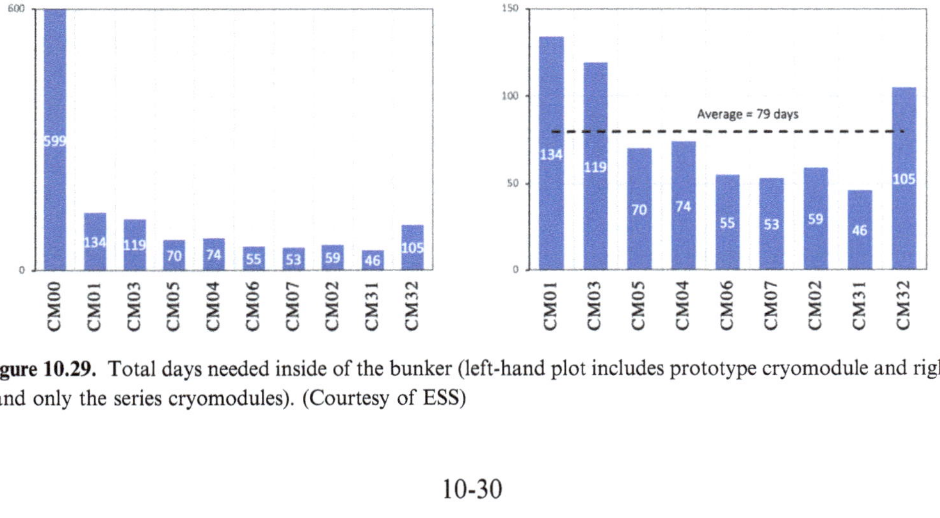

Figure 10.29. Total days needed inside of the bunker (left-hand plot includes prototype cryomodule and right-hand only the series cryomodules). (Courtesy of ESS)

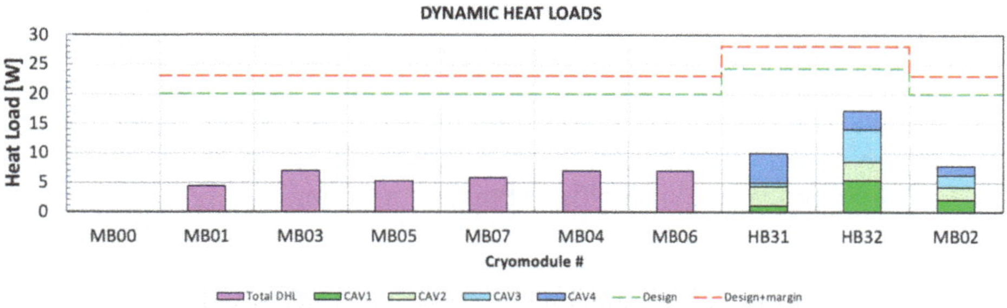

Figure 10.30. Measured DHLs per cryomodule. (Courtesy of ESS)

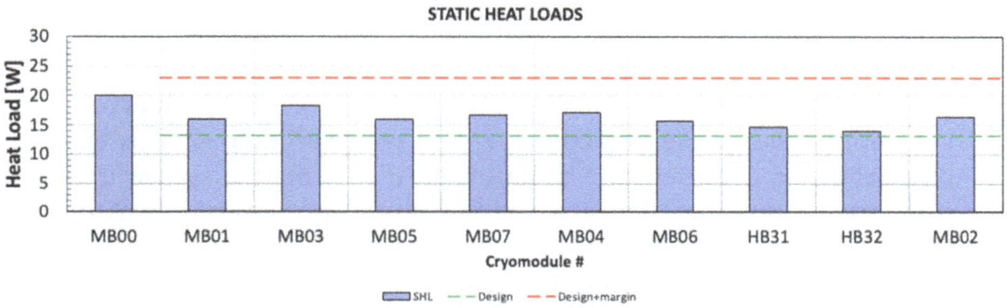

Figure 10.31. Measured SHLs of each qualified cryomodule. (Courtesy of ESS)

the cryogenic system (dynamic + static heat loads), we are confident that the operation in the LINAC can be done easily while keeping some safety margins.

10.2 Lessons learned

TS2 is a test facility for cryomodule acceptance testing at the ESS site.

The project development from the design phase up to steady state operation took several years, and was managed and coordinated by the SRF section within the Accelerator Division, with additional help from a number of stakeholders.

The careful planning of the TS2 workflow was idealized not only from in-house expertise but also with experience from in-kind collaborators (from IFJ-PAN) that are fully integrated in the team.

The test bunker is a cross-section replica of the tunnel with a similar setup of interfaces, allowing timely understanding and mitigation of the shortcomings before tunnel installation.

The operation of the facility has now reached a steady testing pace. Nevertheless, we continuously seek to improve procedures, ways of working, and increase performance. Moreover, we focus on collecting expertise and lessons learned, which will greatly benefit when installing cryomodules in the LINAC tunnel, and fully payoff when commissioning and operating the superconducting LINAC. We list a few key points related to lessons learned at TS2 below:

- Effective Project Management: A key lesson learned is the importance of good project management throughout the entire project. This includes realistic and careful planning, regular communication with stakeholders, effective identification, and mitigation of limitations while keeping the whole team informed of the current status and possible changes.
- Collaboration with Stakeholders: Building a testing facility involves coordination with various stakeholders with deep knowledge of the various systems. Establishing effective communication channels and building strong relationships (formal and informal) with these stakeholders was crucial for successful operation and future improvements.
- Safety: This is a top priority when operating TS2, relying on initially drawn risk assessments which are revisited when needed. All the people with access to TS2 and operating TS2 have to go through regular training (e.g., radiation safety, cryogenics safety, etc). Moreover, when operating the prototype cryomodule we made sure to force (under supervision) the occurrence of the most likely accident scenarios, which allowed us to observe the real behavior of the system and compare it to our expectations. This experience also helped to document the actions needed to bring the system to nominal state and to train personnel for taking those specific actions during such accidental scenarios (i.e., over pressurization of systems, burst of rupture disks, action of safety valves, helium venting and release, and oxygen deficiency scenarios).
- Quality Control and Assurance: Site acceptance testing of cryomodules at TS2 is the gateway for installation of elliptical cryomodules in the tunnel. The careful, precise, and accurate checking of nominal behavior and performance of cavities and cryomodules is the bulk of the work at TS2. We developed robust quality controls, strict inspection protocols, accurate measurement methods, and quality assurance processes in order to ensure accurate and reliable testing results. Deviations from the norm are documented, analyzed, and communicated to the various stakeholders through nonconformity reports. The nonconformities are discussed in coordination with stakeholders and a path forward for mitigation, improvement, amendment, replacement, or repair is proposed depending on the type of nonconformity.
- Scalability and Flexibility: While operating and testing cryomodules at TS2, we keep in mind the ultimate goal of operating the LINAC as a whole, with multiple cryomodules and valve boxes simultaneously cooled down and operated. This allows us to focus and optimize our processes, operator interfaces, and operating procedures.
- Technological Integration: The incorporation of advanced technologies, control systems, and acquisition systems into TS2 greatly improves efficiency and also allows for better data analysis and future performance comparison of cavities and cryomodules at different locations or degradation over time. We use a multitude of available IT tools, and continuously improve our methods and data management systems.

- Staff Training and Development: The staff operating TS2 come from diverse engineering and physics backgrounds. The success of TS2 relies on that diversity and also on the staff members' ability to learn and adapt to the various phases of the project and acquire a multidisciplinary knowledge. Continuous professional development and cross-training are part of the success, relying on a culture of knowledge sharing and collaboration.
- Continuous Improvement: Building and operating the TS2 is an iterative process, a team effort, and requires continuous improvement. This continuous improvement relies on the regular evaluation of operations, openly inviting feedback from staff and stakeholders, while continuously implementing necessary adjustments based on lessons learned.

Acknowledgments

The author would like to thanks the following people for their continuous contribution to the project:

ESS:
Paolo Pierini, Nuno Elias, Cecilia Maiano, Wolfgang Hees, Henry Przybilski, Muyuan Wang, Phlippe Goudket, Felix Schlander, Emilio Asensi, Miklos Boros, Eirikur Magnusson, Peter van Velze, Artur Gevorgyan, Marcelo Juni Ferreira, Morten Jensen, Delphine Ardion, Adalberto Fontoura, and Wojciech Blinczyk

IFJ-PAN:
Michal Sienkiwvicz, Wawrzyniec Gaj, Pawel Halczynski, Marek Skiba, Filip Skalka, and Marcin Wartak

References

[1] Asensi Conejero E *et al* 2017 The cryomodule test stands for the European spallation source *Proc. IPAC'17* (Denmark: Copenhagen) pp 1064–7

[2] Arnold P *et al* 2015 ESS cryogenic system process design *IOP Conf. Ser.: Mater. Sci. Eng.* **101** 012011

[3] Asensi E *et al* 2019 Commissioning the control system for cryomodule cryogenics distribution system in test stand 2 *Proc. ICALEPCS2019* (New York)

[4] Brodrick D P, Brys T, Korhonen T and Sparger J E 2017 EPICS architecture for neutron instrument control at the European spallation source *Proc. ICALEPCS'17* (Spain: Barcelona) pp 1043–7

[5] Ulm G, Bellorini F, Brodrick D P, Fernandes R N, Levchenko N and Piso D P 2017 PLC factory: automating routine tasks in large-scale PLC software development *Proc. ICALEPCS'17* (Spain: Barcelona) pp 495–500

[6] Fernandes R N, Gysin S R, Regnell S, Sah S, Vitorovič M and Vuppala V 2017 Controls configuration database at ESS *Proc. ICALEPCS'17* (Spain: Barcelona) pp 775–9

[7] Kasemir K-U and Grodowitz M L 2017 CS-studio display builder *Proc. ICALEPCS'17* (Spain: Barcelona) pp 1978–81

[8] Maiano C G *et al* 2022 ESS elliptical cryomodules tests at Lund test stand *Proc. IPAC'22* (Thailand: Bangkok) pp 1261–4

[9] Elias N, Gaj W, Halczynski P, Sienkiewicz M, Skalka F D and Su X T 2022 Heat loads measurement methods for the ESS elliptical cryomodules SAT at Lund test stand *Proc. IPAC'22* (Thailand: Bangkok) pp 2819–22

[10] Elias N, Darve C, Fydrych J, Nordt A and Piso D P 2014 Preliminary functional analysis and operating modes of ESS 704 MHz superconducting radio-frequency linac *Proc. IPAC'14* (Germany: Dresden) pp 3317–9

[11] Arnold P *et al* 2020 The ESS test and instruments cryoplant: first test results and operation experiences *IOP Conf. Ser.: Mater. Sci. Eng.* 755.012095

[12] Fydrych J *et al* Cryogenic distribution system for the ESS superconducting proton linac *Phys. Proc.* **67** 828–33

[13] Elias N *et al* 2023 Results of the elliptical cryomodule qualification at the ESS TS2 *Presented at the IPAC'23 (Venice, Italy, May 2023)* Paper WEPA152

IOP Publishing

Cryogenic Technologies at the European Spallation Source
A big science case study
J.G. Weisend II

Chapter 11

Cryomodule testing at Uppsala University

R Ruber and R Santiago Kern

The FREIA laboratory performs the acceptance testing of double-spoke cryomodules before their installation in the ESS tunnel. This chapter describes the available infrastructure at FREIA, the work flow, and the procedures used. Starting from a single cavity, followed by the testing of a prototype cryomodule, and ending with the acceptance testing of the series cryomodules.

11.1 The FREIA laboratory

The FREIA laboratory, which stands for Facility for Research Instrumentation and Accelerators, was established in 2011 and built in 2013 next to the Ångström laboratory at Uppsala University, Uppsala, Sweden [1]. The facility was erected, among other projects, to support ESS in the design of a radio-frequency (RF) power generation and distribution system, the testing of all these components and the testing of superconducting radio-frequency cavities, and later for the testing of double-spoke cryomodules.

With an available space of 1000 m^2, FREIA is divided into different test areas regarding equipment, as shown in figure 11.1. The main areas are the cryogenic equipment and compressors, the RF power stations, their distribution system and controls, and the cryostats.

11.2 Infrastructure at FREIA

A detailed description of the cryogenic systems is given in the following sections.

11.2.1 Helium liquefier

In order to bring equipment into the superconducting state, FREIA has a helium liquefier with liquid nitrogen precooling. The system, with a Linde L140 coldbox and the corresponding recycling equipment, has the characteristics given in table 11.1. Due to an architectural intervention, the coldbox and dewar are on one end of the FREIA

doi:10.1088/978-0-7503-3223-1ch11

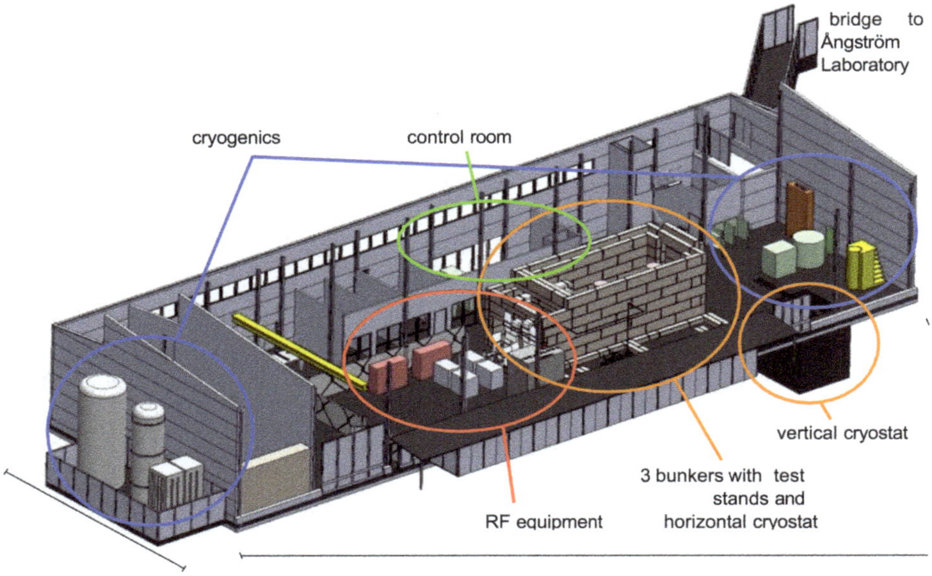

Figure 11.1. The layout of the FREIA laboratory. Reproduced from [1].

Table 11.1. Main characteristics of the helium liquefier system at FREIA. Reprinted from [2].

Liquefier coldbox	Linde L140
– Nominal capacity	140 l h^{-1} at 1.15 bar
– Precooling	150 l h^{-1} at 1.23 bar
– Purier min. inlet pressure	70 l h^{-1} liquid nitrogen
	25 bar
Liquid helium dewar	
– Storage volume	2000 l
– Nominal working pressure	1.15 bar
– Working pressure with HNOSS	1.2–1.25 bar
– Max. test pressure	2.86 bar
– Filling speed mobile dewars	165 l h^{-1} (average)
Recycle compressor	**Kaeser DSD238 SFC**
– Number of units	1
– Mass flow capacity (per unit)	43.5 g s−1
– Nominal discharge pressure	11.5 bar
– Max. discharge pressure	14 bar
Recovery gasbag	
– Storage volume	100 m^3

Recovery compressor	Bauer G18.1–15–5-He	Sauer WP5173L[1]
– Number of units	3	1
– Mass flow capacity (per unit)	25 $m^3 h^{-1}$	180 $m^3 h^{-1}$
– Max. discharge pressure	200 bara	200 bara
Pure gas medium pressure storage		
– Storage volume	30 m^3	
– Working pressure	14.5 bar	
– Max. pressure	23 bar	
Impure gas high pressure storage		
– Storage volume	3480 m^3	
– Max. pressure	200 bar	
Liquid nitrogen tank		
– Storage volume	20 m3	
– Nominal pressure	3 bar	
– Max. pressure	18 bar	
– Filling speed mobile dewars	300 l h^{-1}	

[1] Connected in May 2022.

building while the gasbag, recovery compressors, and the recycle compressor are at the opposite end of the building, next to the outer tanks.

This liquefier has connections to all three of the test stands currently present at FREIA, e.g., the horizontal and vertical cryostats, and a cryomodule. The integration of the liquefier with this equipment is shown in figure 11.2, where the common starting point is the 2000 l dewar. From here the liquid helium can be delivered to each cryostat and the cryomodule. Since the horizontal cryostat and the cryomodule share a bunker, there is one line with one helium and one nitrogen transfer lines inside going into the bunker. It is in this bunker where an intermediate box (called interconnection box) directs the cryogens to one or the other as needed.

The gas recovered from any of the test stands can be sent to the recycle compressor or the gasbag, the main differences being the operating temperature of the test stand and the load on the liquefier at the time. If the test stand is to be operated below atmospheric pressure, a set of vacuum pumps, so-called subatmospheric pumps, are used to decrease the pressure (and thus the temperature) with their output connected to the gasbag. They are also placed in the same room as the gasbag and compressors. Table 11.2 gives an overview of their main parameters. A detailed description of this system and its modeling with Matlab can be found in [2, 3].

11.2.2 Horizontal cryostat 'HNOSS'

The 'HNOSS' horizontal cryostat is a stainless-steel vacuum vessel (figure 11.3) with a room temperature magnetic shield made of μ-metal material, an aluminum

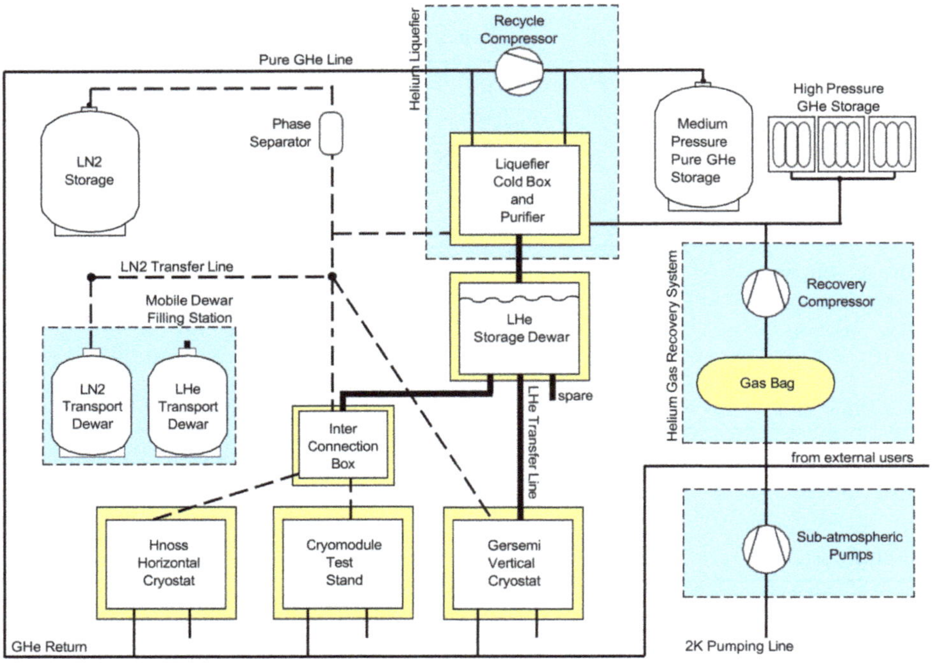

Figure 11.2. Schematic of the helium liquefier and its integration with the different test stands. Reproduced from [1].

Table 11.2. Main characteristics of the subatmospheric pumps at FREIA. Reprinted from [2].

Root pump	Leybold RUVAC WH2500FU
– Quantity	3
Roughing pump	Leybold DRYVAC DV650Helium2.1
– Quantity	2
Mass flow rate (measured)	
– At 10 mbar, 300 K	7250 m^3 h^{-1} (ca. 3.2 g s^{-1})
– At 15 mbar, 300 K	6747 m^3 h^{-1} (ca. 4.3 g s^{-1})
– At 70 mbar, 300 K	2995 m^3 h^{-1} (ca. 9.3 g s^{-1})
Ultimate pressure at inlet	0.19 mbar

thermal shield cooled by liquid nitrogen, and an inner diameter of 1.2 m and inner length of 3.3 m [4, 5]. HNOSS is a research cryostat and, as such, has many additional ports and flanges to accommodate almost any device that is superconducting and has a helium tank. HNOSS can house a maximum of two

Figure 11.3. The horizontal test cryostat HNOSS inside the bunker with the valve box on top. Reproduced from [1].

superconducting cavities with their corresponding helium tanks, while they can be powered with RF from below (ESS 352 MHz double spoke or 704 MHz elliptical cavities) or from the side (TESLA/ILC 1.3 GHz cavities).

Figure 11.4 shows how the cryogens are fed, distributed, and collected in HNOSS and the cavities, while figure 11.5 shows a detailed schematic. The most complex part is the valve box, which is located on top of the cryostat. It contains the necessary piping, cryogenic valves, heat exchangers, and liquid helium pots to store incoming helium from the cryogenic plant at 4.2 K, distribute it to the devices under test, and cool them down further to a minimum temperature of 1.8 K.

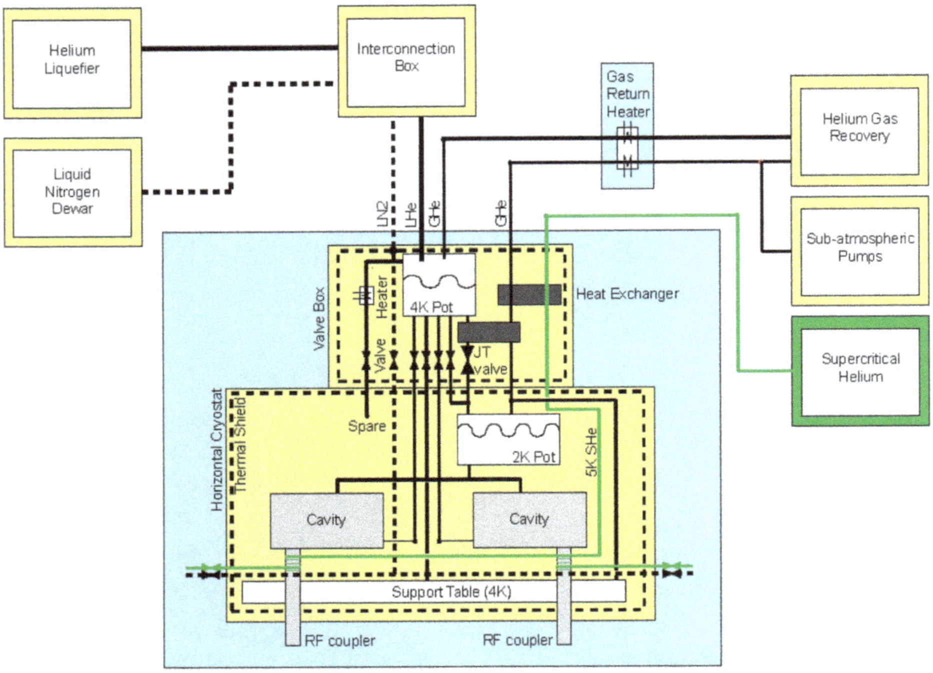

Figure 11.4. Schematic of the horizontal test cryostat HNOSS with the supercritical circuit marked in green.

The liquid helium and nitrogen are fed to the interconnection box and from this box both cryogens enter HNOSS through the valve box, and from there the liquid nitrogen is sent towards the thermal shield and the liquid helium towards a first buffer tank that operates at 4.2 K. Above the cavities there is another buffer tank, called the 2 K tank, that acts as a biphasic volume for the cavities. Depending on the mode of operation, i.e., if it is an initial cool down or if it is keeping cold, the liquid helium from the 4 K tank is sent to the cavities through their lower part or through the 2 K tank, respectively. In addition, if the cavities were not to be hanging by tie-rods, HNOSS operates with a table cooled by liquid helium from the 4 K tank for this purpose. If the cavities are equipped with a fundamental power coupler, it is possible to cool down the antenna by a dedicated supercritical helium circuit, which includes a compressor to operate at 3 bara. In order to use the cooling power of the return gas, there are two heat exchangers in its path: the first is between the outlet from the cavities and the inlet to the Joule–Thomson valve, while the second is between the gas after this heat exchanger and the supercritical helium inlet. Both heat exchangers are combined into one physical unit, of the stack type.

The helium gas return from HNOSS is then taken out of the radiation shielding bunker via a chicane built on top of the labyrinth to enter the bunker and from there it goes through a water reheater, placed halfway between the coldbox and the gasbag. After this reheater, a set of pipes directs the gas towards the recycle

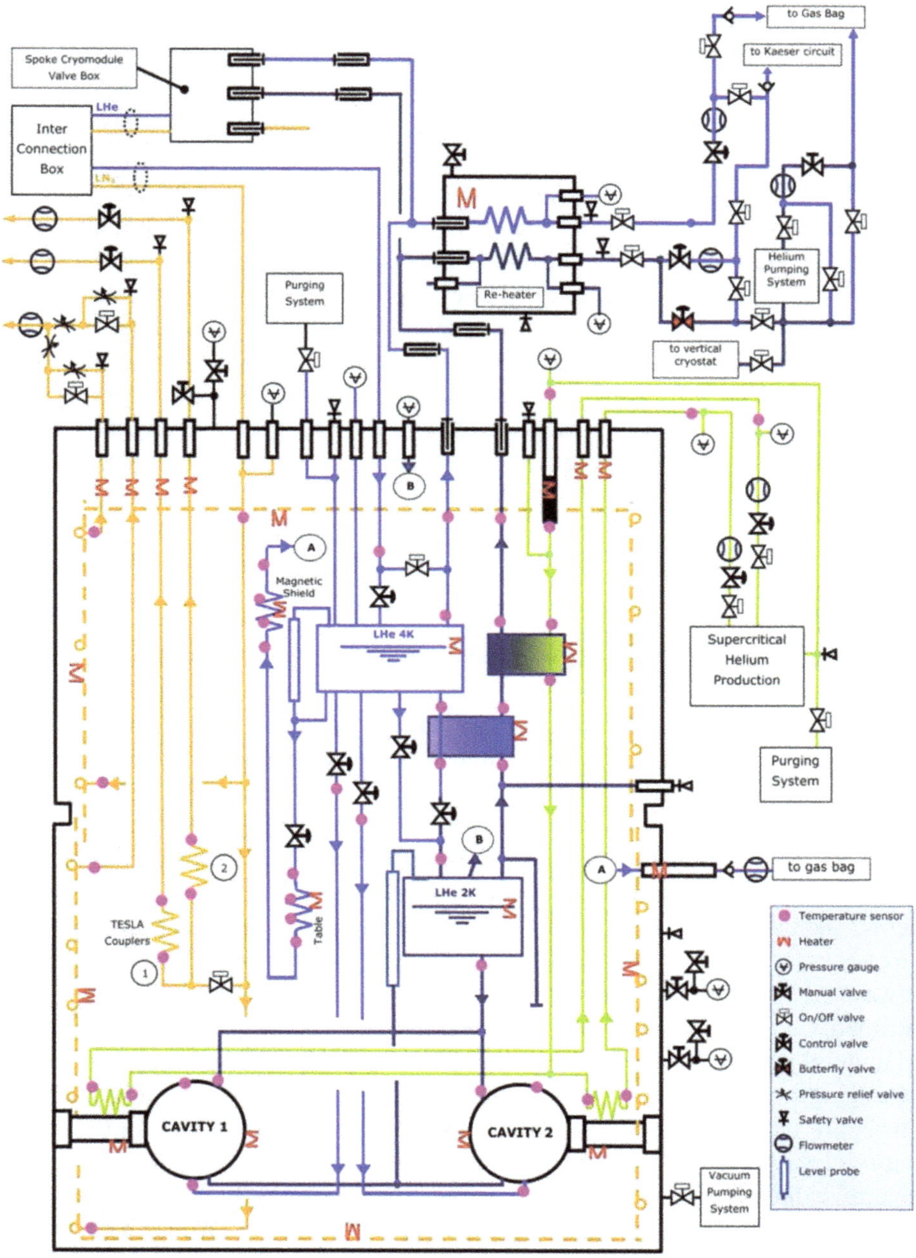

Figure 11.5. Detailed schematic of the horizontal test cryostat HNOSS. Yellow circuit is liquid nitrogen, blue is liquid helium, and green is supercritical helium.

compressor, the subatmospheric pumps, or the gasbag (as necessary). When operating at a temperature lower than 4.2 K, a butterfly valve to keep the pressure constant at the 2 K tank (and thus the device under test) is at the corresponding outlet of this reheater.

11.2.3 Vertical cryostat 'Gersemi'

FREIA also operates with a vertical cryostat called 'Gersemi' (see figure 11.6). Since the underground pit was already excavated during FREIA's construction, the possible dimensions for the cryostat were constrained. In addition, because FREIA is a university facility, the vertical cryostat had to be as versatile as possible to be able to accommodate a wide range of eligible equipment during its lifetime. This led to a cryostat with an effective 1.1 m internal diameter and 4.7 m height, with the possibility to attach a different insert depending on the equipment to be tested and the test conditions: a vacuum insert for test of RF cavities or magnets surrounded by a helium tank, a bath insert for RF cavities or magnets that do not have a helium tank, and a lambda-plate for tests of magnets under pressurized helium [6]. Figure 11.7 shows a schematic representation for each insert, where an external valve box containing most of the valves and heat exchangers direct the cryogens towards the cryostat and the reheater warms up the output gases before sending them to the recycle compressor, to the gasbag, or to the subatmospheric pumps.

11.2.4 Controls

The control system includes slow controls, monitoring of the infrastructure and equipment, and the fast controls for RF power regulation and measurement (or low-level RF system, LLRF). In addition, there are interlock systems for the protection of the superconducting cavities and magnets to be tested, as well as for personnel safety.

The slow control system is based on EPICS (Experimental Physics and Industrial Control System) [7] and LabVIEW [8]. The EPICS software is used at the top level to provide the operator with a common interface to connect together different systems, including systems based on commercial industrial programmable logic controllers or based on home-build laboratory systems, e.g., on Raspberry Pi hardware. The LabVIEW-based controls are typically used for more complex experiments and control programs, such as the conditioning of cavities and fundamental power couplers. It is used for systems that are often modified to adjust to changing and evolving requirements for concrete experiments. The LabVIEW controls are used for the conditioning and testing of superconducting cavities and cryomodules. An archiver is used to store and retrieve data from EPICS. This software, called Archive Appliance, logs all relevant EPICS process variables, continuously, at selected intervals and stores them in a database. The data can later be retrieved in various ways, e.g., plotted using a web interface.

11.3 Preparation procedure

For the ESS project, at the time of publication, FREIA had tested a fully equipped double-spoke cavity, a high-β elliptical cavity, a prototype double-spoke cryomodule, and 13 series cryomodules out of 13 (plus one spare). For all of them, a series of

Figure 11.6. Picture of Gersemi with the valve box on the top left-hand side and the cryostat at the bottom, with the liquid insert in place. Reproduced from [1].

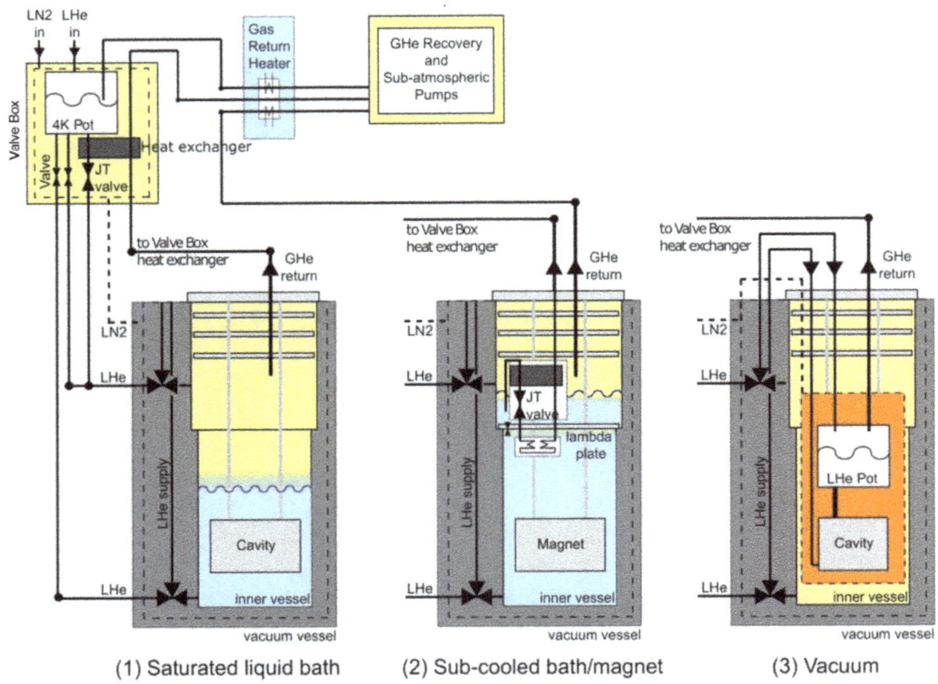

Figure 11.7. Schematic of the vertical cryostat Gersemi with the different inserts. From left to right: liquid, magnet, and vacuum inserts. Reproduced from [1]. © The Author(s). Published by IOP Publishing Ltd. CC BY 4.0.

steps are required before the cavities are filled with liquid helium. These steps, common to all equipment and in chronological order, are:

(a) Install the cryomodule or cavity(ies) in the testing place: This involves placing cavities inside HNOSS or a cryomodule into the bunker and connecting the necessary lines for liquid helium and nitrogen feeding. All connections made are leak tested.

(b) Pump the cavity(ies): Connect the beam vacuum to the corresponding pumping group(s) under clean conditions, leak check the connections before opening to the cavity(ies), connect the volumes, and keep good vacuum levels. The pumping group consists mainly of a turbomolecular pump backed by a dry roughing pump, reaching vacuum levels of ca. 10^{-9} mbar.

(c) Evacuate the cryostat: With the help of a turbomolecular pump backed by a roughing pump, the pressure inside the cryostat is lowered from atmospheric pressure to a pressure lower than 1×10^{-4} mbar, where the effect of heat transfer by conduction and convection is almost nonexistent.

(d) Purge all relevant circuits: In order to avoid having plugs originating from air that has been frozen, all circuits of the cryostat are evacuated to ca. 5 mbar and pressurized to 1200 mbar (one of each per cycle) a number of times to reduce the amount of air present. This sequence is also used to search for possible leaks at both low and high pressures by isolating the circuit for a certain amount of time, typically minutes. Usually there are three cycles performed per circuit.

(e) Condition the fundamental power coupler(s) at room temperature: To get rid of field emitters that might still be present in the power coupler, different RF power values with different pulse lengths are sent to the power coupler for cleaning purposes. During this operation a residual gas analyzer, connected to the beam vacuum of the cavities and power couplers, is used to trace chemical content of the particles released during conditioning.

(f) Cool the thermal shield: The cooling is done with liquid nitrogen at an inlet pressure of 2.5 bara.

(g) Cool the 4 K buffer tank, the 2 K tank, and the cavity/ies with liquid helium at atmospheric pressure: The sequence ends when there is a certain level of liquid helium in all buffer tanks and cavities.

(h) Go to the operating point, usually at 2 K: The temperature of the cavity(ies) is further reduced to the desired temperature (2 K) by reducing the pressure of the liquid helium bath to the corresponding pressure (31 mbar) with the help of the subatmospheric pump system described in section 11.2.1.

(i) Condition the fundamental power coupler(s) and cavity(ies) at cold: The aim is the same as with the conditioning at room temperature, i.e., to get rid of any field emitters left in the couplers and cavities before operation.

All these steps are conducted by automated sequences written in EPICS for the cryogenic processes or LabVIEW for the conditioning. The specifics of each sequence, such as opening and closing of valves, temperature or pressure check and/or regulation, etc, are taken from our and other laboratories' experience and updated when necessary, depending on the equipment connected. After these initial steps, the actual testing can start. Figure 11.8 shows the EPICS interface used for the cryomodule testing. A similar environment was also developed for HNOSS's operation.

11.4 Testing methods at liquid helium temperatures

Depending on the equipment to be tested, more or less tests will be necessary to check the overall performance, but the main tests at cold will be the same. At liquid helium temperatures, the heat load generated by the cavity with no RF power (static) and with RF power applied (dynamic) when at the nominal operating temperature is of great importance. This value is used to calculate the quality Q factor of the cavity and will also show if the actual value agrees with the design value and will provide a first estimation of the total cooling power that might be actually needed. It is worth noting that the testing of any device in any conditions that are not the final ones, such as testing a cavity in a different cryostat or a cryomodule in a different configuration, will incur in the heat loads measured being slightly different than if the cavity were to be in its final configuration.

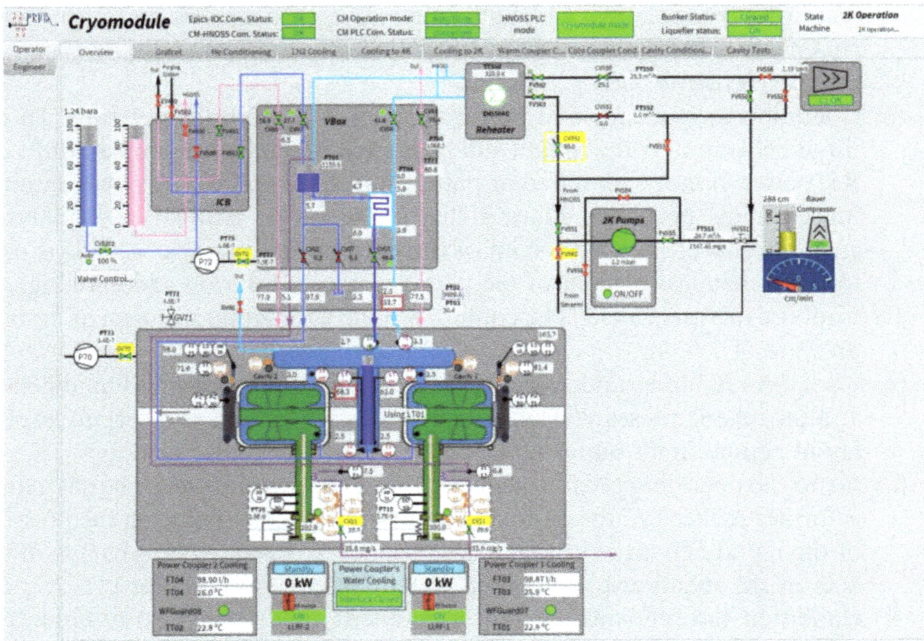

Figure 11.8. Screenshot of the EPICS interface used to test the cryomodules.

The methods used to measure heat loads are:

(a) Flowmeter method: The power dissipated by the cavity(ies) or cryomodule is directly measured by a flowmeter. This method is the simplest, but it is very dependent on the flowmeter's accuracy and requires good pressure stabilization.

(b) Calorimetric method: Heat dissipation is related to the pressure increase in a closed volume. This is done by isolating the cavity(ies), i.e., the inlet (filling) and outlet (exhaust) valves are closed while the rate of pressure increase is measured. This method requires an initial calibration using heaters connected to the cavity(ies) in strategic places. Different heater powers are applied to the cavity(ies) and the rate of pressure increase for that power level is recorded. Once RF power is applied to the cavity(ies), the pressure increase rate can be matched to a certain power level. A valve with a small leak would have an impact on this measurement.

(c) Level decrease method: The heat from the cavity(ies) can be measured via the evaporation rate. In this case the inlet valve is closed and the rate of liquid helium evaporated is estimated by knowing the helium volume in the cavity according to height. Apart from needing good pressure stability for the measurement, this method relies on the accuracy of the level probe used and how close to reality the mapping of the height vs. level is. Thus, knowledge (and accuracy) of the liquid helium volume also affects the measurement.

(d) Heater compensation method: Power dissipation is measured by keeping the level constant with the help of a heater. The filling valve is fixed to a certain opening and the level is kept constant by applying an also constant power to a heater connected to the cavity(ies). When RF power is applied to the cavity(ies), the heater power is changed so as to keep constant level, giving a direct value for the dynamic heat load. The position of the heater and how good the level regulation is will affect this measurement.

Methods a), b), and c) are commonly used at FREIA for static and dynamic measurements, whereas method d) is only applicable for dynamic measurements and is not used.

11.5 Equipment tested at FREIA

11.5.1 Double spoke cavity (ZA-01 Romea)

The first test done at FREIA for ESS was a double-spoke cavity ZA-01 (Romea), equipped with helium tank, cold tuning system, and fundamental power coupler, placed in the horizontal cryostat HNOSS in 2017. This was also the first time that a fully equipped cavity had been tested in HNOSS. A full description of the cryogenic and RF power tests can be found in [9, 10].

Once the cavity was received, it was further equipped with temperature sensors and heaters at different heights (figure 11.9(a)) and afterwards wrapped in multi-layer insulation (MLI). After placing it in HNOSS by hanging rods and making all necessary connections to HNOSS's infrastructure (figure 11.9(b)), the sequences described in section 11.3 were started one after another. After the shields were cold, the cooling with liquid helium at 4.2 K was started. During this time, an important parameter to monitor for a cavity is the cool-down rate through the 100 K temperature barrier. If the cooling through this barrier is too low, i.e., less than 1 K min^{-1}, then the cavity can suffer from the so-called Q-disease [11], which would affect its performance. This rate is usually calculated between 150 K and 50 K, and for the case of Romea a rate of 3.3 K min^{-1} was reached. Meanwhile, at 4.2 K it was observed that for an intermittent filling of the cavities, i.e., the filling valve opens once the level is below the setpoint and closes when the high setpoint has been reached, the level probe would show spikes at intervals, as shown in figure 11.10. Further examinations revealed that these spikes appeared because there was an upward bend in the pipe connecting the cavity to the level probe, as highlighted in figure 11.9(b). Since the line has a disruption, at 4.2 K and with no continuous flow of liquid helium a gas pocket forms, which warms up in time, increasing its size until it moves and is then displaced by incoming LHe, starting the process anew. The reason why it was seen while operating at 4.2 K and not at 2 K was because at 2 K the helium is in the superfluid state, so it filled all the available volume.

After the cavity is filled with liquid helium at atmospheric pressure, the temperature is lowered by pumping with the subatmospheric pumps. The pressure is measured with a pressure gauge connected to the cavities' helium volume. A temperature of 2 K corresponds to a pressure of ca. 31 mbar; however, in our

(a)

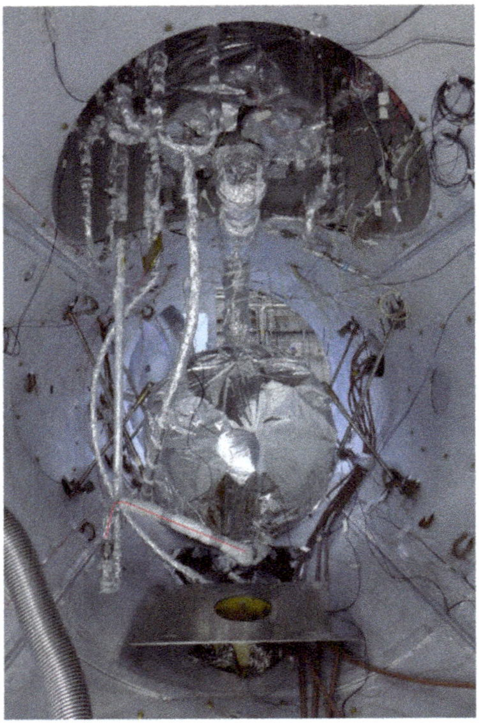

(b)

Figure 11.9. (a) The fully equipped Romea with the fundamental power coupler, the cold tuning system, and some of the temperature sensors (circled in red) and heaters installed. Reprinted from [12]. (b) Romea inside HNOSS, with the pipe connecting the cavity to the liquid helium probe, highlighted in red. Reprinted from [9].

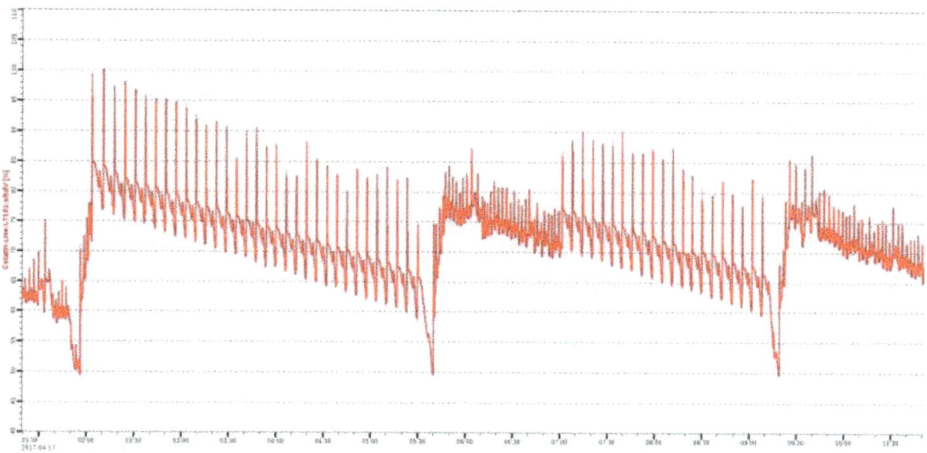

Figure 11.10. Example of instabilities in the level probe vs. time when operating at 4.2 K.

configuration at 31 mbar the temperature sensors placed on the cavity did not show this temperature but higher, so it was decided to lower the pressure until a temperature of ca. 2 K could be read. This pressure ended up being 20 mbar. The pressure was kept stable at 20 mbar via a butterfly valve (see figure 11.5).

As mentioned in section 11.4, if the configuration of the equipment being tested is different from the final one, it will have an impact on the measurements. This is because, for example in the case of HNOSS, first having a thermal shield cooled at 100 K is different from having it cooled with helium gas at 50 K as it will be when the cavity is in the final configuration in its own cryomodule together with another cavity. Second, the presence of the 2 K tank in HNOSS will also add to the heat load, as will the heat from the fundamental power coupler depending on its cooling. As shown in figure 11.5, the cooling of this part is by means of supercritical helium, which uses the return gas from the cavity to further cool down the incoming gas before it enters the coupler. If the filling valve is closed, then the flow of the return gas is reduced and so is the cooling of the power coupler. Due to Romea's design, the outlet of the coupler's cooling circuit was situated outside HNOSS, with a 76 W wired heater located also at the air side. The design cooling settings were a temperature of 5 K at the inlet, room temperature at the outlet, a mass flow of 46 mg s^{-1} and 3 bara during RF on. In reality, the lowest possible temperature reached at the inlet was 5.7 K for a flow of 41.7 mg s^{-1}, at a pressure of 1.7 bara, and at an accelerating gradient of ca. 9 MV m^{-1}. The design pressure of 3 bara proved to be too difficult to achieve and keep at such cold temperatures. In this configuration, it was not possible for the heater to keep the temperature at the outlet above 287 K. Thus, it was agreed to keep the temperature of the coupler higher.

Because of all this, when testing Romea in HNOSS, a rough value for the heat load was estimated. In order to measure the heat load generated by the cavity (and surrounding equipment), the Joule–Thomson (filling) valve, located between the 2 K tank and the heat exchanger (see figure 11.5), was closed and the power dissipated by

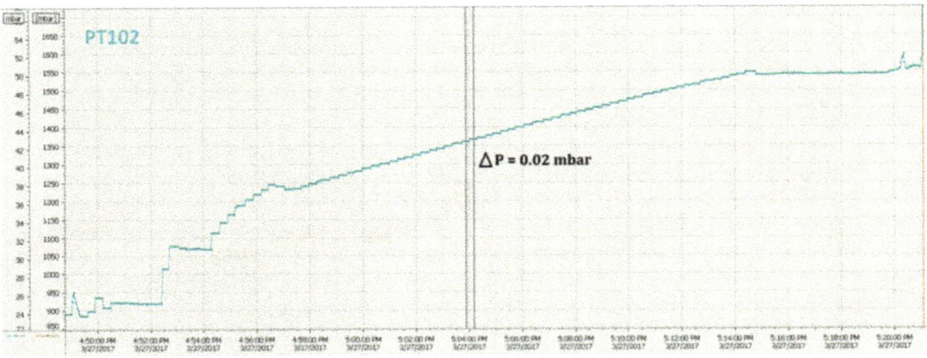

Figure 11.11. Details of the pressure rise vs. time showing a step-like behavior. Reprinted from [9].

the cavity was measured by the volumetric flowmeter located after the subatmospheric pumps, rated at maximum 160 m^3 h^{-1}. The other method used for calculating dynamic heat loads was the calorimetric method, also described in section 11.4. The pressure rise curve obtained via this method turned out to be stepwise instead of linear (see figure 11.11), so the actual value for the static or dynamic heat loads could only be an approximation. The reason for this behavior was found after the tests, and it had to do with the communication between the pressure gauge and the butterfly valve in charge of controlling the pressure.

The outcome of these tests was that with so many variables it was not possible to always have the same conditions, and thus the calculated power would differ between methods. It was considered that, when possible, the calorimetric method should be used because a smaller part of the complete system could affect the measurement.

These tests, which were the first ones ever performed at FREIA and in HNOSS, showed the following:

(a) Special care has to be taken when installing a cavity in HNOSS related to piping. Bends in the piping that would lead to trapped volumes need to be avoided.
(b) Proper sensor installation and thermalization, especially temperature sensors, is key to good operation.
(c) A proper heater rating for the heating of the ceramic window in the fundamental power coupler is crucial for operation. If it is too low, it could damage the window and vent the beam vacuum, rendering the cavity useless.

11.5.2 High-β elliptical cavity (ESS086-P01)

In May 2018, the high-β elliptical cavity ESS086-P01, fitted with fundamental power coupler and cold tuning system, was tested in HNOSS. Detailed descriptions of the tests can be found in [13–15]. As with Romea, this cavity was equipped at FREIA with temperature sensors and heaters, and later wrapped in MLI. Once the tests were

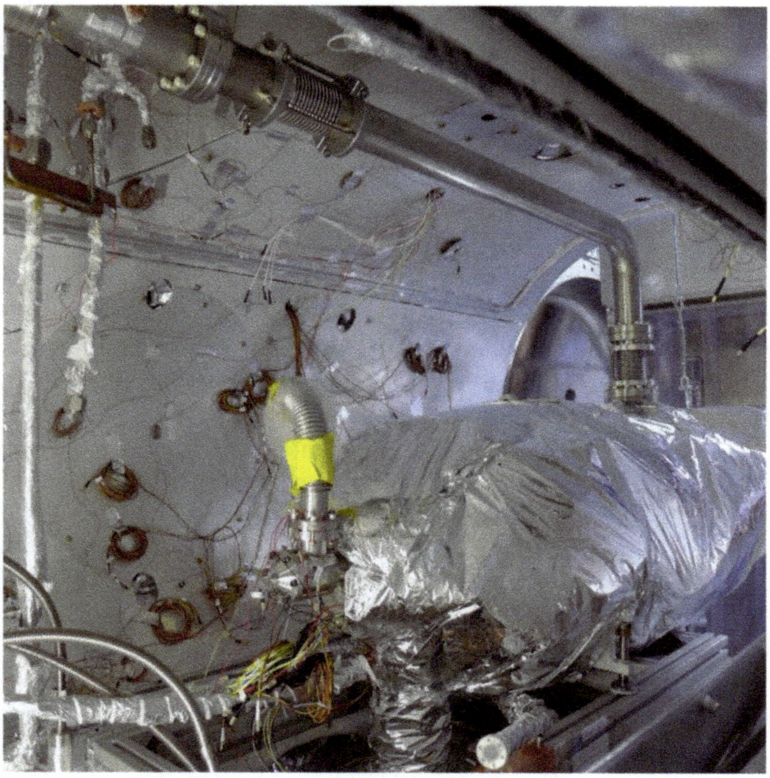

Figure 11.12. Picture of the high-β cavity–cavity installed inside HNOSS. The pipe installed on top of the cavity connects to the 2 K tank and the one in front of the cavity (marked with yellow tape) connects the beam vacuum to the corresponding pumping group. Reprinted from [16].

done and the cavity removed, it could be seen that the temperature sensors were not properly attached to the cavity anymore. This is thought to be due to a new glue used, which although rated for cryogenic operations proved to be unsuitable.

After the sensor installation, the cavity was then placed on top of a table inside HNOSS (see figure 11.12) cooled by liquid helium at 4.2 K. In terms of infrastructure, the main difference between the double-spoke and the elliptical cavity was that in the elliptical the coupler's inlet and outlet are both inside HNOSS. The cavity then followed the same steps as Romea, i.e., the procedure described in section 11.5.1. After some tests at cold, and due to lost connection to the stepper motor, HNOSS had to be warmed up and opened to check the cold tuning system. It turned out that one of the connectors used to link the motor to our control system had slipped away and disconnected. Once this connection was restored and strengthened with Kapton tape, HNOSS was closed, cooled down, and the tests resumed. In both instances. the cooling rate with liquid helium was also above 1 K min^{-1}. Even though special care was put on not having any bends on the connecting pipes, there seemed to still be an upward bend that gave way to spikes in the level probe at 4.2 K.

The heat load measurements at cold focused also on the liquid evaporation method (measured via the flowmeter) and the calorimetric method. However, for the elliptical the pressure reading for the latter method was fixed and, as mentioned in the previous section, the outcome, apart from giving an estimation, was that the dynamic heat loads (relative) agree between methods while the static values (absolute) show a higher deviation.

11.5.3 Double-spoke cryomodules

For the initial section of the superconducting part of the ESS accelerator, 13 cryomodules (plus one as spare), each with two double-spoke cavities, are foreseen. All these cryomodules are to be tested at FREIA, in the bunker where the horizontal cryostat HNOSS is also located (figure 11.13). For the testing, each cryomodule is brought into the bunker and connected to a stationary valve box containing all the necessary valves and a heat exchanger. This valve box is a modification of the series valve box for ESS because at FREIA liquid nitrogen is used for the thermal shield cooling, which required an extra transfer line inlet and outlet. The valve box is connected to the FREIA interconnection box and in turn connects to the cryomodule from the top via a jumper (with a bellow). The cryogens are then sent via the interconnection box (see figure 11.2) and, as with other equipment, we have the possibility to send the recovered helium gas back to the recycle compressor, the subatmospheric pumps, or directly to the gasbag. Due to lack of space in the bunker's walls, these helium outlets are merged with HNOSS's outlets, so when a cryomodule is filled with helium so is HNOSS. This far this has not posed any problem because HNOSS cannot be used at the same time as cryomodules are being tested, but it means more available volume and thus higher need of gaseous helium to fill the circuits. In addition, due to this all the infrastructure after HNOSS, i.e., the reheater, the path to the subatmospheric pumps, the position of the flowmeters and the butterfly for pressure regulation (see figure 11.5) are also common to the cryomodule.

The testing described in this section is divided into the simulator, the prototype, and the 11 series cryomodules tested up to the time of publication.

11.5.3.1 Simulator

During the end of January and beginning of February 2019, a smaller vacuum vessel with a helium tank inside, the so-called simulator, was connected to the valve box, see figure 11.13. This was to first check the performance of the valve box, make a functionality test, and commission the controls. This simulator had, in the helium vessel, a level probe and a heater in direct contact with liquid helium.

After following the relevant steps described in section 11.3, the tests focused on reaching 2 K, checking that no cold leaks appeared, that all instrumentation was working, and the effect of heat dissipation in the system, together with the corresponding sequence debugging. What was observed during these tests was that, even though we could reach 2 K without major problems, the system showed thermoacoustic oscillations, seen in the valve box by a frozen surface on the outlet's

Figure 11.13. Picture of the simulator connected to the valve box via the jumper and HNOSS in the background.

Figure 11.14. Picture of the exhaust valve CV04 showing frost on the housing when opened at 100%. Reprinted from [17].

valve housing, as shown in figure 11.14. In order to modify the acoustic impedance of the line and try to get rid of the oscillations, a circuit consisting of a small volume, a valve, and a pipe (RLC circuit) was added to the same circuit where the oscillations were originating from. The idea was to modify the resistance and/or inductance by changing the opening of the valve and/or changing the diameter and length of the pipe, respectively. The inductance could not be changed because the volume was kept the same. This exercise helped to reduce the oscillations but unfortunately did not suppress them. In any case, since the impedance of the system will be changed once a real cryomodule is connected, there was no point in spending more time looking into it because they might disappear on their own. Another important observation was that the temperature profile of the heat exchanger present in the valve box showed a possible contact with the vacuum barrier, rendering this exchanger ineffective for this and future tests because the valve box remains in place while the cryomodules are swapped for testing. Still, given that it was possible to reach 2 K, the sequences could be validated and the system's weaknesses could be identified, the tests with the simulator were considered successful, and the green light was given to move onto cryomodule testing.

11.5.3.2 Prototype cryomodule

The prototype cryomodule was tested at FREIA between the end of February and the middle of October 2019, see figure 11.15. This cryomodule housed the cavities ZA-01 (Romea) and ZA-02 (Giulietta), with their corresponding fundamental power couplers and cold tunning systems. Note that Romea had already been tested at FREIA in HNOSS, see above. The goal of these tests was to validate the

Figure 11.15. The cryomodule (center), the valve box (left-hand side) and the interconnection box (top right-hand side). Reproduced from [1].

cryomodule's design (cryogenics and RF), further debug the sequences, and find the best ways to get useful information for future cryomodule testing, such as which methods, as described in section 11.4, worked best to measure heat loads and how to implement and interpret them. A full description of these tests is given in [17, 18].

Before the actual testing can start, prior to the points defined in section 11.3 and common also to the series testing, the following extra steps are taken:

(a) Unpack the cryomodule and check that the shock sensor attached to the box has not been triggered.
(b) Visually check for possible damage.
(c) Verify electrical connections on the cryomodule flanges.
(d) Mount the doorknobs (RF adapter between the waveguide and the fundamental power coupler).
(e) Move the cryomodule into the bunker.
(f) Connect the relief and safety valves to the cryomodule's helium circuit (shared among all cryomodules).

Once these steps have been done, the actual testing can start. As a summary, the prototype went through three full thermal cycles up to 293 K due to different circumstances, and one partial thermal cycle only up to 20 K for the Easter holidays. The tests done within each of these thermal cycles, except for the first, comprised all these tests:

(a) Cool-down rate of the thermal shield.
(b) Cool-down rate of the cavities between 150 K and 50 K.
(c) Static heat load measurement at 2 K.
(d) Dynamic heat load measurement at 2 K and at different accelerating gradients.

In general, the cool down of the valve box's and cryomodule's shields from room temperature to ca. 85 K took only a couple of hours and the cooling of the cavities to 4.2 K less than double the time. Almost right at the beginning, after the cavities were cold and power was being applied to each individually, there was an incident with a ruptured safety relief disk. The issue was that the interlock was not properly set while one of the cavities was being powered [19]. During a cavity quench the evaporated helium increased the pressure in the cavity to 1.7 bara, which was above the limit for the safety relief disk. After this event, the system was warmed up to room temperature, the disk replaced, and the cool down restarted because there was fortunately no damage to the cavities.

The tests performed in this cryomodule highlighted the continuous presence of thermoacoustic oscillations, which could not be damped by the RLC circuit that was still in place from the simulator tests. Because of this, the outlet valve CV04 (see figure 11.8) also showed frost when the cryomodule was at or below 4.2 K (figure 11.14) and vibrations could be felt on the outlet transfer line, which reduced or almost disappeared when this valve was set to an opening of ca. 40%, but it still could be seen in the high flows from the cryomodule when measuring heat loads. This disturbance also affected the proportional-integral-derivative (PID) fine-tuning of the filling and the butterfly valves, making it difficult to find the right parameters that would work well under any change of conditions. Even when the system seemed to be in a stable state, unfortunately no good valve settings were found during the testing with the prototype cryomodule after long time-consuming efforts. As found when running the simulator, the performance of the heat exchanger in the valve box was suboptimal, but in general this did not pose a big problem with the actual testing of the cavities below atmospheric pressure because it was still possible to reach the superfluid helium state, maintain it, and do measurements, all while having some margin on the liquefier's cooling power and recovery system's capacity.

Regarding the tests pertaining to cryomodule performance at cold, all methods described in section 11.4 except the level decrease and heater compensation methods were carried out. The reason for not attempting the level decrease method was because at the time the mapping of the helium volume in terms of height according to the level probe was unavailable. The heat compensation method was not used due to needing a very fine and stable level regulation and, for fast and visible effect, heaters placed in direct contact with liquid helium, which was not the case.

It turned out that the absolute values obtained for the dynamic heat loads were, at times independent of the method chosen as measurement, even lower than the values found for the static heat loads, which of course has no physical meaning. One thing to consider in this case is that the flowmeter's maximum range is 160 m^3 h^{-1}, which compared to ca. 11 m^3 h^{-1} of flow from the cavity, might not be accurate enough as to distinguish 1–2 m^3 h^{-1} steps, as expected for the relative dynamic heat load value.

The heat load measurement then proved to be nontrivial. As mentioned earlier, since good PID parameters for the filling valve could not be found, at the beginning of the measurements the filling valve would be manually fixed to a certain opening that would keep the level as constant as possible. However, depending on the RF power applied to the cavities, this opening might not be adequate and the level would decrease or increase faster than expected affecting the heat load because it also depends on the liquid helium level. If the valve opening was then changed, then the gas flashing through the valve would also change, which affects the measurement. After these problems were encountered, it was decided to keep the filling valve closed to avoid this extra flashing. However, since the fundamental power coupler is cooled with flow taken from the exhaust of the cavities (see figure 11.8), this cooling is very much reduced and the coupler warms up, giving more heat to the cavity and thus to the helium bath. Even though this is the case, it was considered that the time of the measurement would not be long enough to have too big of an impact in the helium bath.

It is worth mentioning that apart from the filling valve, there is also a butterfly valve that regulates the pressure of the helium bath. The PID regulation of this valve also proved to be difficult, mostly due to the presence of thermoacoustic oscillations, adding uncertainty to the heat load measurements because the regulation would not be smooth, as it would if the PID parameters were good and the system stable.

Under these circumstances, the calorimetric method then took an important role because in that case the system does not depend on any valve regulation and it was already decided that the lack of power cooling during a short time would not have a big impact on the heat loads. This method gave a maximum of 1 W in dynamic heat load, agreeing with other methods in not being discernible enough by other means.

11.5.3.3 Series cryomodule

After the testing of the prototype cryomodule, it was obvious that there would still be a learning curve when testing the series cryomodules because not all issues had been solved or understood when testing the prototype. It then still took some time to streamline the tests and pinpoint the weak points. Even after 13 tested cryomodules, this process is still ongoing. Since the start of the cryomodule tests in October 2020, at the time of writing (May 2023) a total of 13 cryomodules had been checked. Of these 13 cryomodules, three of them had issues with the stepper motor, one had a leak in the double wall of the fundamental power coupler, and another had both issues: a leak in the double wall tube of the fundamental power coupler and a problem with the stepper motor. Because of these problems these cryomodules had, or will have, to be re-tested after being repaired at either FREIA (when possible), at ESS's premises or at Laboratoire Irène Joliot-Curie (IJCLab) in Orsay, France. A small issue that was observed for some was that one of the level probes (there are two in total) was short-circuited when checked at reception. The reason was found by IJCLab to be the cable insulation: the isolating tape around the cables would thin out from rubbing against the edge of the flange's opening inside the cryomodule during transport, creating a short circuit to ground. This problem was solved by adding a PEEK insulator around the cables, and this is currently being implemented in all cryomodules.

Regarding installation, once a cryomodule was received at FREIA, the procedure followed until it was inside the bunker connected to the liquefier and the recovery system is the same as it was for the prototype cryomodule, as detailed in sections 11.3 and 11.5.3.2. Even though all cryomodules are supposedly equal in terms of dimensions, there were a handful of times where the valve box needed to be minimally displaced to be able to connect the lines between the cryomodule and the valve box at the jumper. Being able to move the valve box and the cryomodule in the *xyz* plane proved to be a very useful feature. Since some equipment is installed from cryomodule to cryomodule (i.e., insulation and beam vacuum pumping groups, the corresponding pressure gauges, and the safety and relief valves), for this and other obvious reasons, a detailed operation manual explaining the steps to follow since the arrival of the cryomodule at FREIA all the way to cool down, warmup, and disconnection proved to be very useful in avoiding minor mistakes that would otherwise take time from testing.

The methods used to calculate heat loads were the same as with the prototype cryomodule, encountering the same problems for the same type of measurements. One thing that has been changed after the testing of a few cryomodules is the possibility to regulate the pressure in the cavities with the exhaust valve (CV04, see figure 11.8) instead of the butterfly valve, common also to HNOSS (see figure 11.5).

This change alone reduced the flows measured by the flowmeter by 33%, which clearly indicated that the flows measured before might have been conditioned by the thermoacoustic oscillations. The regulation of the pressure with this valve has been used since then.

For one of the cryomodules the level drop method to calculate heat loads was also checked, thanks to the mapping of liquid helium volume vs. height in the cavities that we received from ESS. In this case, the value found for the dynamic heat load was in agreement with the flowmeter method, although the absolute value for the static heat load was 30% higher than when measured with the flowmeter. A possible explanation for this behavior might be that, as in all measurements, it has uncertainties. When the flow is measured via the flowmeter, it is assumed that the temperature of the cold circuit is independent of the flow, which means that the quantity of the cold gas stored in the cryogenic circuit does not change with the flow. This hypothesis might not be necessarily true but if we would be able to reach a steady state then the temperature of the cold circuit would stabilize and the flow measured by the flowmeter should reach a value close to that obtained by the level variation. One thing to be taken into account while using this method is that the variation should not be too big otherwise the surface wetted by the liquid helium would decrease as the level decreases, and thus the liquid helium evaporation would also decrease.

Once the cryomodules have met ESS's requirements (e.g., achieved accelerating gradient, dynamic heat loads, the cold tuning systems have performed without issues, and no leaks have been found at either room or liquid helium temperatures) they were transferred to ESS for installation. At the time of publication, nine cryomodules out of 13 (with ongoing discussions about testing of a spare cryomodule) have been successfully tested.

11.6 Lessons learned

Apart from what has been already mentioned in the corresponding sections, one thing that had to be generally accepted was that cryogenics take time. At times we wanted to speed things up by cooling the shields with liquid nitrogen one day and start the cool down with helium on the next. Although possible, this would have increased the amount of helium and time needed because the cavities, and especially the cold tuning systems, would be warmer. As an example, when only the thermal shields are being cooled with liquid nitrogen, the cold tuning system takes 66 h to reach 180 K from room temperature, and an extra 107 h to reach 90 K once the cooling with liquid helium has started. So, if possible, the thermal shield would be started on a Friday to have the weekend for the cryomodule to cool down, or at least left for more than a day before starting the cavities' cool down with liquid helium if the schedule permits.

In terms of infrastructure, if one could change part of the existing installation, the best option to measure heat loads would be to use a flowmeter as close to the outlet of the cryomodules as possible, which would mean using a cold gas flowmeter and adding a pressure and temperature sensor to be able to calculate the actual flow rate. Unfortunately, this is not possible in the case of FREIA, but another reasonable and simpler solution would be to add a bypass line to the current flowmeter that would include a flowmeter with a lower operating range, e.g., 0–20 $m^3\ h^{-1}$. In this case the high range flowmeter could be used during cool down and the low range flowmeter during measurements. Another infrastructure point that proved to be a bottleneck during the cool down of the cavities was the low capacity of the recovery compressors to empty the gasbag. As stated in table 11.1, the combined capacity of the recovery compressors was less than the size of the gasbag, which unbeknown to us turned out to be a problem during liquid helium cool down because the cooling would need to be stopped or very much reduced midway to avoid venting gas helium to atmosphere. This might lead to Q-disease. This changed in May 2022 when a recovery compressor with a much higher capacity was put in place, operating in parallel to the existing recovery compressors. In terms of controls, it is obvious that a good set of regulation parameters for all valves involved is crucial for good measurements, and time must be spent on this task for systems that are in their final position. In addition, in the presence of thermoacoustic oscillations, time should be taken to find ways to dampen the oscillations whenever possible.

In general, cryogenic problems coming from the testing of cryomodules at FREIA can be divided into two main sources:

(a) The liquefier: Lack of liquid helium production, problems with the main compressor, and full gasbag.
(b) The cryomodule itself: Cold leaks, equipment malfunction, and high field emission because then the heat loads would be higher.

The problems that have been, so far, dominating the schedule during all stages of the testing had to do with the cryomodules. However, at the beginning of the tests problems with the RF stations were also predominant, whereas the liquefier is now the dominant problem after more than half of the tests have been done. The reason why this is so might be because the RF stations' performance correlated to the cavities is better understood with more testing, and fine-tuning of the corresponding parameters is possible. In the case of the liquefier, the performance degrades with time if the extra 80 K adsorber installed inside the liquefier's coldbox is not used. This leads to contaminants being accumulated after many days of running the liquefier instead of being efficiently removed, which reduces the output's mass flow and incurs time-consuming regeneration processes. Apart from this, proper maintenance has to be done. If there is a tight schedule, as in the case of this cryomodule testing at FREIA, then it is easy to keep postponing maintenance until it is not possible to delay it anymore because issues affecting the cryomodule testing start to appear, incurring delays.

After so many cryomodules had been tested, we found that two weeks spent for installation, two weeks at cold (from cooling of the thermal shields to operating at 31 mbar), and one week to warmup and disconnect was a good estimation if no major problems arise during testing, such as a cold leak or issues with the liquefier.

The work done at the FREIA laboratory for the ESS involves many disciplines, e.g., cryogenics, radio frequency, controls, etc. As a result, knowledge and experience have been gathered in all of these disciplines and we find ourselves in a better position to understand, plan, and carry out future research needed for the development of the accelerator facilities. Because of our specialized and versatile infrastructure and competence, our options are multiple and diverse.

References

[1] Ruber R *et al* 2021 Accelerator development at the FREIA laboratory *J. Inst.* **16** P07039

[2] Ruber R *et al* 2015 *The Cryogenic System at the FREIA Laboratory* (Uppsala University FREIA) Report 2015/03

[3] Waagaard E 2020 Benchmarking a cryogenic code for the FREIA helium liquefier *BSc Thesis* Uppsala University, Uppsala

[4] Chevalier N R *et al* 2014 Design of a horizontal test cryostat for superconducting RF cavities for the FREIA facility at Uppsala University *AIP Conf. Proc.* **1573** 1277–84

[5] Santiago Kern R *et al* 2014 The HNOSS horizontal cryostat and the helium liquefaction plant at FREIA *5th Int. Particle Accelerator Conf. (IPAC) Proc. (Dresden, Germany)* pp 2759–61

[6] Santiago Kern R *et al* 2017 The Gersemi vertical cryostat at FREIA *Proc. 14th Int. Institute of Refrigeration Conf. (Dresden, Germany)* pp 441–6

[7] Experimental Physics and Industrial Control System (EPICS) https://ref11x07-controls.org/ (accessed 7 July 2022)

[8] LabVIEW https://www.ni.com/ref11x08/ (accessed 7 July 2022)

[9] Santiago Kern R *et al* 2017 *Cryogenic Synopsis from the Testing of the Fully Equipped ESS' Double Spoke Cavity Romea* (Uppsala University FREIA) Report 2017/12

[10] Li H *et al* 2017 High power testing of the first ESS spoke cavity package *Proc. 18th International Conf. on RF Superconductivity* ed V Schaa, Y He, L Li and N Zhao (China: Lanzhou) pp 817–20

[11] Knobloch J 2003 The 'Q disease' in superconducting niobium RF cavities *AIP Conf. Proc.* **671** 133–50

[12] Santiago-Kern R, Eriksson J, Hermansson L and Jönsson Å 2017 *Cryogenic Settings for Testing of the Fully Equipped ESS' Double Spoke Cavity Romea* Internal Uppsala University FREIA Report 2017/11

[13] Santiago Kern R *et al* 2018 *Cryogenic Synopsis from the Testing of the Fully Equipped ESS' High Beta Cavity ESS086-P01 (Part I)* (Uppsala University FREIA) Report 2018/04

[14] Santiago Kern R *et al* 2018 *Cryogenic Synopsis from the Testing of the Fully Equipped ESS' High Beta Cavity ESS086-P01 (Part II)* (Uppsala University FREIA) Report 2018/06

[15] Li H *et al* 2018 *First High Power Test of the ESS High Beta Elliptical Cavity Package* (Uppsala University FREIA) Report 2018/07

[16] Santiago-Kern R, Eriksson J, Hermansson L and Jönsson Å 2018 *Cryogenic Settings for Testing of the Fully Equipped ESS' Double Spoke Cavity Romea* Internal Uppsala University FREIA Report 2018/03

[17] Santiago Kern R *et al* 2019 *Cryogenic Performance of the Spoke Prototype Cryomodule for ESS* (Uppsala University FREIA) Report 2019/05

[18] Li H *et al* 2019 *RF Performance of the Spoke Prototype Cryomodule for ESS* (Uppsala University FREIA) Report 2019/08

[19] Burst Disc Review https://indico.uu.se/event/627/ (accessed 7 July 2022)

Chapter 12

Lessons learned and final comments

J.G. Weisend II

This chapter discusses the high-level lessons learned from the development of the ESS cryogenic system. These lessons are particularly valuable for other big science projects that will use the model of in-kind contributions in project execution. This chapter also make some final comments on the state of the European Spallation Source (ESS) cryogenics system at the time of publication and the expected next steps.

One of the most important features of ESS, as stated in chapter 1, is the use of in-kind contributions to provide equipment, staffing, and expertise to the project. This use of in-kind contributions was not restricted to the cryogenic system and is used extensively throughout the accelerator, target, and scientific instruments at ESS. The initial goal of the project was to have up to 50% of the project cost provided by in-kind contributions. Due to the increasing size and cost of big science projects today, the use of the in-kind funding model is seen more and more. Current examples include the ITER fusion project in Cadarache, France, the PIP-II accelerator project at Fermilab, USA, and the FAIR accelerator project in Darmstadt, Germany. Of course, space, telescope, and particle detector projects have long used this model at a somewhat smaller scale.

One of the contributions of ESS to the broader scientific community is the lessons that we learned using in-kind contributions. Chapters 3–11 all end with a list of lessons learned due the production of the systems covered in the chapter. These lessons tend to be both technical and organizational. This chapter will mainly discuss lessons learned that affect the entire ESS cryogenics systems. All the comments below are the opinion of the author and do not necessarily represent any sort of official viewpoint of the ESS laboratory.

With that said, the following were found to be valuable to our success and should be considered by future projects.

doi:10.1088/978-0-7503-3223-1ch12

- A good working relationship with clear communication is key when working with in-kind partners. While formal documentation describing each party's responsibilities, content to be provided, and schedule are a necessary part of this, possibly more important is developing strong personal relationships between the staff of the in-kind partner and the laboratory. This can be done via regular meetings, visits to each other's facilities, and joint participation in reviews. The in-kind partner should be viewed and treated as a scientific collaborator and not as a subcontractor. These relationships should be developed at all levels of the organizations and not just at the senior management levels. Clear communications and a strong relationship mean that if a problem arises, it will be quickly raised and solved to everyone's benefit.
- People make the difference in projects such as this. It was very important that we had sufficient, qualified staff at both ESS and the in-kind partners to carry out the work. These people were recruited early on in the project. It is particularly important to have technicians and operators in place early enough so that they can participate in the installation and commissioning of the equipment. This will provide them with the experience needed when the regular operation of the cryogenic system starts.
- Design reviews at all stages of the project are invaluable. Such reviews should always include discussions of safety issues and interfaces. Reviews should, whenever possible, include people from outside the project.
- Early collaboration with the people designing the conventional buildings and utilities is also vital. Clear detailed requirements on building size, access, crane size, and utility needs, including such things as water quality, need to be prepared and discussed with the conventional facilities team very early in the project. Be prepared to insist that your needs are met.
- ESS took advantage of early prototyping and testing whenever possible. This included testing cryomodule prototypes, commissioning cryomodule test facilities as soon as possible, and planned testing of the cryogenic moderator system (CMS) with helium before the start of hydrogen operations. Discovering issues earlier allows corrections to be made before the schedule is affected. The discovery and remediation of thermoacoustic oscillations in the cryogenic distribution system (CDS; chapter 4) is a good example of this approach.

At the time of publication (early 2024), the ESS cryogenics system is near completion. All the cryoplants have been commissioned and accepted, almost all the cryomodules required for the first stage of the linear accelerator's operations have been delivered to ESS, tested, and placed in the tunnel. Later this summer, the CMS will be commissioned using helium with hydrogen operations planned for the following year. The recent second commissioning test of the accelerator cryoplant and CDS (chapters 3 and 4) included the installation of one spoke cavity cryomodule and one elliptical cavity cryomodule in their final positions in

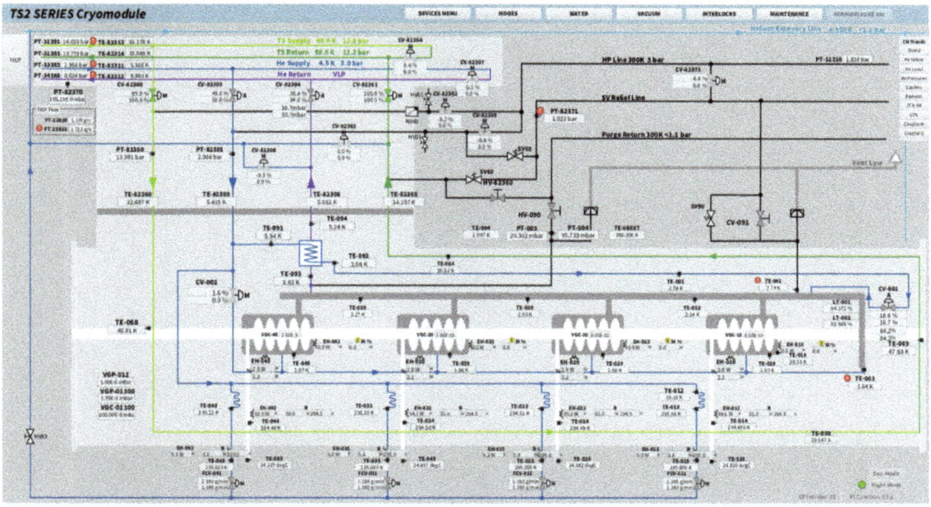

Figure 12.1. 2 K operation within a cryomodule at Test Stand 2. Similar results have been seen in tunnel operations. (Courtesy of ESS.)

the tunnel. During the test, 2 K helium was established in the tunnel for the first time (see figure 12.1). The current status of the ESS cryogenics system is something that we can all be proud of and helps set up the ESS as a major facility for future scientific discoveries.

www.ingramcontent.com/pod-product-compliance
Ingram Content Group UK Ltd.
Pitfield, Milton Keynes, MK11 3LW, UK
UKHW051907100926
8556IPUK00005B/798

9 780750 332217